KAIST Research Series

The KAIST Research Series has been established to enable the wide dissemination of KAIST's outstanding research achievements among researchers at an international level. It provides a framework for KAIST researchers to publish monographs synthesizing the key results of their extended research programmes and setting these in a global context. Volumes published in the KAIST Research Series include single- and joint-authored monographs written by KAIST researchers, and edited volumes arranged by KAIST academics but including an international authorship. The KAIST Research Series also provides a forum for undergraduate-level textbooks and graduate-level textbooks based around lecture courses given at KAIST. It is coordinated by a team of Series Editors, representing most academic departments within KAIST.

About KAIST KAIST was established by the government in 1971 as the nation's first graduate school specializing in science and engineering education and research. By 2006, KAIST has produced over 30,000 graduates. A large number of KAIST graduates have become researchers, venture entrepreneurs and technological bureaucrats, contributing to Korea's economic growth and development of science and technology. KAIST ranked seventh in Asia in a ranking compiled jointly by the Chosun Ilbo, a major Korean daily, and global university evaluation institute QS of Britain, and was included in the "2012 Top 100 Global Innovative Organizations" published by Thomson Reuters, the world's leading media and financial-data firm. KAIST is committed to becoming one of the world's leading universities focused on science and technology with a vision to create knowledge for human society.

About KAIST Press KAIST Press is a university press established to disseminate KAIST's outstanding research achievements and popularize science & technology. KAIST Press has published research papers, reports and teaching materials, as well as books aimed at public understanding of science and technology. It has recently expanded its operations into organizing a series of guest lectures designed to promote public understanding of science and technology.

SooJean Han

Linear Systems

Theory and Applications

Springer

SooJean Han
Department of Electrical Engineering
Korea Advanced Institute of Technology
(KAIST)
Daejeon, Korea (Republic of)

ISSN 2214-2541 ISSN 2214-255X (electronic)
KAIST Research Series
ISBN 978-981-96-9773-1 ISBN 978-981-96-9774-8 (eBook)
https://doi.org/10.1007/978-981-96-9774-8

This Springer imprint is published by the registered company Springer Nature Singapore Pte Ltd.
The registered company address is: 152 Beach Road, #21-01/04 Gateway East, Singapore 189721, Singapore

Preface

Linear Systems: Theory and Applications is one of the most fundamental and important prerequisites necessary to study control engineering. The overall field of control engineering studies how systems respond to various inputs, and how to design controllers that ensure desired behaviors. Such a system is said to be *linear* if its governing model is linear, which means the following two properties. First, scaling the input of a linear system means that the output is also scaled proportionally. Second, adding two inputs of a linear system means that the output is also the sum of the outputs to each input individually. While many real-world control systems are affected by nonlinearities and uncertainties, it is often appealing to simplify them to the linear domain due to these convenient properties. In fact, the analysis of linear systems primarily uses only basic tools from linear algebra, linear differential equations, and frequency domain analysis.

This textbook was conceived from the lecture notes I created and used while teaching the graduate-level linear systems course at KAIST in Spring 2024. While teaching the course, my notes were inspired by the many excellent references used by the linear systems courses taught at leading institutions like Caltech, Berkeley, MIT, Stanford, and others. This textbook differs from these courses in two main ways. First, it introduces foundational mathematics background (e.g., Jordan canonical form, $\mathcal{L}_2$-space) as the topics progress, providing a more seamless transition to advanced linear systems topics. Second, it provides natural progressions towards more advanced subfields of control systems engineering, namely optimal control, state-estimation, and robust control, and their connections to the main linear systems topics are emphasized.

This book is organized into four main parts: (1) *Linear System Properties*, (2) *Linear Stability*, (3) *Linear Control and Estimation*, and (4) *Linear Optimal Control and Estimation*. Each part is itself composed of multiple chapters that cover all aspects of the focused topic.

- Part I introduces the basic classifications of signals and systems, important properties such as time invariance and causality, as well as their mathematical descriptions in both time and frequency domains. The notion of what it means to "control" and how "uncontrolled" systems differ from "controlled" ones is explained by way of block-diagrams.

- Part II introduces stability, a particularly important property of general systems. Various different types of stability are introduced, such as input-output stability and internal stability. When a system is linear, its stability can be analyzed by discerning some properties of its eigenvalues. Lyapunov sense internal stability is especially crucial in more general systems which are not necessarily linear, and this book dedicates a few chapters towards the stability analysis of such nonlinear systems.
- Part III finally builds upon the notion of "control" introduced in Part I. Two key preliminary questions are addressed: (1) the "what extent" can a system be controlled, which is characterized its *controllability*, and (2) to "what extent" can the system's full state be observed, which is characterized by its *observability*. The duality between controllability and observability for linear systems is also explored. Methods for controlling a linear system are also introduced, including feedback control via pole-placement.
- Part IV extends the methods of control and state-observation discussed in Part III. Pole-placement requires a predetermined collection of desired poles in order to perform control design, but it is often difficult to choose them in practice. Instead, the subfield of optimal control is concerned with the optimization of a specific cost functional, which quantifies properties of the state and control effort directly. The linear quadratic regulator (LQR) and Kalman filter, two hallmark methods in control theory beyond linear systems, are introduced and explored. In particular, the connection among LQR control and $\mathcal{H}_2$ and $\mathcal{H}_1$ robust control is also emphasized in one chapter.

SooJean Han
KAIST
Daejeon, Korea (Republic of)

Acknowledgements

As optional extra credit, students in my course were asked to LaTeX the lecture notes. Many thanks to those who participated. Special thanks also go to our members in the KAIST Autonomous Control of Stochastic Systems (ACSS) research lab, who spent time proofreading the manuscript and figures: Jeongyong Yang, Eunwoo Sung, Seunghwan Jang, Hojin Ju, and Sanghun Park. Finally, additional thanks go to our teaching assistants for the course—Minwoo Kim, Suhwan Sung, and Je In Yu—for completing homework solutions and carrying out recitation sessions, many of which have been adapted into this textbook as exercise problems.

Contents

Part I Linear System Properties

1 Introduction and System Definitions ... 3
1.1 State-Space Model Description ... 4
1.2 Transfer Function Description ... 8
1.3 Classification of Systems ... 12
1.4 Mathematical Reviews ... 15
1.4.1 Matrix Properties, Matrix Algebra ... 15
1.4.2 Spaces and Basis ... 17
1.4.3 Diagonalization and Jordan Form ... 22
1.4.4 Functions of Square Matrices ... 24
References ... 27

2 Characteristic Modes ... 29
2.1 The Matrix Exponential ... 29
2.1.1 Computing e^A via Power Series Expansion ... 29
2.1.2 Computing e^A via Cayley-Hamilton Theorem ... 30
2.1.3 Computing e^A via Jordan Canonical Form ... 30
2.2 The Time-Indexed Matrix Exponential ... 32
2.3 Equivalent Systems ... 34
2.4 Solutions to Autonomous Systems ... 36
2.4.1 Connection to System Stability ... 37
2.5 Discretization ... 38

3 State-Transition Matrix ... 41
3.1 STM for the DT LTI System ... 41
3.2 STM for CT LTV Systems ... 42
3.3 Fundamental Matrix for CT LTV Systems ... 45
3.4 Fundamental Matrix for DT LTV Systems ... 46
3.5 Uniqueness of STM ... 46

3.6 Special Topic: Periodic LTV Systems 47
3.6.1 Fundamental Matrices Under Equivalence Transformations 48
3.6.2 Introductory Floquet Theory 49

4 Problems and Exercises 51
References 61

Part II Linear Stability

5 Input-Output Stability 65
5.1 BIBO Stability for LTI Systems 66
5.2 BIBO Stability for LTV Systems 69

6 Internal Stability 71
6.1 Linearization 71
6.2 Determining Stability via Eigenvalues 73
6.3 Classifying Types of Equilibria 74
6.3.1 Node Equilibria 75
6.3.2 Saddle Equilibria 77
6.3.3 Focus Equilibria 78
6.3.4 Center Equilibria 79

7 Lyapunov Stability 83
7.1 A Motivating Example 83
7.2 Lyapunov-Sense Stability 84
7.2.1 Additional Definitions 85
7.2.2 Preliminary Mathematical Reviews 86
7.3 Lyapunov Indirect Method 88
7.4 Lyapunov Direct Method 89
7.5 Exponential Stability 90
7.6 Lyapunov Stability for LTI Systems 91
7.7 Invariant Set Theorem 94
References 97

8 Uniform Stability 99
8.1 Lyapunov Direct Method for Nonautonomous Systems 101
8.1.1 LTV Case 101
8.2 Comparison Functions 102
8.3 Proof of Stability Using STM 103
8.4 Converse Theorem for LTV Systems 105
8.5 Lyapunov Indirect Method for Nonautonomous Systems 106

9 Problems and Exercises 109
References 115

Part III Linear Control and Estimation

10 Canonical Forms 119
10.1 System Realizations 119
10.2 Three Main Canonical Forms 120
10.3 Controllable Canonical Form 121
10.4 Observable Canonical Form 123
10.5 Modal Canonical Form 125
10.5.1 Case: $D(s)$ has Distinct Real Roots 125
10.5.2 Case: $D(s)$ has Some Repeating Real Roots 127

11 Minimum-Energy Input 129
11.1 Preliminary Reviews and Background 129
11.1.1 Least-Squares Methods 129
11.1.2 $\mathcal{L}^2$ Space and Norm 132
11.1.3 Linear Mapping 134
11.1.4 Adjoint Operators 135
11.2 Minimum Energy Input: Continuous-Time Case 136
11.3 Minimum Energy Input: Discrete-Time Case 140

12 Controllability 143
12.1 Controllability and Reachability 143
12.2 Proofs of Equivalence of the Controllability Tests 147
12.2.1 Proof: (CTRB1.) and (CTRB2.) are Equivalent to (CTRB3.) 147
12.2.2 Proof: (CTRB1.) and (CTRB2.) are Equivalent to (CTRB4.) 153
12.2.3 Proof: (CTRB1.) and (CTRB2.) are Equivalent to (CTRB5.) 154

13 Observability 157

14 Minimal Realizations 161
14.1 The Kalman Decomposition 161
14.2 Duality Principle 162
14.3 Controllability and Observability Properties in Realizations 165
14.4 Minimal Realizations 166
14.5 Gilbert Realization 169

15 Controller and Observer Design 171
15.1 Pole-Placement 171
15.1.1 State-Feedback Control 171
15.1.2 Ackermann's Formula 173
15.2 General Feedback Interconnection System 174
15.3 State Observers (State Estimators) 175
15.3.1 Full-Order Observers 176

15.3.2 Reduced-Order Observers ... 178
15.4 Observer-Based Controllers ... 179
References ... 181

16 Problems and Exercises ... 183
References ... 194

Part IV Linear Optimal Control and Estimation

17 Feedback Stabilization ... 197
17.1 Parametrizing Stabilizing Controllers ... 197
17.2 Youla Parametrization ... 202
17.2.1 Case 1: Stable Plant ... 202
17.2.2 Case 2: General, Possibly Unstable Plant ... 205

18 The Linear Quadratic Regulator ... 209
18.1 Dynamic Programming ... 209
18.2 Basic Derivations of the Linear Quadratic Regulator ... 211
18.2.1 Discrete-Time Dynamics ... 211
18.2.2 Continuous-Time Dynamics ... 215
18.2.3 LQR via Lagrange Multipliers ... 217
18.2.4 LQR is a Stabilizing Controller ... 219
18.2.5 Output-Feedback Case ... 221
18.3 Solving Riccati Equations ... 222
18.3.1 With the Hamiltonian Matrix ... 223

19 Linear Robust and Stochastic Control ... 229
19.1 Stabilization Using LMIs ... 230
19.2 Hardy Spaces and System Norms ... 232
19.2.1 Signal Norms Review ... 232
19.2.2 System Norms ... 233
19.3 $\mathcal{H}_\infty$ Optimal Control ... 236
19.4 Stochastic Linear Systems ... 240
19.4.1 System Dynamics ... 240
19.4.2 Interpretation of the $\mathcal{H}_2$ System Norm ... 240
19.4.3 Signal Processing Background ... 241
19.4.4 Stochastic Processes Background ... 242
19.4.5 The White Noise Terminology ... 244
19.5 The Linear Quadratic Gaussian ... 245
19.6 $\mathcal{H}_2$ Optimal Control ... 247
19.7 Relationship Between $\mathcal{H}_2$ and LQG ... 249
References ... 251

20 Linear State Estimation 253
20.1 Preliminaries: Minimum Mean-Squared Estimator 253
20.1.1 Stochastic Linear Least-Squares 254
20.1.2 The Orthogonality Principle 255
20.2 Sequential Estimation Over Time 257
20.2.1 The Linear Gaussian Case 257
20.2.2 The Innovations Sequence 258
20.3 The General Bayesian Filtering Problem 259
20.4 The Discrete-Time Kalman Filter 260
20.5 The Continuous-Time Kalman Filter 262
References 264

21 Problems and Exercises 265
References 278

Part I
Linear System Properties

Chapter 1
Introduction and System Definitions

A *system* $\mathcal{H}$ is a signal processor which transforms the some input signal into an output signal. In relation to terminologies from mathematics, signals are *functions* and systems are *operators* which take those functions as input and returns another function as output. Systems can be visually represented using *block diagrams*, where arrows and lines represent signals, and blocks represent system components. Some common simple block diagrams of systems are shown in Fig. 1.1. Arrows indicate the signals $\mathbf{u}$, $\mathbf{y}$, $\mathbf{w}$, and $\overline{\mathbf{z}}$ which are involved with the system. The main signals are the *control input* $\mathbf{u}: \mathbb{R}^{\geq 0} \to \mathbb{R}^m$ and the *output* signal $\mathbf{y}: \mathbb{R}^{\geq 0} \to \mathbb{R}^k$. Additional signals include the following. *Exogenous inputs* $\mathbf{w}: \mathbb{R}^{\geq 0} \to \mathbb{R}^n$ model external inputs such as noise and external disturbances. *Auxiliary outputs* $\overline{\mathbf{z}}: \mathbb{R}^{\geq 0} \to \mathbb{R}^\ell$ are outputs which are separate from $\mathbf{y}$ in that it is typically used to measure some performance criteria. Here, $m, k, \ell \in \mathbb{N}$ all represent their respective signal dimensions. Often, $\mathcal{H}$ is called the *open-loop system* or the *plant*, and we can also write the system as $\mathbf{y}(t) = \mathcal{H}\{\mathbf{u}\}(t)$ for all time t. Exogenous inputs and auxiliary outputs will mainly influence our discussion of more general linear systems in later chapters (see Part III and Part IV).

When a *controller* is included, another block, typically labeled $\mathcal{K}$, is included beneath the plant. The overall system is then called a *closed-loop system*, or a *feedback interconnection*, in reference to the fact that the output of the controller's signal is being *fed back* to the plant. When a controller is involved, it is often useful to categorize the input and output signals according to what is being used by the controller and what is not. $\overline{\mathbf{z}}$ is called the *regulated* or *auxiliary output* while $\mathbf{w}$ is an *external disturbance*. The main function of a controller is to change the natural behavior of a plant. There are many applications, including output regulation, reference tracking, disturbance rejection.

SJ. Han, *Linear Systems*, KAIST Research Series,
https://doi.org/10.1007/978-981-96-9774-8_1

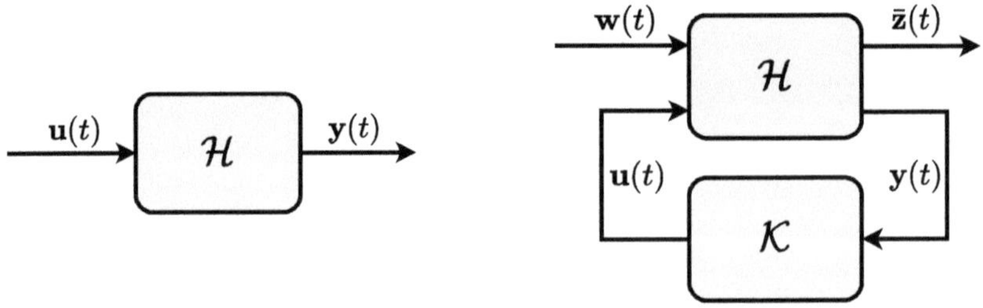

Fig. 1.1 Simplified system block diagrams describing only the input and output signals. [Left] $\mathcal{H}$ admits only one input **u** and one output **y**. [Right] $\mathcal{H}$ with external disturbance **w** and auxiliary output $\bar{\mathbf{z}}$, and a feedback controller $\mathcal{K}$ that uses output signal **y** to construct input signal **u**

1.1 State-Space Model Description

Mathematical descriptions of systems include differential equations (or difference equations, in the case of discrete-time systems), including ordinary differential equations (ODEs), partial differential equations (PDEs), and stochastic differential equations (SDEs). Systems representation based on differential equations are often referred to as the *state-space representation*, where the system is characterized by an *internal state* signal (or simply, *state*) **x**.

$$\mathcal{H} \triangleq \begin{cases} \dot{\mathbf{x}}(t) = f(\mathbf{x}(t), \mathbf{w}(t), \mathbf{u}(t)), \\ \bar{\mathbf{z}}(t) = g(\mathbf{x}(t), \mathbf{w}(t), \mathbf{u}(t)), \\ \mathbf{y}(t) = h(\mathbf{x}(t), \mathbf{w}(t), \mathbf{u}(t)), \\ \mathbf{x}(0) = \mathbf{x}_0 \text{ initial condition} \end{cases} \tag{1.1}$$

where the following variables are defined (same as before):

- $\mathbf{x} : \mathbb{R} \geq 0 \to \mathbb{R}^n$ with initial condition $\mathbf{x}_0 \in \mathbb{R}^n$ is the *(internal) state* of $\mathcal{H}$, of dimension $n \in \mathbb{N}$
- $\mathbf{u} : \mathbb{R} \geq 0 \to \mathbb{R}^m$ is the *control input* of $\mathcal{H}$, of dimension $m \in \mathbb{N}$
- $\mathbf{y} : \mathbb{R} \geq 0 \to \mathbb{R}^k$ is the *output* or *measurement* of $\mathcal{H}$, of dimension $k \in \mathbb{N}$
- $\mathbf{w} : \mathbb{R} \geq 0 \to \mathbb{R}^d$ is another input, called the *exogenous input*, of $\mathcal{H}$, of dimension $d \in \mathbb{N}$
- $\bar{\mathbf{z}} : \mathbb{R} \geq 0 \to \mathbb{R}^\ell$ is another output, called the *auxiliary output*, of $\mathcal{H}$, of dimension $\ell \in \mathbb{N}$

Here, f, g, and h are possibly nonlinear functions. In this book, however, we are primarily interested in *linear systems*, in which the state-space model is expressed as

$$\mathcal{H} \triangleq \begin{cases} \dot{\mathbf{x}}(t) = A\mathbf{x}(t) + B_1\mathbf{w}(t) + B_2\mathbf{u}(t), \\ \bar{\mathbf{z}}(t) = C_1\mathbf{x}(t) + D_{11}\mathbf{w}(t) + D_{12}\mathbf{u}(t), \\ \mathbf{y}(t) = C_2\mathbf{x}(t) + D_{21}\mathbf{w}(t) + D_{22}\mathbf{u}(t), \\ \mathbf{x}(0) = \mathbf{x}_0 \text{ initial condition} \end{cases} \tag{1.2}$$

Here, $A \in \mathbb{R}^{n\times n}$, $B_2 \in \mathbb{R}^{n\times m}$, and the other coefficient matrices (defined with the appropriate dimensions) are called the *system matrices* of (1.2).

In the following, we present a few examples of linear systems and their system matrices. Although many systems in the real world are nonlinear, one can convert a nonlinear system (1.1) to a linear description (1.2) using a technique called *linearization*, which will be discussed in detail in Chap. 2.

Example 1.1 (*Simple Pendulum*) Consider a simple pendulum suspended from a point in space. The bob has mass m and the string attached to it has length L. Let $\theta \in \mathbb{R}$ represent the angle that the string of the pendulum makes with the vertical line pointing downwards from the suspension point. See Fig. 1.2 for a visualization. Applying Newton's second law along the x-axis of this system gives us the system dynamics:

$$-mg\sin\theta \triangleq F_x \triangleq ma = mL\ddot{\theta} \implies \ddot{\theta} = -\frac{g}{L}\sin\theta \tag{1.3}$$

Note that this is a nonlinear system due to the sinusoidal term, which requires us to perform linearization to obtain a linear description. Define the state as $\mathbf{x}(t) \triangleq \left[\theta(t)\ \dot{\theta}(t)\right]^\top$. We use the small-angle approximation $\sin\theta \approx \theta$ and get

$$\dot{\mathbf{x}}(t) = \begin{bmatrix} 0 & 1 \\ -\frac{g}{L} & 0 \end{bmatrix} \mathbf{x}(t) \tag{1.4}$$

which is in the form of $\dot{\mathbf{x}}(t) = A\mathbf{x}(t)$, with the other matrices equal to 0. In particular, since there is no control input required to drive the system, this system is said to be an *uncontrolled system.* □

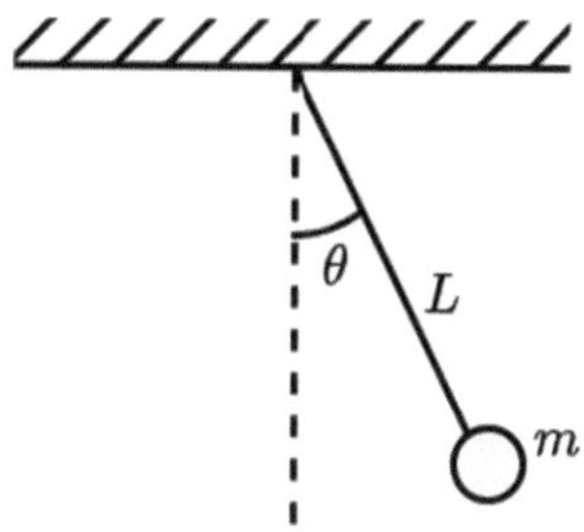

Fig. 1.2 The simple pendulum of Example 1.1

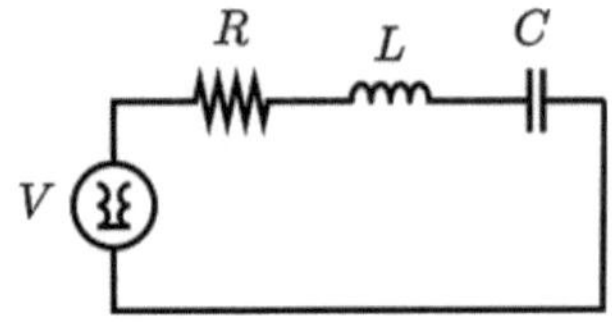

Fig. 1.3 The simple RLC series circuit of Example 1.2

Example 1.2 (*RLC Series Circuit*) Consider a standard simple RLC series circuit composed of a resistor with resistance R, inductor with inductance L, and capacitor with capacitance C connected in series to a constant voltage source V. See Fig. 1.3 for a visualization.

By Kirchoff's voltage law, we know that the sum of the voltages across each component in the circuit must equal the total voltage:

$$V = V_R(t) + V_L(t) + V_C(t) = RI(t) + L\dot{I}(t) + V_C(0) + \frac{1}{C}\int_0^t I(s)ds$$

$$\implies RI(t) + L\dot{I}(t) + V_C(0) + \frac{1}{C}\int_0^t I(s)ds$$

Taking the derivative across the entire equation yields $0 = R\dot{I}(t) + L\ddot{I}(t) + (1/C)I(t)$. Define the state to be $\mathbf{x}(t) = \begin{bmatrix} I(t) & \dot{I}(t) \end{bmatrix}^\top$. Then the dynamics can be rewritten as

$$\dot{\mathbf{x}}(t) = \begin{bmatrix} 0 & 1 \\ -\frac{1}{LC} & -\frac{R}{L} \end{bmatrix} \mathbf{x}(t) \tag{1.5}$$

which is again in the form of $\dot{\mathbf{x}}(t) = A\mathbf{x}(t)$, and uncontrolled system. Note that since this circuit is already linear, no linearization is needed. □

Why are linear systems like Examples 1.1 and 1.2 so appealing to study? There are two main, related reasons.

1. They are easy to solve directly.
2. They mostly require only linear algebra and differential equations background to analyze.

To understand this simplicity, let us consider the scalar uncontrolled system $\dot{x}(t) = ax(t)$ and $x(0) = x_0$, where $x(t), a \in \mathbb{R}$ for all t. We can solve this ODE directly by separation of variables:

$$\frac{dx(t)}{dt} = ax(t) \implies \frac{dx(t)}{x(t)} = adt \implies \ln x(t) - \ln x_0 = at \implies x(t) = e^{at}x_0$$

and this gives an explicit solution trajectory $x(t)$ over time t. We can conduct a simple analysis of this solution form: the "stability" of the system depends on the sign of a. Namely, if $a > 0$, then $x(t) \to \infty$ as $t \to \infty$ but if $a < 0$, then $x(t) \to 0$. We discuss stability of linear systems more formally in Part II.

When a control term is introduced, we are still able to solve the scalar linear system explicitly. Consider the following *controlled* system

$$\dot{x}(t) = ax(t) + bu(t), \quad x(0) = x_0$$

where $x(t), u(t), a, b \in \mathbb{R}$ for all t. Using integrating factor $\exp(-\int_0^t adt) = e^{-at}$:

$$\begin{aligned}\frac{dx(t)}{dt} = ax(t) + bu(t) \quad &\Longrightarrow \quad \left(\frac{dx(t)}{x(t)} - ax(t)\right) e^{-at} = bu(t)e^{-at} \\ &\Longrightarrow \quad x(t) = e^{at}x_0 + \int_0^t e^{a(t-s)}bu(s)ds\end{aligned}$$

Note that compared to the uncontrolled case, the controlled system has an additional convolution integral term $\int_0^t e^{a(t-s)}bu(s)ds$. Essentially, this term accumulates the entire past history of the inputs $\{u(s) : 0 \leq s \leq t\}$ and stores this information in the state $x(t)$. This gives us a nice conceptual interpretation of the *state* $x(t)$:

> The *state* is an internal variable whose values $\{\mathbf{x}(s) : 0 \leq s \leq t\}$ together with "future" input $\mathbf{u}(t)$ are enough to determine the system output $\mathbf{y}(t)$.

Even in the vector case, the system can be solved directly via the integrating factor technique. For the uncontrolled system, we have

$$\dot{\mathbf{x}}(t) = A\mathbf{x}(t) \Longrightarrow e^{-At}\dot{\mathbf{x}}(t) - e^{-At}A\mathbf{x}(t) = 0 \Longrightarrow \mathbf{x}(t) = e^{At}\mathbf{x}_0$$

and for controlled systems (with $B_2 \equiv B$ and the other matrices in (1.2) still equal to 0), we have

$$\begin{aligned}\dot{\mathbf{x}}(t) = A\mathbf{x}(t) + B\mathbf{u}(t) &\Longrightarrow e^{-At}(\dot{\mathbf{x}}(t) - A\mathbf{x}(t)) = e^{-At}B\mathbf{u}(t) \\ &\Longrightarrow \mathbf{x}(t) = e^{At}\mathbf{x}_0 + \int_0^t e^{A(t-s)}B\mathbf{u}(s)ds\end{aligned}$$

The *matrix exponential* e^{At} is a special type of square matrix we will investigate in the subsequent Chap. 2.

Remark 1.1 Every concept discussed throughout this subsection can be extended to the discrete-time (DT) setting as well. The main difference is that the differential equations become *difference* equations.

For the majority of this book until Part IV, we will focus on linear systems without noise $\mathbf{w}$ and without considering the auxiliary output $\bar{\mathbf{z}}$. Consequently, we will remove the subscripts in our system matrices in order to simplify notation.

$$\mathcal{H} \triangleq \begin{cases} \dot{\mathbf{x}}(t) = A\mathbf{x}(t) + B\mathbf{u}(t), \\ \mathbf{y}(t) = C\mathbf{x}(t) + D\mathbf{u}(t), \\ \mathbf{x}(0) = \mathbf{x}_0 \text{ initial condition} \end{cases} \tag{1.6}$$

A fundamental question for any linear system (and nonlinear systems too) is its *realizability*, which refers to the ability to implement a given state-space model using physical components or a practical control system. It is important because it ensures that the theoretical models we use to describe and design systems can actually be constructed in the real world.

Definition 1.1 *(Realization)* A *realization* of a CT (linear) system is the representation of its input-output behavior in terms of a state-space model $\{A(t), B(t), C(t), D(t)\}$, meaning

$$\dot{\mathbf{x}}(t) = A(t)\mathbf{x}(t) + B(t)\mathbf{u}(t)$$
$$\mathbf{y}(t) = C(t)\mathbf{x}(t) + D(t)\mathbf{u}(t)$$

Note that a realization does not have to be linear. Realizations also exist for DT systems, and are defined similarly to their CT counterparts.

Definition 1.2 *(Realizability)* Input-to-output transfer function $H(s)$ for a CT LTI system is said to be *realizable* if there exists a finite-dimensional state-space $\{A, B, C, D\}$ such that $H(s) = C(sI - A)^{-1}B + D$. We have a similar definition for DT LTI systems, i.e., $H(z) = C(zI - A)^{-1}B + D$.

Loosely-speaking, a system is realizable if there exists a physical or computational setup (such as a set of actuators, sensors, or algorithms) that can reproduce the behavior described by the model.

Remark 1.2 We mention one caveat regarding the notion of *realizability* is the literature. Some may argue that improper transfer functions still have a valid realization. For example, the system realization $y(t) = \dot{u}(t)$ has transfer function $y(s)/u(s) = s$, which is clearly improper. However, it is difficult to implement a perfect derivative in reality, and it is this practicability that influences our choice to focus only on realizations for proper transfer matrices.

1.2 Transfer Function Description

Although most of our discussion in this book will be focused on the state-space model representation, it is worth mentioning a few remarks about the concepts that are widely used in the study of transfer function representations. Transfer functions are greatly used in the study of preliminary topics related control theory, such as signals and systems, the frequency domain, and various common transforms (e.g., Fourier, Laplace, z).

We first define a few common signals.

Definition 1.3 *(Continuous-time Impulse)* The *continuous-time impulse function* (also called the *Dirac Delta*) is the function $\delta(t)$ can be defined as the following limit:

$$\delta(t) \triangleq \lim_{\Delta \to 0} h_\Delta(t) = \lim_{\Delta \to 0} g_\Delta(t) = \cdots$$

where

$$h_\Delta(t) \triangleq \begin{cases} \frac{1}{\Delta} & (\text{if } -\frac{\Delta}{2} < t < \frac{\Delta}{2}) \\ 0 & (\text{otherwise}) \end{cases}, \quad g_\Delta(t) \triangleq \begin{cases} \frac{1}{\Delta} - \frac{1}{\Delta^2}t & \text{if } 0 < t < \Delta \\ \frac{1}{\Delta} + \frac{1}{\Delta^2}t & \text{if } -\Delta < t < 0 \\ 0 & \text{otherwise} \end{cases}$$

are a few common limiting functions that are used in the construction of the Dirac delta. Note that this means the method of constructing δ is not unique; we can use any sequence curve such that the area underneath is always 1. See Fig. 1.4 for visualization.

The Dirac delta has a few useful properties:

1. the *sifting property*: $\int_{-\infty}^{\infty} f(t)\delta(t-\tau)dt \triangleq \lim_{\Delta \Longrightarrow 0} \int_{-\infty}^{\infty} f(t)h_\delta(t-\tau)dt = f(\tau)$
2. the *Laplace transform is constant*: $\mathcal{L}\{\delta(t)\} \triangleq \int_{0-}^{\infty} \delta(t)e^{-st}dt = 1$

Definition 1.4 *(Unit Step)* The *unit step function* is defined as

$$\theta(t) = \begin{cases} 0 & \text{if } t < 0 \\ 1 & \text{if } t \geq 0 \end{cases}$$

Note that $\theta(t)$, which is a discontinuous function, can be constructed from a limit of continuous functions

$$g_\Delta(t) \triangleq \begin{cases} 0 & \text{if } t < -\Delta \\ \frac{1}{\Delta}(t+\Delta) & \text{if } -\Delta \leq t < 0 \\ 1 & \text{if } t \geq 0 \end{cases},$$

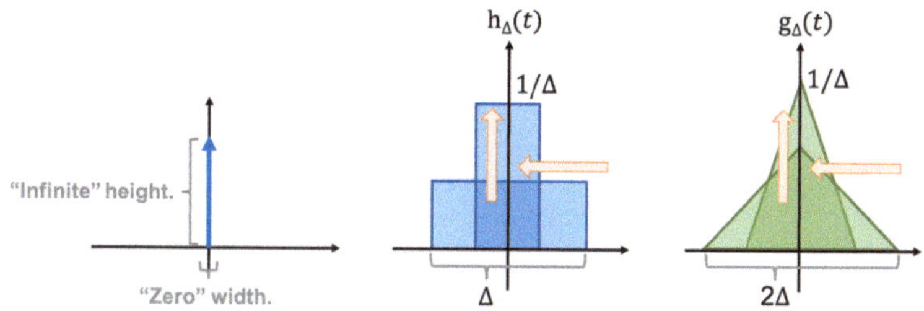

Fig. 1.4 The continuous-time impulse function (i.e., Dirac delta). [Left] Dirac delta. [Middle] The construction of the Dirac delta as the limit of rectangles. [Right] The construction of the Dirac delta as the limit of triangles, which further emphasizes that the construction of the Dirac delta is not unique

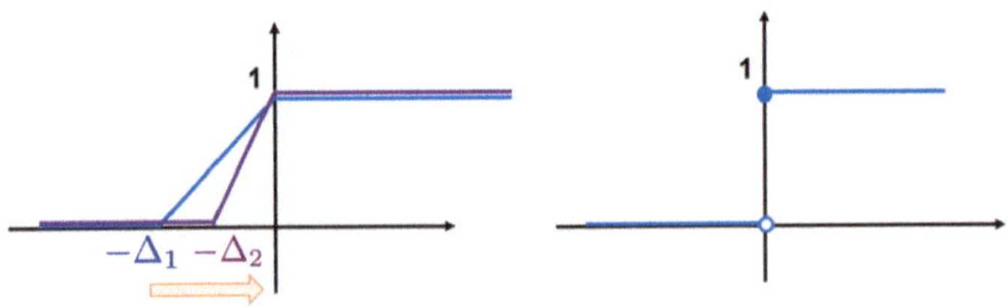

Fig. 1.5 The continuous-time unit step. [Left] The limit of a sequence of continuous functions g_Δ. [Right] The unit step function, which g_Δ approaches as $\Delta \to 0$

as $\theta(t) \triangleq \lim_{\Delta \to 0} g_\Delta(t)$. A visualization is shown in Fig. 1.5.

Definition 1.5 *(Impulse and Step Responses)* The *impulse response* (typically denoted $h(t)$) of a system $\mathcal{H}$ is the output signal when input $u(t) = \delta(t)$. Likewise, the *step response* is the output when the input is the step function $\theta(t)$.

If impulse response $h(t)$ is known for system $\mathcal{H}$, then the output of any general input $u(t)$ can be determined via convolution

$$(f * g)(t) := \int_{-\infty}^{\infty} f(\tau)g(t-\tau)d\tau.$$

This can be seen as follows:

$$u(t) = \int_{-\infty}^{\infty} u(\tau)\delta(t-\tau)d\tau,$$

$$y(t) = \int_{-\infty}^{\infty} u(\tau)\mathcal{H}(\delta(t-\tau))d\tau = \int_{-\infty}^{\infty} u(\tau)h(t-\tau)d\tau.$$

Here, $\int_{-\infty}^{\infty} u(\tau)\delta(t-\tau)d\tau$ can be viewed as a superposition of shifted impulse functions, where $\delta(t-\tau)$ is weighted by $u(\tau)$.

We also note the following relationship by recalling that the Laplace transform of a convolution is multiplication:

$$y(t) = \int_{-\infty}^{\infty} u(\tau)h(t-\tau)d\tau \implies Y(s) = U(s)H(s), \quad \therefore H(s) = \frac{Y(s)}{U(s)}$$

Here, Y and U are the Laplace transforms of y and u, respectively, and $H(s)$ is called the *input-to-output transfer function*. It is exactly the Laplace transform of the impulse response.

Let's compute the impulse response of our system by substituting the shifted impulse function $u(t) = \delta(t-\tau)$, for some time shift $\tau \in \mathbb{R}$. By taking the Laplace transform across our system dynamics (1.6), we get the transfer function representation:

$$X(s) = (sI - A)^{-1}(x_0 + Bu(s)), \quad Y(s) = C(sI - A)^{-1}(x_0 + Bu(s)) + Du(s). \tag{1.7}$$

By taking the inverse Laplace transform across (1.7), we can obtain the solution of $\mathbf{x}(t)$ and $\mathbf{y}(t)$ as functions of the input $\mathbf{u}(t)$ and initial condition $\mathbf{x}_0$.

Let's return to the time domain and derive the solution to (1.6) that way. Using the integrating factor method similar to what we did before in the scalar case, we get the solution of $\dot{\mathbf{x}}(t) = A\mathbf{x}(t) + B\mathbf{u}(t)$:

$$\mathbf{x}(t) = \Phi(t, t_0)\mathbf{x}_0 + \int_{t_0}^{t} \Phi(t, s)B\mathbf{u}(s)ds. \tag{1.8}$$

Here, we usually call $\Phi(t, \tau) \triangleq e^{A(t-\tau)}$ as the *state-transition matrix*, which propagates initial state $\mathbf{x}_0$ along forward in time.

Substituting in (1.8), the measurement equation can be expressed as:

$$\begin{aligned} \mathbf{y}(t) &= C\mathbf{x}(t) + D\mathbf{u}(t) \\ &= C\Phi(t, t_0)\mathbf{x}_0 + \int_{t_0}^{t} C\Phi(t, s)B\mathbf{u}(s)ds + D\mathbf{u}(t) \\ &= C\Phi(t, t_0)\mathbf{x}_0 + \int_{t_0}^{t} (C\Phi(t, s)B + D\delta(t-s))\,\mathbf{u}(s)ds \end{aligned} \tag{1.9}$$

where $\delta(t-s)$ is the shifted Dirac delta. Thus, if we recall that $\mathcal{L}\{e^{At}\} = (sI - A)^{-1}$, the expressions in (1.7) make more sense.

Substituting in (1.8) and (1.9) in the time-domain yields:

$$h(t) := \int_{t_0}^{t} (C\Phi(t, s)B + D\delta(t-s))\,\delta(s-\tau)ds = \begin{cases} C\Phi(t, s)B & \text{if } t \neq \tau \\ C\Phi(t, s)B + D & \text{if } t = \tau \end{cases}$$

Definition 1.6 *(Zero-Input and Zero-State Responses)* $\Phi(t, t_0)x_0$, which captures the system response without any control inputs, is called the *zero-input response*, and $\int_{t_0}^{t} \Phi(t, s)Bu(s)ds$, which captures how the input affects the system, is called the *zero-state response*.

Remark 1.3 Discrete-time systems can also be characterized by its impulse response. The discrete-time impulse function δt is simply a stem of height 1 positioned at $t = 0$, and so it is obvious that any discrete-time signal can be written as a linear combination of shifted impulses, $\mathbf{u}_t = \sum_{k=0}^{\infty} \mathbf{u}_k \delta_{t-k}$. Moreover, the output to any input can be written as a convolution $y_t = \sum_{k=0}^{\infty} \mathbf{u}_k h_{t-k}$, and after applying the z-transform, $Y(z) = U(z)H(z)$.

1.3 Classification of Systems

Systems can be categorized in many different ways.

Definition 1.7 *(CT versus DT)* A system is *continuous-time (CT)* if all its signals are CT. Likewise, a system is *discrete-time (DT)* if all its signals are DT.

We will henceforth abbreviate continuous-time as CT and discrete-time as DT.

As a brief digression, there are also *hybrid systems*, where both CT and DT signals are involved in the system. An example of this is a thermostat with two discrete modes (`ON`/`OFF`), trying to regulate a room around the temperature $C \in [\underline{C}, \overline{C}]$ degrees. Here, $\underline{C}$ and $\overline{C}$ are the minimum and maximum possible temperatures that the thermostat can take, respectively.

Let $x(t)$ be the temperature of the room at time t. When the thermostat is `OFF`, the temperature evolves according to the dynamics $\dot{x}(t) = -\alpha(x(t) - \underline{C})$, as it starts to decrease towards $\underline{C}$. When the thermostat is `ON`, the temperature evolves according to the dynamics $\dot{x}(t) = -\alpha(x(t) - \overline{C})$, as it starts to increase towards $\overline{C}$. Here, $\alpha > 0$ is the rate of the temperature's change.

The state of the overall system can be represented in two parts: $x(t) \in \mathbb{R}^{\geq 0}$ is the continuous temperature of the system, while $q(t) \in \{\texttt{ON}, \texttt{OFF}\}$ is the discrete mode. In fact, $q(t)$ can still be interpreted as a CT signal, but it can be treated like a discrete object; this is shown in Fig. 1.6.

Definition 1.8 *(SISO versus MIMO)* A system is *single-input, single-output (SISO)* if it only has one input $u(t)$ and one output $y(t)$. Likewise, it is *multi-input, multi-output (MIMO)* if it only has multiple inputs $u_1(t), \ldots, u_M(t)$ and multiple outputs $y_1(t), \ldots, y_N(t)$.

Remark 1.4 Both SISO and MIMO systems can be related to each other by a vector representation. For example, a vector state $\mathbf{x}(t) \in \mathbb{R}^n$ can be thought of as a single vector input, or multiple scalar inputs $\mathbf{x}(t) \triangleq [x_1(t), \ldots, x_n(t)]^\top$.

Given Remark 1.4, one might wonder how to tell the difference between a SISO and a MIMO system. In general, it depends on the application of interest. For example, wireless communication involving multiple transmitters and receivers requires different hardware than wireless communication involving a single transmitter and receiver pair. Written mathematically, however, the representation of each system may seem the same.

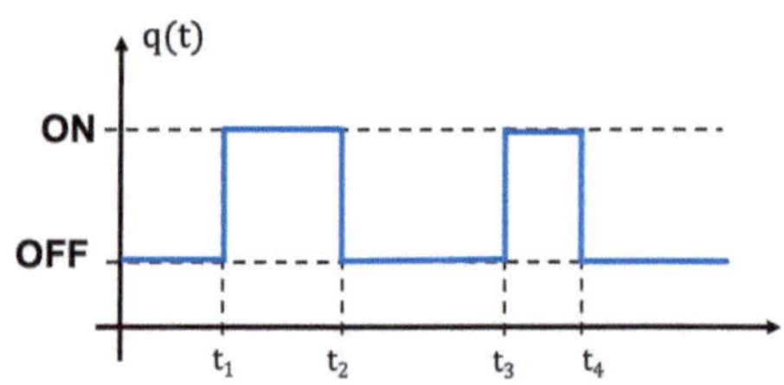

Fig. 1.6 A sample trajectory of $q(t)$, a CT signal that can be interpreted in DT by a choice of "switching times" $\{t_i\}$

Definition 1.9 *(Linear)* System $\mathcal{H}$ is said to be *linear* if it satisfies the following two properties:

1. *additivity*: if $y_1(t) = \mathcal{H}\{x_1\}(t)$ and $y_2(t) = \mathcal{H}\{x_2\}(t)$, then $y_1(t) + y_2(t) = \mathcal{H}\{x_1 + x_2\}(t)$.
2. *homogeneity*: if $y(t) = \mathcal{H}\{x\}(t)$, then for all scalars $\alpha \in \mathbb{R}$, $\alpha y(t) = \mathcal{H}\{\alpha x\}(t)$.

Combined together, a system is linear iff $\alpha y_1(t) + \beta y_2(t) = \mathcal{H}\{\alpha x_1 + \beta x_2\}(t)$ for any scalars $\alpha, \beta \in \mathbb{R}$ and any inputs $x_1, x_2 : \mathbb{R} \to \mathbb{R}$ with respective outputs $y_i = \mathcal{H}\{x_i\}, \ \forall\, i = 1, 2$.

Definition 1.10 *(TI versus TV)* For any signal f, define the time-shift operator as $\mathcal{S}_\tau\{f\}(t) \triangleq f(t-\tau) \ \ \forall t \in \mathbb{R}$. System $\mathcal{H}$ is *time-invariant (TI)* if for any input signal x, it commutes with $\mathcal{S}_\tau$ for any τ:

$$y(t) = \mathcal{H}\{x\}(t) \implies \mathcal{S}_\tau\{y\}(t) = \mathcal{S}_\tau\{\mathcal{H}\{x\}\}(t) = \mathcal{H}\{\mathcal{S}_\tau\{x\}\}(t), \ \forall\, \tau, t \in \mathbb{R}$$

and it is *time-varying (TV)* if this property is not satisfied.

Example 1.3 (*TI versus TV*) The squaring system, defined by $y(t) = x^2(t)$, is TI, while the system $y(t) = tx^2(t)$ clearly isn't. □

Definition 1.11 *(Causal)* System $\mathcal{H}$ is *causal* if its output depends only on the past and present input values.

$$\forall\, \tau \in \mathbb{R}, \quad \begin{cases} y_1(t) = \mathcal{H}(x_1(t)) \\ y_2(t) = \mathcal{H}(x_2(t)) \\ \text{where } x_1(t) = x_2(t) \ \ \forall\, t \le \tau \end{cases} \implies y_1(\tau) = y_2(\tau)$$

Example 1.4 The following two systems are respectively classified as

$$y(t) = \int_{t-1}^{t+1} u(s)\, ds \quad \text{is noncausal,} \qquad y(t) = \begin{cases} 2u(t) + 3 & \text{if } t \ge 1 \\ 0 & \text{else} \end{cases} \quad \text{is causal}$$

□

Definition 1.12 *(Memoryless)* CT system $\mathcal{H}$ is *memoryless* if its output $y(t)$ depends only on the input $x(t)$ at the same time $t \in \mathbb{R}$.

Note that memoryless systems are always causal, but causal systems are not always memoryless since they might also depend on past inputs.

Another way of defining a memoryless system is whether it requires "memory" of "any variables" in order to influence the current output y(t). For example, $y(t) = u(t)^2$ is a memoryless system, while $y(t) = u(t) + u(t-1)$ is not. Pure functions of time (e.g., $y(t) = t$) are not memoryless because it requires you to store the time variable t into memory. However, $y(t) = t$ "is" memoryless if your input signal itself is $u(t) = t$. For the same reason, $y(t) = tu(t)$ is also not memoryless for general inputs $u(t)$.

Definition 1.13 *(Lumped versus Distributed)* System $\mathcal{H}$ is *lumped* (or *lumped-parameter*) if its state-space is finite-dimensional. On the other hand, $\mathcal{H}$ is *distributed (or distributed-parameter)* if its state-space is infinite-dimensional.

Lumped systems are typically modeled where dependent variables of interest are a function of *time alone*. For instance, the pendulum and RLC circuit examples examined in Examples 1.1 and 1.2 are both lumped systems. Distributed systems have dependent variables which are functions of time and one or more spatial variables (e.g., partial differential equations). However, lumped approximations to distributed systems can be made.

Example 1.5 The following systems have the respective classifications

$$y(t) = \int_{t-1}^{t+1} u(s)\, ds \quad \text{is distributed,}$$

$$\frac{\partial y(t,x)}{\partial t} = \frac{\partial^2 y(t,x)}{\partial x^2} + u(t,x) \quad \text{is also distributed}$$

The second system is a well-known PDE called the Heat Equation. It can have an approximate lumped representation via sampling the infinite-dimensional, continuous state $y(t,x)$ at specific points in space $\{x_1, \ldots, x_N\}$. See Fig. 1.7 for a visualization. □

Throughout this textbook, we will mostly focus on lumped systems.

As a brief remark, many concepts in control systems engineering use the terminology "distributed" to mean "distributed control", which differs from "centralized control" in the following way. *Centralized control* refers to a system where a single central unit or authority makes decisions and manages the entire operation, with all control functions concentrated in one location. In mathematical terms, we are designing a single $\mathbf{u}$ signal for entire system $\dot{\mathbf{x}} = A\mathbf{x} + B\mathbf{u}$. Many introductory control systems courses (including linear systems) focus primarily on the centralized control setting. In contrast, *distributed control* involves multiple independent units or controllers that work collaboratively, with decision-making and control tasks distributed across different nodes or agents. In mathematical terms, we are designing

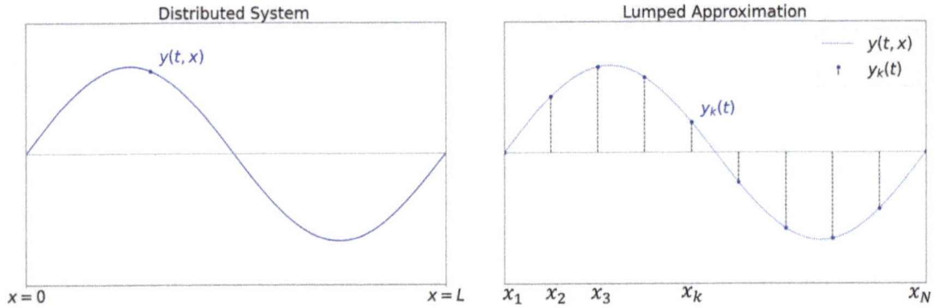

Fig. 1.7 A lumped approximation [Right] to a distributed system [Left]

a $\mathbf{u}_i$ for local subsystems, represented by $\dot{\mathbf{x}}_i = A_i\mathbf{x} + B_i\mathbf{u}_i$, where the full state is given by $\mathbf{x} \triangleq (\mathbf{x}_1^\top, \ldots, \mathbf{x}_N^\top)$ and $N \in \mathbb{N}$ is the number of subsystems.

While centralized control offers simplicity and easier coordination, it may suffer from bottlenecks and single points of failure. Distributed control, on the other hand, is more scalable and flexible, especially for larger systems. However, it can be more complicated to implement due to the need for communication and coordination among the decentralized units.

The terminology "distributed" as described above is more relevant to decentralized control, multi-agent systems, and other related subfields. However in the context of Definition 1.13, "distributed" means something else. note that both centralized and decentralized/distributed systems are lumped-parameter systems, because their state admits a finite-dimensional vector representation.

Definition 1.14 *(Internal versus External)* Another way to categorize system models is *internal* versus *external models*.

- Internal models…
  - …describe the input, output signals like $\mathbf{u}$ and $\mathbf{y}$, as well as all relevant internal state variables $\mathbf{x}$.
  - …are more suitable for representing complex systems, including nonlinear, time-varying, and more. Thus, it is often used in modern control approaches. For example, state-space models are one of the most common types of internal models.
- External models…
  - …describe only how the input $\mathbf{u}$ affects the output $\mathbf{y}$. The system itself is viewed as a black box functions.
  - …is typically only valid for LTI systems (although some research is being done towards extending external model representation beyond LTI systems). For example, transfer function models are one of the most common types of external models.

1.4 Mathematical Reviews

To conclude this introductory chapter, we provide a brief review of some of the linear algebra and analysis background that will be useful for the rest of the textbook.

1.4.1 Matrix Properties, Matrix Algebra

We start with some elementary properties and operations on matrices.

Definition 1.15 *(Row Echelon Form (REF))* A matrix is in *row echelon form* if the followings hold.

- Rows with zero entries must be below any row with entries.
- The leading entry (that is, the left-most nonzero entry) of every nonzero row, called the pivot, is on the right of the leading entry of every row above.

Many properties of matrices may be easily deduced from their row echelon form, such as the rank and the kernel.

Definition 1.16 *(Reduced Row Echelon Form (RREF))* A matrix is in *reduced row echelon form* if the followings hold.

- It is in row echelon form.
- The leading entry in each nonzero row is 1 (called a leading one).
- A leading one is the only non-zero entry in its column.

Definition 1.17 *(Elementary Row Operations)* There are three types of elementary row operations.

- Swap two rows of a matrix.
- Multiply a nonzero scalar to a row of a matrix.
- Given two different rows r_1 and r_2, change r_2 into $r_2 + cr_1$ where c is a scalar.

Fact: One can use elementary row operations to reduce any matrix into a reduced row echelon form. While a matrix may have several row echelon forms, it can have only one unique reduced row echelon form.
Solving linear equations of the form $A\mathbf{x} = \mathbf{y}$: Form the augmented matrix $[A|y]$, and reduce this augmented matrix into a row-reduced echelon form; after being reduced, the system is trivial to solve.

Definition 1.18 Let A be an m-by-n matrix with entries in the set $\mathbb{F}$ where $\mathbb{F} = \mathbb{R}$ or $\mathbb{F} = \mathbb{C}$. The column space or the range space of A is defined as the set of all possible linear combinations of the columns of A. The dimension of the column space is called the rank of the matrix A. The kernel or nullspace of A is defined as $\ker(A) \ = \{v \in \mathbb{F}^n : Av = 0\}$. The dimension of $\ker(A)$ is called the nullity of A.

Definition 1.19 The row space of A is defined as the set of all possible linear combinations of the rows of A.

Theorem 1.1 ("row rank" equals "(column) rank") *The dimension of the row space of A equals the rank of A.*

Theorem 1.2 (Rank-Nullity Theorem) *Let A be an m-by-n matrix with entries in the set $\mathbb{F}$ where $\mathbb{F} = \mathbb{R}$ or $\mathbb{F} = \mathbb{C}$. Then, we have*

$$\text{rank}(A) + \text{nullity}(A) = n. \tag{1.10}$$

Definition 1.20 *(Determinant)* Let A be an n-by-n square matrix with complex entries.

$$\det(A) = \sum_{\sigma \in S_n} \operatorname{sgn}(\sigma) \prod_{i=1}^{n} A_{i,\sigma(i)} \tag{1.11}$$

where S_n is the set of all permutations on the set $\{1, 2, \ldots, n\}$, i.e., set of all bijections from the set $\{1, 2, \ldots, n\}$ onto itself. Here, $\operatorname{sgn}(\sigma) = 1$ for even permutations $\sigma \in S_n$ and $\operatorname{sgn}(\sigma) = -1$ for odd permutations $\sigma \in S_n$.

Theorem 1.3 *The function* det *is multilinear as a function on the columns of the input matrix, i.e., as* $\det : \mathbb{C}^n \times \cdots \times \mathbb{C}^n \to \mathbb{C}$. *Also,* $\det(A)$ *is zero whenever* A *has two identical columns. Finally,* $\det(\mathrm{id}_n) = 1$.[1]

Definition 1.21 *(Trace)* Let A be an n-by-n square matrix with complex entries. The *trace* of A is defined as $\operatorname{tr}(A) = \sum_{i=1}^{n} A_{ii}$.

Proposition 1.1 (Cyclic invariance of trace) *Let A_i be n-by-n square matrices with complex entries. Then, the following equality holds.*

$$\operatorname{tr}(A_1 A_2 A_3 \cdots A_n) = \operatorname{tr}(A_2 A_3 \cdots A_n A_1) = \operatorname{tr}(A_3 A_4 \cdots A_n A_1 A_2) = \cdots \tag{1.12}$$

Definition 1.22 *(Inverse matrices)* Let A be an n-by-n square matrix with complex entries. A complex matrix B with the same size is said to be the inverse of A if $AB = BA = \mathrm{id}_n$.

Proposition 1.2 *The following hold.*

- *If an inverse matrix exists, it is unique. The unique inverse of A (if exists) is denoted as A^{-1}.*
- *If a square matrix A has real entries and A^{-1} exists, then A^{-1} has real entries, too.*
- *If A, B are n-by-n matrices and $AB = \mathrm{id}$, then A and B are inverses to each other.*

1.4.2 Spaces and Basis

Several types of mathematical spaces will be useful in our study of linear systems.

Definition 1.23 *(Field)* Define $\mathbb{F} \triangleq (F, +, \cdot)$ such that F is a set containing at least two elements, $+ : F \times F \to F$ is an addition operator, and $\cdot : F \times F \to F$ is a multiplication operator. Then $\mathbb{F}$ is called a *field* if the following axioms hold:

(A) • uniqueness: $\forall\, a, b \in F$, $a + b$ is uniquely defined.
• commutativity: $a + b = b + a$.

---

[1] Actually, these three properties uniquely determine the function $\det : \mathbb{C}^n \times \cdots \times \mathbb{C}^n \to \mathbb{C}$.

- associativity: $(a + b) + c = a + (b + c)$, where $c \in F$ also.
- additive identity: $\exists\, 0 \in F$ s.t. $\forall\, a \in F, 0 + a = a$.
- additive inverse: $\forall\, a \in F,\ \exists\, (-a) \in F$ s.t. $a + (-a) = 0$.

(M)
- uniqueness: $\forall\, a, b \in F, a \cdot b (\equiv ab$ for simplicity)) is uniquely defined.
- commutativity: $ab = ba$.
- associativity: $(ab)c = a(bc)$, where $c \in F$ also.
- multiplicative identity: $\exists\, 1 \in F$ s.t. $\forall\, a \in F, 1a = a$.
- multiplicative inverse: $\forall\, a \in F,\ \exists\, a^{-1} \in F$ s.t. $aa^{-1} = 1$.
- "non-degenerate": $0 \neq 1$.

(D) distributivity: $a(b + c) = ab + ac$.

Remarks It can be shown that the identity elements 0 and 1 are unique. Furthermore, if $ab = 0$, then $a = 0$ or $b = 0$.

Definition 1.24 Define $\mathcal{V} \triangleq (V, \mathbb{F}, +, \cdot)$ such that V is a set of elements (vectors), $+ : V \times V \to V$ is an addition operator, and $\cdot : F \times V \to V$ is a scalar multiplication operator. Then $\mathcal{V}$ is called a *vector space* if the following axioms hold:

(A)
- commutativity: $\mathbf{v} + \mathbf{w} = \mathbf{w} + \mathbf{v}$.
- associativity: $(\mathbf{v} + \mathbf{w}) + \mathbf{u} = \mathbf{v} + (\mathbf{w} + \mathbf{u})$, where $\mathbf{u} \in V$ also.
- additive identity: $\exists\, 0 \in V$ s.t. $\forall\, \mathbf{v} \in V, 0 + \mathbf{v} = \mathbf{v}$.
- additive inverse: $\forall\, \mathbf{v} \in V,\ \exists\, (-\mathbf{v}) \in V$ s.t. $\mathbf{v} + (-\mathbf{v}) = 0$.

(SM)
- associativity: $(ab)\mathbf{v} = a(b\mathbf{v})$, where $b \in \mathbb{F}$ also.
- distributivity 1: $a(\mathbf{v} + \mathbf{w}) = a\mathbf{v} + a\mathbf{w}$.
- distributivity 2: $(a + b)\mathbf{v} = a\mathbf{v} + b\mathbf{v}$.
- multiplicative identity: $\exists\, 1 \in \mathbb{F}$ such that $1\mathbf{v} = \mathbf{v}, \forall\, \mathbf{v} \in V$.

Definition 1.25 *(Subspaces)* A *subspace* $\mathcal{W}$ of a vector space $\mathcal{V}$ is a subset of vectors that itself forms a vector space. A necessary and sufficient condition for a nonempty subset to form a subspace is that it must be closed w.r.t. vector addition and scalar multiplication.

Definition 1.26 *(Linear (In)dependence)* Let $\mathcal{V}$ be a vector space over field $\mathbb{F}$ and let $\mathcal{W} \triangleq \{\mathbf{v}_1, \ldots, \mathbf{v}_n\} \subset \mathcal{V}$ be a subset of vectors. $\mathcal{W}$ is *linearly dependent* if there exist $c_1, \ldots, c_n \in \mathbb{F}$ not all 0 such that

$$c_1\mathbf{v}_1 + \cdots + c_n\mathbf{v}_n = 0$$

$\mathcal{W}$ is *linearly independent* if the only scalars which satisfy the condition above is $c_1 = \cdots = c_n = 0$.

Definition 1.27 *(Dimension)* The maximal number of linearly independent vectors in a vector space $\mathcal{V}$ is the *dimension* of $\mathcal{V}$ (i.e., dim $\mathcal{V}$).

Definition 1.28 *(Basis and Span)* A set $\mathcal{W} \subset \mathcal{V}$ of vector space $\mathcal{V}$ is a *basis* if every vector in $\mathcal{V}$ can be expressed as a unique linear combination of the vectors in $\mathcal{W}$:

$$\mathcal{W} \triangleq \{\mathbf{v}_1, \ldots, \mathbf{v}_N\} \Longrightarrow \mathcal{V} \triangleq \{c_1\mathbf{v}_1 + \cdots + \mathbf{c}_N\mathbf{v}_N : c_1, \ldots, c_N \in \mathbb{F}\}$$

For example, $\{\mathbf{e}_!, \ldots, \mathbf{e}_n\} \triangleq \{[1, 0, \ldots, 0]^\top, [0, 1, \ldots, 0]^\top, \ldots, [0, 0, \ldots, 1]^\top\} \subset \mathbb{R}^n$ is often called the standard basis of $\mathbb{R}^n$. The *(linear) span* of a set of vectors $\mathcal{S}$ is the set of all linear combinations of the vectors in $\mathcal{S}$. For example, $\mathcal{V} = \text{span}(\mathcal{W})$.

Lemma 1.1 *Let $\mathcal{V}$ over field $\mathbb{F}$ be a vector space such that dim $\mathcal{V} = n \in \mathbb{N}$. Then any subset of $\mathcal{V}$ which contains n linearly independent vectors forms a basis for $\mathcal{V}$. (Consequently, N in the notation of $\mathcal{W}$ in Definition 1.28 is always equal to n.)*

Definition 1.29 *(Linear Operator)* Let $\mathcal{V}$ and $\mathcal{W}$ be two vector spaces over the same field $\mathbb{F}$, and define a function $T : V \to W$ such that $\mathbf{w}_i = f(\mathbf{v}_i)$ for $i = 1, 2$. Then T is said to be a *linear mapping* (or *linear transformation*) iff $T(\alpha_1\mathbf{v}_1 + \alpha_2\mathbf{v}_2) = \alpha_1\mathbf{w}_1 + \alpha_2\mathbf{w}_2$ for any $\alpha_1, \alpha_2 \in \mathbb{F}$. Furthermore, T is a *linear operator* when $\mathcal{W} = \mathcal{V}$ (a linear map which maps to the same vector space as its domain).

Definition 1.30 *(Invariant)* Let $\mathcal{V}$ be a vector space and $T : \mathcal{V} \to \mathcal{V}$ be a linear operator. A subspace $\mathcal{W} \subseteq \mathcal{V}$ is called an *invariant* subspace under T if

$$\mathbf{v} \in \mathcal{W} \Longrightarrow T(\mathbf{v}) \in \mathcal{W}$$

or equivalently, $T\mathcal{W} \subseteq \mathcal{W}$.

Theorem 1.4 (Rank-Nullity Theorem, Revisited) *Let $T : \mathcal{V} \to \mathcal{W}$ be a linear mapping between vector spaces $\mathcal{V}$ and $\mathcal{W}$ over the same field $\mathbb{F}$, and let $\mathcal{V}$ be finite-dimensional. Then Im(T) and Ker(T) are linear subspaces of $\mathcal{V}$ which are invariant under T, and*

$$\underbrace{rank(T)}_{\triangleq dimIm(T)} + \underbrace{nullity(T)}_{\triangleq dimKer(T)} = dim\mathcal{V}$$

(Because a matrix is a type of linear mapping, the same property holds–see Theorem 1.2.)

Theorem 1.5 (Range-Nullspace Decomposition) *Suppose we have the same setup as in Theorem 1.4. If, in addition, $Im(\mathcal{V}) \cap Ker(\mathcal{V}) = \{0\}$, then*

$$\mathcal{V} = Im(\mathcal{V}) \oplus Ker(\mathcal{V})$$

is a decomposition of $\mathcal{V}$ as a direct sun of subspaces invariant under T.

Lemma 1.2 (Range-Nullspace Decomposition of $\mathbb{C}^n$) *For $A \in \mathbb{C}^n$, there exists a smallest $k \in \mathbb{N}$ such that*

$$\mathbb{C}^n = Im(A^k) \oplus Ker(A^k)$$

Definition 1.31 *(Vector norms)* Let V be a vector space over $\mathbb{F} = \mathbb{R}$ or $\mathbb{F} = \mathbb{C}$. A function $\|\cdot\| : V \to \mathbb{R}$ is called a *(vector) norm* on V if the followings hold.

- $\|v\| \geq 0$ for all $v \in V$
- $\|v\| = 0$ if and only if $v = 0$
- $\|cv\| = |c| \, \|v\|$ for all $c \in \mathbb{F}$ and $v \in V$
- $\|v_1 + v_2\| \leq \|v_1\| + \|v_2\|$ for all $v_1, v_2 \in V$ (triangle inequality)

Definition 1.32 *(Inner products)* Let V be a vector space over $\mathbb{F} = \mathbb{R}$ or $\mathbb{F} = \mathbb{C}$. A function $\langle \cdot, \cdot \rangle : V \times V \to \mathbb{F}$ is called an *inner product* on V if the followings hold.

- $\langle v, w \rangle^* = \langle w, v \rangle$ where * denotes complex conjugate
- $\langle c_1 v_1 + c_2 v_2, w \rangle = c_1 \langle v_1, w \rangle + c_2 \langle v_2, w \rangle$
- $\langle v, v \rangle \geq 0$
- $\langle v, v \rangle = 0$ if and only if $v = 0$

A pair $(V, \|\cdot\|)$ is called a normed space if $\|\cdot\| : V \to \mathbb{R}$ is a norm on V. In an analogous manner, we can define inner product spaces as pairs $(V, \langle \cdot, \cdot \rangle)$ for which $\langle \cdot, \cdot \rangle : V \to \mathbb{F}$ is an inner product on V. We often omit the inner product/norm symbols and simply say that V is an inner product/normed space.

Theorem 1.6 (Inner product spaces are normed spaces) *Let $(V, \langle \cdot, \cdot \rangle)$ be an inner product space. Define $\|v\| = \sqrt{\langle v, v \rangle}$. Then, $(V, \|\cdot\|)$ is a normed space.*

Theorem 1.7 (Cauchy-Schwarz inequality) *If V is an inner product space, then for any vectors $v, w \in V$, we have*

$$|\langle v, w \rangle| \leq \|v\| \, \|w\| \tag{1.13}$$

where $\|\cdot\|$ is defined as in Theorem 1.6.

Definition 1.33 *(Induced norms)* Let V, W be vector spaces over $\mathbb{F} \in \{\mathbb{R}, \mathbb{C}\}$. Assume moreover that V is finite dimensional.[2] Let $\mathcal{L}(V, W)$ be the set of all linear transformations from V to W. Then, this set is a vector space over $\mathbb{F}$ with the definitions $(cT)(v) = c(T(v))$ and $(T_1 + T_2)(v) = T_1(v) + T_2(v)$. Moreover, each linear transformation $T \in \mathcal{L}(V, W)$ can be assigned a norm, called the *induced norm*, which is defined as follows.[3]

$$\|T\| = \sup_{v \neq 0} \frac{\|T(v)\|}{\|v\|} \left(= \max_{\|v\|=1} \frac{\|T(v)\|}{\|v\|} \right). \tag{1.14}$$

Of course, the induced norm is indeed a norm.

---

[2] This condition is needed to ensure that the definition for the induced norm results in a finite value; i.e., the supremum is not equal to ∞.

[3] The maximum is attained because V is finite dimensional; this is a consequence of the Heine-Borel theorem.

Definition 1.34 *(Completeness of normed spaces)* Let $(V, \|\cdot\|)$ be a normed space. If for every sequence $v_1, v_2, \cdots$ in V such that $\lim_{m,n\to\infty} \|v_m - v_n\| = 0$[4] there exists a $v \in V$ such that $\lim_{n\to\infty} \|v_n - v\| = 0$, the normed space V is said to be *complete*.

Theorem 1.8 *A normed space* $(V, \|\cdot\|)$ *is complete if and only if every absolutely convergent series in V converges, i.e., if and only if* $\sum_{n=1}^{\infty} \|v_n\| < \infty$ *implies the convergence of* $\sum_{n=1}^{\infty} v_n$ *to an element in V for any sequence* $\{v_n\}$.

Definition 1.35 *(Equivalence of norms)* Let V be a vector space over $\mathbb{R}$ or $\mathbb{C}$, and let $\|\cdot\|_1$ and $\|\cdot\|_2$ be norms on V. These two norms are said to be *equivalent* if there exists two *positive* constants C_1, C_2 such that $\|v\|_2 \le C_1 \|v\|_1$ and $\|v\|_1 \le C_2 \|v\|_2$ for all $v \in V$.

Theorem 1.9 *Let V be a* finite dimensional *vector space over* $\mathbb{R}$ *or* $\mathbb{C}$. *Then, it is complete under any norm on V. Also, any two norms on V are equivalent.*

Definition 1.36 *(Banach spaces and Hilbert spaces)* A complete normed space is called a *Banach space*. A complete inner product space (i.e., an inner product space which is complete as a normed space) is called a *Hilbert space*.

If a Banach (Hilbert) space's field of scalars $\mathbb{F}$ equals $\mathbb{R}$ ($\mathbb{C}$), it is said to be a real (complex) Banach (Hilbert) space.

Definition 1.37 *(Vector projection)* Let V be an inner product space, and let $u, v \in V$. Suppose $u \neq 0$. Define the projection of v onto u as

$$\text{proj}_u(v) = \frac{\langle v, u\rangle}{\|u\|^2} u. \tag{1.15}$$

Note that this is a linear transformation from V into itself.

Theorem 1.10 (Gram-Schmidt orthogonalization) *Let* $v_1, v_2, \ldots, v_n \in V$ *be linearly independent vectors in an inner product space V. Define* $u_1, u_2, \ldots, u_n \in V$ *as follows.*

$$u_1 = v_1 \tag{1.16a}$$
$$u_2 = v_2 - \text{proj}_{u_1}(v_2) \tag{1.16b}$$
$$u_3 = v_3 - \text{proj}_{u_1}(v_3) - \text{proj}_{u_2}(v_3) \tag{1.16c}$$
$$\vdots \tag{1.16d}$$

Then, $u_i \neq 0$ *for all i*, $\text{span}(u_1, u_2, \ldots, u_k) = \text{span}(v_1, v_2, \ldots, v_k)$ *for all* $1 \le k \le n$, *and* u_i *are mutually orthogonal.*

Remark 1.5 If we define $e_i = \frac{u_i}{\|u_i\|}$ for all i, $\text{span}(e_1, e_2, \ldots, e_k) = \text{span}(v_1, v_2, \ldots, v_k)$ for all $1 \le k \le n$, and e_i are orthonormal. With this additional step, the whole procedure is called the *Gram-Schmidt orthonormalization*.

[4] Such sequences are called *Cauchy sequences*.

1.4.3 Diagonalization and Jordan Form

A matrix $A \in \mathbb{R}^{n \times n}$ can be viewed as a linear function which maps $\mathbb{R}^n$ to itself: if $A\mathbf{x} = \mathbf{y}$ for some $\mathbf{x}, \mathbf{y} \in \mathbb{R}^n$, then $\mathbf{x}$ is mapped to $\mathbf{y}$. Recall that $\mathbb{R}^n$ has the standard basis $\{\mathbf{e}_!, \ldots, \mathbf{e}_n\}$. Then $A\mathbf{e}_k$ contains the coefficients use to represent vector $A\mathbf{e}_k$ with respect to the standard basis.

$$\begin{bmatrix} | & & | \\ A\mathbf{e}_1 & \cdots & A\mathbf{e}_n \\ | & & | \end{bmatrix} = A \begin{bmatrix} | & & | \\ \mathbf{e}_1 & \cdots & \mathbf{e}_n \\ | & & | \end{bmatrix}$$

Consider a new basis $\{\mathbf{v}_1, \ldots, \mathbf{v}_n\}$ of $\mathbb{R}^n$ such that A has a different representation $\tilde{A} \in \mathbb{R}^{n \times n}$. Note that $\tilde{A}$ also maps $\mathbb{R}^n$ to itself: $\tilde{A}\tilde{\mathbf{x}} = \tilde{\mathbf{y}}$. Here, $\tilde{\mathbf{x}}, \tilde{\mathbf{y}} \in \mathbb{R}^n$ are the representations of $\mathbf{x}, \mathbf{y}$ with respect to this new basis:

$$\tilde{\mathbf{x}} = V^{-1}\mathbf{x}, \quad \tilde{\mathbf{y}} = V^{-1}\mathbf{y}, \quad \text{where } V \triangleq \begin{bmatrix} | & & | \\ \mathbf{v}_1 & \cdots & \mathbf{v}_n \\ | & & | \end{bmatrix}$$

This transformation procedure is often called a *change-of-basis*. (**Exercise:** V is a nonsingular matrix. Why?) We can relate the two matrices in the following way:

$$\tilde{A}\tilde{\mathbf{x}} = \tilde{\mathbf{y}} \Longrightarrow \tilde{A}V^{-1}\mathbf{x} = V^{-1}\mathbf{y} \Longrightarrow \underbrace{V\tilde{A}V^{-1}}_{\equiv A}\mathbf{x} = \mathbf{y}$$

The relationship $A = V\tilde{A}V^{-1}$, equivalently $\tilde{A} = V^{-1}AV$ or $V\tilde{A} = AV$, is called a *similarity transformation*. Note that

$$A \begin{bmatrix} | & & | \\ \mathbf{v}_1 & \cdots & \mathbf{v}_n \\ | & & | \end{bmatrix} = \begin{bmatrix} | & & | \\ A\mathbf{v}_1 & \cdots & A\mathbf{v}_n \\ | & & | \end{bmatrix} = \begin{bmatrix} | & & | \\ \mathbf{v}_1 & \cdots & \mathbf{v}_n \\ | & & | \end{bmatrix} \tilde{A},$$

which means that the kth column of $\tilde{A}$ is the representation of $A\mathbf{v}_k$ with respect to the new basis.

Similarity transformations are useful for transforming a matrix to a simpler form. Before discussing various common forms, we recall a few more definitions.

Definition 1.38 *(Eigenvalues and Eigenvectors)* Scalar $\lambda \in \mathbb{R}$ (or $\mathbb{C}$) is called an *eigenvalue* of square matrix $A \in \mathbb{R}^{n \times n}$ if there exists a vector $\mathbf{v} \in \mathbb{R}^n$ (or $\mathbb{C}^n$), $\mathbf{v} \neq 0$ such that $A\mathbf{v} = \lambda\mathbf{v}$. Vector $\mathbf{v}$ is the corresponding *(right) eigenvector* of A associated with λ. (The *left eigenvector* $\mathbf{w}^\top \in \mathbb{R}^n$ (or $\mathbb{C}^n$), $\mathbf{w} \neq 0$ satisfies the equation $\mathbf{w}A = \lambda\mathbf{w}$.)

Eigenvalues are the solutions to the *characteristic equation* $\chi(\lambda) = 0$, where $\chi(\lambda) \triangleq \det(A - \lambda I)$ is the *characteristic polynomial*. The idea is that if $(A - \lambda I)$ is nonsingular, the only solution to $(A - \lambda I)\mathbf{v} = 0$ is $\mathbf{v} = 0$, and so eigenvalues λ are the places where $(A - \lambda I)$ are singular, i.e., have determinant zero. Because $\chi(\lambda)$ is a polynomial of degree n, there are always n eigenvalues associated with A.

Definition 1.39 *(Companion Matrices)* Matrices of the form

$$\begin{bmatrix} 0 & 0 & \cdots & 0 & -\alpha_n \\ 1 & 0 & \cdots & 0 & -\alpha_{n-1} \\ 0 & 1 & \cdots & 0 & -\alpha_{n-2} \\ \vdots & \vdots & \ddots & \vdots & \vdots \\ 0 & 0 & \cdots & 1 & -\alpha_1 \end{bmatrix}, \quad \begin{bmatrix} -\alpha_1 & 1 & 0 & \cdots & 0 \\ -\alpha_2 & 0 & 1 & \cdots & 0 \\ \vdots & \vdots & \vdots & \ddots & \vdots \\ -\alpha_{n-1} & 0 & 0 & \cdots & 1 \\ -\alpha_n & 0 & 0 & \cdots & 0 \end{bmatrix}$$

and their transposes are called *companion form matrices*. They all have characteristic polynomials of the form:

$$\chi(\lambda) = \lambda^n + \alpha_1 \lambda^{n-1} + \alpha_2 \lambda^{n-2} + \cdots \alpha_{n-1}\lambda + \alpha_n,$$

with coefficients that can be easily determined from the matrix entries.

Now, we are ready to review some common similarity transformations of square matrices. There are two cases depending on the structure of the eigenvalues.

1. **All distinct eigenvalues:** $\lambda_i \neq \lambda_j$ **for all** $i, j = 1, \ldots, n$ and $i \neq j$.
   Let $\{\mathbf{v}_1, \ldots, \mathbf{v}_n\}$ denote the eigenvectors corresponding to the eigenvalues $\{\lambda_1, \ldots, \lambda_n\}$. Clearly, $\{\mathbf{v}_1, \ldots, \mathbf{v}_n\}$ are linearly independent, and can be used to form a basis for $\mathbb{R}^n$ (Lemma 1.1). This gives rise to a similarity transformation

$$\Lambda = V^{-1}AV, \text{ where } V \triangleq \begin{bmatrix} | & & | \\ \mathbf{v}_1 & \cdots & \mathbf{v}_n \\ | & & | \end{bmatrix} \text{ and } \Lambda \triangleq \begin{bmatrix} \lambda_1 & & & \\ & \lambda_2 & & \\ & & \ddots & \\ & & & \lambda_n \end{bmatrix}$$

   This particular similarity transformation $A \to \Lambda$ is called *diagonalization*. (**Exercise**: Is this diagonalized form Λ unique?)
2. **Some eigenvalues are repeating: there exist some** $i, j = 1, \ldots, n$ **and** $i \neq j$ **such that** $\lambda_i = \lambda_j$.
   The similarity transformation in this case is known as the *Jordan canonical form*:

$$J = V^{-1}AV, \text{ where } V \triangleq \begin{bmatrix} | & & | \\ V_1 & \cdots & V_r \\ | & & | \end{bmatrix} \text{ and } J \triangleq \begin{bmatrix} J_1 & & & \\ & J_2 & & \\ & & \ddots & \\ & & & J_r \end{bmatrix}, \; J_k \triangleq \begin{bmatrix} \lambda_k & 1 & 0 & \cdots & 0 \\ 0 & \lambda_k & 1 & \cdots & 0 \\ \vdots & \vdots & \vdots & \ddots & \vdots \\ 0 & 0 & 0 & \cdots & 1 \\ 0 & 0 & 0 & \cdots & \lambda_k \end{bmatrix}$$

Here, $r < n$, $k = 1, \ldots, r$, and each $J_k \in \mathbb{R}^{n_k \times n_k}$ is called a *Jordan block*, where $\sum_k n_k = n$. Each submatrix $V_k \in \mathbb{R}^{n \times n_k}$ contains the *generalized eigenvectors* corresponding to eigenvalue λ_k:

$$V_k \triangleq \begin{bmatrix} | & & | \\ \mathbf{v}_1^{(k)} & \cdots & \mathbf{v}_{n_k}^{(k)} \\ | & & | \end{bmatrix} \text{ such that } \ker(A - \lambda_k I)^{n_k} = \operatorname{span}\{\mathbf{v}_1^{(k)}, \ldots, \mathbf{v}_{n_k}^{(k)}\}$$

and are generated by

$$\mathbf{v}_{n_k}^{(k)} \triangleq \mathbf{v}^{(k)} \to \mathbf{v}_{n_k-1}^{(k)} \triangleq (A - \lambda_k I)\mathbf{v}_{n_k}^{(k)} \to \mathbf{v}_{n_k-2}^{(k)} \triangleq (A - \lambda_k I)\mathbf{v}_{n_k-1}^{(k)} \to \cdots \to \mathbf{v}_1^{(k)} \triangleq (A - \lambda_k I)\mathbf{v}_2^{(k)}$$

where $\mathbf{v}^{(k)} \neq 0$ is the vector satisfying $(A - \lambda_k I)v^{(k)} = 0$. Note that these generalized eigenvectors are still linearly independent, and so can be used as a basis.

1.4.4 Functions of Square Matrices

The easiest class of functions to study are *polynomials*. Recall that polynomials $f : \mathbb{R} \to \mathbb{R}$ of scalar variables $x \in \mathbb{R}$ are defined

$$p(x) = c_0 x^k + c_1 x^{k-1} + \cdots + c_{n-1} x + c_n \tag{1.17}$$

To extend polynomials to matrices $A \in \mathbb{R}^{n \times n}$, we first define the *matrix power* A^m for any $m \in \mathbb{N}$ to be $\prod_{i=1}^m A$ (i.e., A multiplied m times) and $A^0 \triangleq I$. If A is invertible, negative powers of A can be defined $A^{-m} \triangleq (A^{-1})^m$ for any $m \in \mathbb{Z}^{\geq 0}$. Some properties of the matrix power include $A^m A^k = A^{m+k}$ for any $m, k \in \mathbb{Z}$.

The matrix version of (1.17) can then be expressed

$$p(A) = c_0 A^k + c_1 A^{k-1} + \cdots + c_{n-1} A + c_n I \tag{1.18}$$

Consider the similarity transform $\tilde{A} \triangleq V^{-1} A V$. Then for any matrix polynomial of the form (1.18), we have $p(\tilde{A}) = V^{-1} p(A) V$. Furthermore, for any block-diagonal A, it follows that

$$A = \begin{bmatrix} A_1 & & & \\ & A_2 & & \\ & & \ddots & \\ & & & A_r \end{bmatrix} \Longrightarrow p(A) = \begin{bmatrix} p(A_1) & & & \\ & p(A_2) & & \\ & & \ddots & \\ & & & p(A_r) \end{bmatrix}$$

This gives us an easy way of computing polynomials of A matrices which can be diagonalized or decomposed into Jordan form.

Definition 1.40 *(Rational Functions)* Scalar-valued functions f is a *rational function* iff it can be written as $p(x)/q(x)$ for two polynomials p and q. In the square matrix case:

$$f(A) = \frac{p(A)}{q(A)} \triangleq p(A)q(A)^{-1} = q^{-1}(A)p(A)$$

can be defined as long as $q(A)$ is invertible.

Theorem 1.11 (Cayley-Hamilton) *Every complex-valued square matrix $A \in \mathbb{C}^{n\times n}$ satisfies its own characteristic equation:*

$$\chi(A) \triangleq A^n + c_1 A^{n-1} + c_2 A^{n-2} + \cdots + c_{n-1}A + c_n I = 0$$

A consequence of this is that the matrix power A^n can be expressed as a linear combination of all the lower powers $\{A^k : 0 \le k \le n-1\}$.

Proof This is easy to check for upper triangular matrices. From the fact that every complex-valued square matrix is similar to an upper triangular matrix, it can be generalized to arbitrary square matrices. ■

Definition 1.41 *(Power Series)* A *power series* $f(x)$ in scalar variable x is an infinite series of the form

$$f(x) = \sum_{k=0}^{\infty} c_k x^k$$

with scalar coefficients $\{c_k\}$. Likewise, power series of matrices can be written as

$$f(A) = \sum_{k=0}^{\infty} c_k A^k$$

A nice property is that if the scalar power series $f(x)$ is convergent for all x, then the matrix equivalent $f(A)$ is convergent too, for any square A.

One especially common matrix function is the matrix exponential e^A, which can be defined by its power series[5]

$$e^A \triangleq I + A + \frac{1}{2}A^2 + \frac{1}{3!}A^3 + \cdots = \sum_{k=0}^{\infty} \frac{1}{k!}A^k \tag{1.19}$$

[5] If two elements in a Banach space V can be multiplied in a "nice" way, and the norm satisfies $\|xy\| \le \|x\|\,\|y\|$ for all vectors x and y, V is called a *Banach algebra*. If the multiplication has an identity element, the Banach algebra is said to be *unital*. In unital Banach algebras, elements can be exponentiated by an analogous power series definition; the matrix exponential is a special case in which $V = \mathbb{R}^{n\times n}$ or $V = \mathbb{C}^{n\times n}$.

Properties of the matrix exponential include[6]:

1. $e^0 = I$
2. $A\mathbf{v} = \lambda\mathbf{v} \Longrightarrow e^A\mathbf{v} = e^\lambda\mathbf{v}$
3. $\left(e^A\right)^\top = e^{A^\top}$
4. $\det(e^A) = e^{\mathrm{tr}(A)}$
5. if $X, Y \in \mathbb{R}^{n\times n}$ commute, then $e^{X+Y} = e^X e^Y = e^Y e^X$
6. $\left(e^A\right)^{-1} = e^{-A}$

There are several ways we can compute the matrix exponential. We will see these discussed in class.

1. direct computation (easy only if A has special properties, e.g., nilpotent, idempotent, diagonal, etc.)
2. via Cayley-Hamilton theorem
3. via Jordan decomposition

Some Recommended References

Note that this section is just a brief summary to inform the reader about which topics from linear algebra and analysis are the most relevant in the study of linear systems. To the student who wishes to recover their full knowledge on these preliminaries, we recommend the following references, among many others. A standard undergraduate-level treatment of linear algebra is Axler's *Linear Algebra Done Right* [1] and Strang's *Introduction to Linear Algebra* [2]. The standard reference for real analysis is Rudin's *Principles of Mathematical Analysis* [3]. References about matrix computations, properties, and related equations include Golub and van Loan's *Matrix Computations* [4] and Horn and Johnson's *Matrix Analysis* [5]. A more advanced and rigorous discussion of linear algebra topics is Hoffman and Kunze's *Linear Algebra* [6]. And a more advanced discussion of real analysis subject, such as infinite dimensional Banach and Hilbert spaces, is Brezis' *Functional Analysis, Sobolev Spaces and Partial Differential Equations* [7].

[6] Property 5: if X and Y do not commute, there is a formula called the Baker-Campbell-Hausdorff formula which computes a solution Z to $e^X e^Y = e^Z$.

References

1. Sheldon Jay Axler. *Linear Algebra Done Right*. Undergraduate Texts in Mathematics. New York: Springer, 1997.
2. Gilbert Strang. *Introduction to Linear Algebra*. Fourth. Wellesley, MA: Wellesley-Cambridge Press, 2009.
3. Walter Rudin. *Principles of Mathematical Analysis, 3rd Ed.* USA: McGraw-Hill, Inc., 1976.
4. Gene H. Golub and Charles F. van Loan. *Matrix Computations*. Fourth. JHU Press, 2013.
5. Roger A. Horn and Charles R. Johnson. *Matrix Analysis*. Cambridge University Press, 1985.
6. Kenneth Hoffman and Ray A. Kunze. *Linear Algebra*. Second. PHI Learning, 2004.
7. Haim Brezis. *Functional Analysis, Sobolev Spaces and Partial Differential Equations*. Universitext. Springer New York, 2010.

Chapter 2
Characteristic Modes

2.1 The Matrix Exponential

In the previous chapter, we considered the controlled system dynamics $\dot{\mathbf{x}}(t) = A\mathbf{x}(t) + B\mathbf{u}(t)$ and obtained a solution $\mathbf{x}(t)$ in terms of the *matrix exponential* $e^{A(t-s)}$. The question we will address in the present chapter is as follows: how do we actually compute the matrix exponential?

First, we will consider the matrix exponential without a *time-index* t, i.e., e^A. Here, there are several ways to compute the matrix exponential.

1. Direct calculation by truncation (using the properties of A)
2. the Cayley-Hamilton theorem
3. the Jordan canonical form

Let's investigate each of these methods individually.

2.1.1 *Computing e^A via Power Series Expansion*

Just as how the scalar exponential function $f(x) = e^x$, $x \in \mathbb{R}$ admits a power series representation, the matrix exponential also has a power series representation too. Given square matrix $A \in \mathbb{R}^{n\times n}$, we have

$$e^A \triangleq \sum_{k=0}^{\infty} \frac{1}{k!} A^k = I + A + \frac{A^2}{2} + \frac{A^3}{3!} + \cdots \tag{2.1}$$

This expression can be simplified depending on specific properties of A:

- If A is *nilpotent*, i.e., $\exists\ n \in \mathbb{N}$ such that $A^n = 0$, then the sum in (2.1) simply gets truncated:

SJ. Han, *Linear Systems*, KAIST Research Series,
https://doi.org/10.1007/978-981-96-9774-8_2

$$e^A \triangleq \sum_{k=0}^{n-1} \frac{1}{k!} A^k.$$

- if A is *idempotent*, i.e., $A^2 = A$, then $e^A = I + \sum_{k=1}^{\infty} \frac{1}{k!} A^k = I + (\sum_{k=1}^{\infty} \frac{1}{k!})A = I + (e-1)A$.
- if $A \triangleq \text{diag}(a_1, \ldots, a_n)$ is diagonal, then e^A is diagonal too:

$$e^A = \begin{pmatrix} e^{a_1} & & 0 \\ & \ddots & \\ 0 & & e^{a_n} \end{pmatrix}$$

2.1.2 Computing e^A via Cayley-Hamilton Theorem

For more general square matrices, direct computation of the matrix exponential is difficult. Instead, we can use one of the other two techniques, the Cayley-Hamilton theorem and the Jordan canonical form.

Theorem 2.12 (Cayley-Hamilton) *Every real-valued square matrix* $A \in \mathbb{R}^{n \times n}$ *satisfies its own characteristic equation* $\det(A - \lambda I) = 0$, *i.e.,*

$$A^n = -c_{n-1}A^{n-1} - \cdots - c_2 A^2 - c_1 A - c_0 I$$

Theorem 2.12 implies that all powers of A, i.e., A^m for $m \geq n$, can be written as linear combinations of $\{A^k, 0 \leq k \leq n-1\}$. For the matrix exponential, we can use Theorem 2.12 to simplify

$$e^A = \sum_{k=0}^{\infty} \frac{A^k}{k!} = \alpha_0 I + \alpha_1 A + \alpha_2 A^2 + \cdots + \alpha_{n-1} A^{n-1}$$

where α_i are functions of $\{c_1, \ldots, c_{n-1}\}$. Thus, the infinite sum becomes truncated into a finite sum, and we can compute each term individually.

2.1.3 Computing e^A via Jordan Canonical Form

For a square matrix $A \in \mathbb{R}^{n \times n}$, we recall the *Jordan canonical form (JCF)* (also called the *Jordan decomposition*) as $A = VJV^{-1}$, where $J \triangleq \text{diag}(J_1, \ldots, J_r)$ is a block-diagonal matrix, with *Jordan block*

$$J_k \triangleq \begin{bmatrix} \lambda_k & 1 & & \\ & \lambda_k & \ddots & \\ & & \ddots & 1 \\ & & & \lambda_k \end{bmatrix} \in \mathbb{R}^{n_k \times n_k},$$

that has dimension $n_k \in \mathbb{N}$ such that $\sum_{k=1}^{r} n_k = n$, and

$$V \triangleq \begin{bmatrix} | & & | \\ V_1 & \cdots & V_r \\ | & & | \end{bmatrix}$$

is the stack of *generalized eigenvectors*, where $V_k \in \mathbb{R}^{n \times n_j}$ is the submatrix of generalized eigenvectors corresponding specifically to eigenvalue λ_k.

$$V_k \triangleq \begin{bmatrix} | & & | \\ v_1^{(k)} & \cdots & v_r^{(k)} \\ | & & | \end{bmatrix} \text{ such that } \ker(A - \lambda_k I)^{n_k} = \text{span}\{v_1^{(k)}, \ldots, v_r^{(k)}\}.$$

Note that the Jordan decomposition gives us a block-diagonal structure J. Thus, we can extend and apply the form of the matrix exponential for diagonal matrices to compute e^A for general A.

$$e^A = Ve^J V^{-1} = V \begin{bmatrix} e^{J_1} & & \\ & \ddots & \\ & & e^{J_r} \end{bmatrix} V^{-1} \text{ and } e^{J_k} \triangleq \begin{bmatrix} e^{\lambda_k} & e^{\lambda_k} & \frac{1}{2}e^{\lambda_k} & \cdots & \frac{1}{(n_k-1)!}e^{\lambda_k} \\ & e^{\lambda_k} & e^{\lambda_k} & \cdots & \\ & & e^{\lambda_k} & \cdots & \vdots \\ & & & \ddots & \\ & & & & e^{\lambda_k} \end{bmatrix}. \tag{2.2}$$

The derivation of (2.2) follows from rote calculation. Note that we can write each Jordan block J_k as a sum of a diagonal matrix and a nilpotent matrix.

$$J_k = \underbrace{\begin{bmatrix} \lambda_1 & & \\ & \ddots & \\ & & \lambda_k \end{bmatrix}}_{=\lambda_k I} + \underbrace{\begin{bmatrix} 0 & 1 & & \\ & 0 & \ddots & \\ & & \ddots & 1 \\ & & & 0 \end{bmatrix}}_{\triangleq N_k}$$

Therefore, we can use the methods in Sect. 2.1.1 to compute the matrix exponential of each Jordan block, and subsequently compute te matrix exponential of A. Consider a generic, smooth (i.e., infinitely-differentiable) real-valued function f. By Taylor-expanding $f(\lambda)$ at $\lambda = \lambda_k$, we get

$$f(\lambda) = f(\lambda_k) + f'(\lambda_k)(\lambda - \lambda_k) + \frac{1}{2} f''(\lambda_k)(\lambda - \lambda_k)^2 + \cdots$$

and replacing λ with A (while treating f as a matrix-valued function now) leads to

$$\begin{aligned}
f(A) &= f(\lambda_k)I + f'(\lambda_k)(A - \lambda_k I) + \frac{1}{2} f''(\lambda_k)(A - \lambda_k I)^2 + \cdots \\
&= f(\lambda_k)I + f'(\lambda_k)(VJV^{-1} - \lambda_k I) + \frac{1}{2} f''(\lambda_k)(VJV^{-1} - \lambda_k I)^2 + \cdots \\
&= V\left\{ f(\lambda_k)I + f'(\lambda_k)(J - \lambda_k I) + \frac{1}{2} f''(\lambda_k)(J - \lambda_k I)^2 + \cdots \right\} V^{-1} \\
&= V\left\{ f(\lambda_k)I + f'(\lambda_k)N_k + \frac{1}{2} f''(\lambda_k)N_k^2 + \cdots \right\} V^{-1} = V f(J_k) V^{-1}.
\end{aligned}$$

Letting $f(A) = e^A$, $f(\lambda_k) = e^{\lambda_k}$ and $f(J_k) = e^{J_k}$, we get

$$\begin{aligned}
e^{J_k} &= e^{\lambda_k} I + e^{\lambda_k} N_k + \frac{1}{2} e^{\lambda_k} N_k^2 + \cdots \\
&= e^{\lambda_k} I + e^{\lambda_k} N_k + \frac{1}{2} e^{\lambda_k} N_k^2 + \cdots + \frac{1}{(n_k - 1)} e^{\lambda_k} N_k^{n_k - 1},
\end{aligned}$$

where the second equality comes from the fact that $(N_k)^{n_k} = 0$ (due to nilpotence). Then, we have

$$\begin{aligned}
e^A &= I + A + \frac{1}{2} A^2 + \cdots \\
&= V \underbrace{\left(I + J + \frac{1}{2} J^2 + \cdots\right)}_{=e^J} V^{-1} \\
&= V \begin{bmatrix} e^{J_1} & & \\ & \ddots & \\ & & e^{J_r} \end{bmatrix} V^{-1},
\end{aligned}$$

with e^{J_k} defined as above in (2.2).

2.2 The Time-Indexed Matrix Exponential

Now we are ready to address the original *time-indexed matrix exponential* e^{At}. The power series expansion still holds:

$$e^{At} \triangleq \sum_{k=0}^{\infty} \frac{1}{k!}(At)^k = I + At + \frac{1}{2}A^2t^2 + \frac{1}{3}A^3t^3 + \cdots$$

and can be viewed as a matrix-valued function of time t.

There are several notable properties of the time-indexed matrix exponential:

1. $Av = \lambda v \implies e^{At}v = e^{\lambda t}v$
2. $(e^{At})^\top = e^{A^\top t}$
3. $\det(e^{At}) = e^{tr(A)t}$
4. if $X, Y \in \mathbb{R}^{n\times n}$ commute, then $e^{(X+Y)t} = e^{Xt}e^{Yt} = e^{Yt}e^{Xt}$
5. $e^{A(t_1+t_2)} = e^{At_1}e^{At_2} = e^{At_2}e^{At_1}$
6. $(e^{At})^{-1} = e^{-At}$
7. If A is skew symmetric, then e^{At} is orthogonal for all t: $e^{At}(e^{At})^\top = e^{At}e^{A^\top t} = e^{At}e^{-At} = I$

We can use similar methods to explicitly compute the time-indexed matrix exponential. For example, with the JCF, we can compute e^{At} as

$$e^{At} = V\begin{bmatrix} e^{J_1 t} & & \\ & \ddots & \\ & & e^{J_r t}\end{bmatrix}V^{-1} \text{ where } e^{J_k t} \triangleq \begin{bmatrix} e^{\lambda_k t} & te^{\lambda_k t} & \frac{1}{2}t^2e^{\lambda_k t} & \cdots & \frac{1}{(n_k-1)!}t^{n_k-1}e^{\lambda_k t} \\ & e^{\lambda_k t} & te^{\lambda_k t} & \cdots & \\ & & e^{\lambda_k t} & \cdots & \vdots \\ & & & \ddots & \\ & & & & e^{\lambda_k t}\end{bmatrix}.$$

Alternatively, we can also convert the matrix exponential to the frequency domain via *Laplace transform*. Recall that

$$\mathcal{L}\{t\} = \int_0^\infty te^{-st}dt = -\frac{1}{s}te^{-st}\Big|_0^\infty + \frac{1}{s}\int_0^\infty e^{-st}dt = \frac{1}{s^2},$$

and more generally, we can calculate via integration-by-parts:

$$\begin{aligned}\mathcal{L}\left\{\frac{1}{k!}t^k\right\} &= \int_0^\infty \frac{1}{k!}t^k e^{-st}dt = -\frac{1}{k!}\frac{1}{s}t^k e^{-st}\Big|_0^\infty + \frac{1}{s}\int_0^\infty \frac{1}{(k-1)!}t^{k-1}e^{-st}dt \\ &= \frac{1}{s^{k-1}}\int_0^\infty te^{-st}dt = \frac{1}{s^{k+1}}.\end{aligned}$$

Using the power series representation of the time-indexed matrix exponential, and taking the Laplace transform yields

$$\mathcal{L}\{e^{At}\} = \sum_{k=0}^{\infty} A^k \mathcal{L}\left\{\frac{1}{k!}t^k\right\} = \frac{1}{s}\sum_{k=0}^{\infty}(s^{-1}A)^k.$$

Note that the scalar infinite series $\sum_{k=0}^{\infty}(s^{-1}x)^k$ converges to $\frac{1}{1-s^{-1}x}$ if $|s^{-1}x| < 1$ due to the Geometric series. Therefore,

$$\mathcal{L}\{e^{At}\} = \frac{1}{s}\sum_{k=0}^{\infty}(s^{-1}A)^k = s^{-1}(I - s^{-1}A)^{-1} = (sI - A)^{-1} \tag{2.3}$$

Deriving (2.3) in the way described above requires s to be large enough for all eigenvalues of $s^{-1}A$ to have magnitude less than 1. But $\mathcal{L}\{e^{At}\} = (sI - A)^{-1}$ actually holds *for all s* except at eigenvalues of A. This can be more easily seen using the following alternative derivation.

Recall the derivative property of the Laplace transform: $\mathcal{L}\{\frac{d}{dt}f(t)\} = sF(s) - f(0)$ property. The time derivative of e^{At} can be derived from the power series:

$$\frac{d}{dt}\{e^{At}\} = \sum_{k=0}^{\infty}\frac{1}{k!}A^k\frac{d}{dt}\{t^k\} = \sum_{k=1}^{\infty}\frac{1}{(k-1)!}A^k t^{k-1} = \left(\sum_{m=0}^{\infty}\frac{1}{m!}A^m t^m\right)A = e^{At}A.$$

Furthermore, it can also be expressed as:

$$\frac{d}{dt}\{e^{At}\} = \sum_{k=0}^{\infty}\frac{1}{k!}A^k\frac{d}{dt}\{t^k\} = \sum_{k=1}^{\infty}\frac{1}{(k-1)!}A^k t^{k-1} = A\sum_{m=0}^{\infty}\frac{1}{m!}A^m t^m = Ae^{At}.$$

$$\therefore Ae^{At} = e^{At}A$$

It shows that the matrix A commutes with its matrix exponential. From there, the rest of the derivation follows straightforwardly:

$$A\mathcal{L}\{e^{At}\} = \mathcal{L}\left\{\frac{d}{dt}e^{At}\right\} = s\mathcal{L}\{e^{At}\} - e^0 = s\mathcal{L}\{e^{At}\} - I.$$

$$\implies (sI - A)\mathcal{L}\{e^{At}\} = I.$$

$$\therefore \mathcal{L}\{e^{At}\} = (sI - A)^{-1}$$

2.3 Equivalent Systems

Following the Jordan decomposition method of computing the matrix exponential, we can simplify te analysis of linear systems by simply reviewing the change of coordinates principle from linear algebra. Define $\tilde{x} := P^{-1}x$ where

$$P := \begin{bmatrix} | & & | \\ P_1 & \cdots & P_r \\ | & & | \end{bmatrix} \in \mathbb{R}^{n\times n}$$

is a nonsingular transformation, and $\tilde{x}$ is the new coordinate of x in the new basis. Then the LTI system $\dot{x} = Ax$ in the new $\tilde{x}$-coordinate can be represented as:

$$P\dot{\tilde{x}} = AP\tilde{x} \implies \dot{\tilde{x}} = P^{-1}AP\tilde{x} = \tilde{A}\tilde{x}$$

with $\tilde{x}(0) = P^{-1}\mathbf{x}_0$. The basis of the state space changes from $\{\mathbf{e}_1, \cdots, \mathbf{e}_n\}$ to $\{P_1, \ldots, P_n\}$.

In the case of diagonalizable A, i.e., $A = V\Lambda V^{-1}$, we can change coordinates using V.

$$\dot{\tilde{x}} = V^{-1}AV\tilde{x} = \Lambda\tilde{x} \implies \begin{cases} \dot{\tilde{x}}_1 = \lambda_1\tilde{x}_1 \\ \dot{\tilde{x}}_2 = \lambda_2\tilde{x}_2 \\ \vdots \\ \dot{\tilde{x}}_n = \lambda_n\tilde{x}_n \end{cases}$$

The coefficients c_i in (2.5) can be determined explicitly.

$$\begin{aligned} A = V\Lambda V^{-1} &= \begin{bmatrix} | & & | \\ V_1 & \cdots & V_n \\ | & & | \end{bmatrix} \begin{bmatrix} \lambda_1 & & \\ & \ddots & \\ & & \lambda_n \end{bmatrix} \begin{bmatrix} -w_1^\top- \\ \vdots \\ -w_n^\top- \end{bmatrix} \\ &= \begin{bmatrix} | & & | \\ V_1 & \cdots & V_n \\ | & & | \end{bmatrix} \begin{bmatrix} -\lambda_1 w_1^\top- \\ \vdots \\ -\lambda_n w_n^\top- \end{bmatrix} \\ &= \lambda_1\mathbf{v}_1 w_1^\top + \lambda_2\mathbf{v}_2 w_2^\top + \cdots + \lambda_n\mathbf{v}_n w_n^\top, \end{aligned}$$

Substituting this back into the JCF method of computing the matrix exponential yields

$$\begin{aligned} e^{At} = Ve^{\Lambda t}V^{-1} &= \begin{bmatrix} | & & | \\ V_1 & \cdots & V_n \\ | & & | \end{bmatrix} \begin{bmatrix} e^{\lambda_1 t} & & \\ & \ddots & \\ & & e^{\lambda_n t} \end{bmatrix} \begin{bmatrix} -w_1^\top- \\ \vdots \\ -w_n^\top- \end{bmatrix} \\ &= e^{\lambda_1 t}\mathbf{v}_1 w_1^\top + e^{\lambda_2 t}\mathbf{v}_2 w_2^\top + \cdots + e^{\lambda_n t}\mathbf{v}_n w_n^\top. \end{aligned}$$

Therefore,

$$\mathbf{x}(t) = e^{At}\mathbf{x}_0 = e^{\lambda_1 t}\mathbf{v}_1 c_1 + e^{\lambda_2 t}\mathbf{v}_2 c_2 + \cdots + e^{\lambda_n t}\mathbf{v}_n c_n$$

$$\text{with } c_1 = w_1^\top\mathbf{x}_0,\ c_2 = w_2^\top\mathbf{x}_0,\ \ldots,\ c_n = w_n^\top\mathbf{x}_0.$$

In the most general case, suppose matrix A has JCF, i.e., $A = VJV^{-1}$. Then we can change coordinates using V, i.e., $\tilde{x} = V^{-1}x$.

$$\mathbf{x}(t) = e^{At}\mathbf{x}_0 = Ve^{Jt}V^{-1}\mathbf{x}_0 = \begin{bmatrix} | & & | \\ V_1 & \cdots & V_r \\ | & & | \end{bmatrix} \begin{bmatrix} e^{J_1 t} & & \\ & \ddots & \\ & & e^{J_n t} \end{bmatrix} \begin{bmatrix} -w_1^\top- \\ \vdots \\ -w_r^\top- \end{bmatrix} \mathbf{x}_0 = \sum_{k=1}^{r} V_k e^{J_k t} W_k^\top \mathbf{x}_0 \tag{2.4}$$

In conclusion, a change of coordinates yields $r \in \mathbb{N}$ total independent LTI systems, with the kth subsystem having state $\tilde{\mathbf{x}}_k(t) \in \mathbb{R}^{n_k}$.

$$\begin{cases} \dot{\tilde{x}}_1 = J_1\tilde{x}_1 \\ \dot{\tilde{x}}_2 = J_2\tilde{x}_2 \\ \quad\vdots \\ \dot{\tilde{x}}_n = J_r\tilde{x}_r \end{cases}$$

The fact that a collection of independent subsystems are revealed through an eigendecomposition suggests that a change of coordinates (in the basis for the solution space $\mathcal{X}$) can allow us to study the original system more easily. We will analyze this in the following section, on determining the solutions the autonomous linear systems.

2.4 Solutions to Autonomous Systems

Now we are ready to analyze the solution trajectories of the zero-input response (i.e., $B = 0$ uncontrolled case).

Definition 2.42 (*Autonomous*) A general (possibly nonlinear) system $\dot{\mathbf{x}}(t) = f(t, \mathbf{x}(t))$ is said to be *autonomous* if it has no explicit time-dependence aside from the state variable $\mathbf{x}(t)$, i.e., $\dot{\mathbf{x}}(t) = f(t, \mathbf{x}(t)) = f(\mathbf{x}(t))$.

Definition 2.43 (*Solution Space*) The *solution space* $\mathcal{X}$ of uncontrolled LTI system $\dot{\mathbf{x}}(t) = A\mathbf{x}(t)$ is the set of all its possible solutions:

$$\mathcal{X} := \{\mathbf{x} : \mathbb{R} \geq 0 \Longrightarrow \mathbb{R}^n \mid \dot{\mathbf{x}}(t) = A\mathbf{x}(t), \mathbf{x}(0) = \mathbf{x}_0 \in \mathbb{R}^n\}.$$

Definition 2.44 (*Characteristic Modes*) Suppose $A \in \mathbb{R}^{n\times n}$ is a square matrix. If it is diagonalizable, $A = V\Lambda V^{-1}$, where

$$\Lambda := \begin{bmatrix} \lambda_1 & & \\ & \ddots & \\ & & \lambda_n \end{bmatrix} \text{ and } V := \begin{bmatrix} | & & | \\ V_1 & \cdots & V_n \\ | & & | \end{bmatrix}, V^{-1} := \begin{bmatrix} | & & | \\ w_1 & \cdots & w_n \\ | & & | \end{bmatrix}^\top.$$

The *characteristic modes* of LTI system $\dot{\mathbf{x}}(t) = A\mathbf{x}(t)$ are the solutions from initial condition $\mathbf{x}_0 \equiv V_k$.

$$\mathbf{x}(t) = e^{At}V_k = e^{\lambda_k t}V_k, \ k = 1, \ldots, n$$

Alternatively, if A is not diagonalizable, and the Jordan form must be used, each characteristic mode corresponds to each Jordan block.

More intuitively, the characteristic modes of a linear system represent the natural solutions to the system's dynamics, which typically arise from inherent phenomena of the system, such as vibrations or oscillations. They can be thought of as the "building blocks" of the system's response.

Since the eigenvectors $\{\mathbf{v}_1, \ldots, \mathbf{v}_n\}$ form a basis of $\mathbb{R}^n$, we can write

$$\mathbf{x}_0 = c_1\mathbf{v}_1 + \cdots + c_n\mathbf{v}_n \text{ for any } \mathbf{x}_0 \in \mathbb{R}^n, \text{ scalars } c_i.$$

Thus, the n modes $\{e^{\lambda_1 t}\mathbf{v}_1, \ldots, e^{\lambda_n t}\mathbf{v}_n\}$ form a basis of the solution space $\mathcal{X}$ ($\dim(\mathcal{X}) = n$), so any solution $\mathbf{x}(t)$ of $\dot{\mathbf{x}}(t) = A\mathbf{x}(t)$ can be written as a linear combination of the modes:

$$\begin{aligned} \mathbf{x}(t) = e^{At}\mathbf{x}_0 &= c_1 e^{At}\mathbf{v}_1 + c_2 e^{At}\mathbf{v}_2 + \cdots + c_n e^{At}\mathbf{v}_n \\ &= c_1 e^{\lambda_1 t}\mathbf{v}_1 + c_2 e^{\lambda_2 t}\mathbf{v}_2 + \cdots + c_n e^{\lambda_n t}\mathbf{v}_n \end{aligned} \tag{2.5}$$

with the property $Av = \lambda v \implies e^{At}v = e^{\lambda t}v$.

2.4.1 Connection to System Stability

There are several types of characteristic modes that we can categorize. These different types of modes determine the stability of the system. We will formally describe linear stability in the following Part II

1. If $\lambda_k \in \mathbb{R}$, the mode $e^{\lambda_k t}V_k$ is
   (a) stable if $\lambda_k < 0$ ($\lim_{t \Longrightarrow 0} e^{\lambda_k t}V_k = 0$)
   (b) unstable if $\lambda_k > 0$ ($\lim_{t \Longrightarrow 0} e^{\lambda_k t}V_k = \infty$)
   (c) marginally stable if $\lambda_k = 0$ ($\lim_{t \Longrightarrow 0} e^{\lambda_k t}V_k = V_k$, mode is stationary)
2. If $\sigma_k + j\omega_k =: \lambda_k \in \mathbb{C}$ and $V_k := V_k^{(R)} + jV_k^{(C)}$,
   (a) there is another mode $e^{\bar{\lambda}_k t}\bar{V}_k$, since the mode is complex
   (b) a real solution in $\mathcal{X}$ is $2\,\mathrm{Re}(e^{\lambda_k t}V_k)$

   Then, the stability type of $e^{\lambda_k t} = e^{\sigma_k t}e^{j\omega_k t} = e^{\sigma_k t}(\cos(\omega_k t) + j\sin(\omega_k t))$ can be determined by the real part of λ_k:

   (a) stable if $\sigma_k < 0$
   (b) unstable if $\sigma_k > 0$
   (c) marginally stable if $\sigma_k = 0$

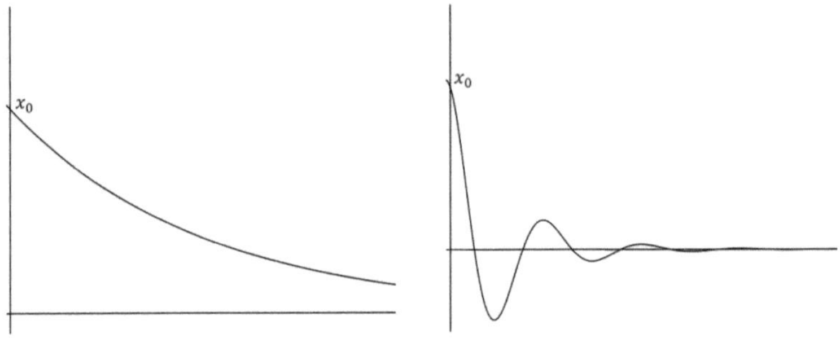

Fig. 2.1 Sample scalar state trajectories versus time for stable systems, with and without oscillations

However, due to the $\cos(\omega_k t) + j \sin(\omega_k t)$ term, there is oscillation in the system. Sample trajectories for stable systems, with and without oscillations, is shown in Fig. 2.1.

2.5 Discretization

So far, our discussion has been focused on continuous-time (CT) systems rather than discrete-time (DT) systems. *Discretization* is a common procedure used to convert a CT system into a DT approximation. Discretization is often necessary in order to simulate such systems on a digital computer. We will present three methods.

Start from the CT LTI system $\dot{\mathbf{x}}(t) = A\mathbf{x}(t) + B\mathbf{u}(t)$. The approximation is based on the definition of derivative from calculus:

$$\dot{\mathbf{x}}(t) = \lim_{\Delta t \to 0} \frac{\mathbf{x}(t + \Delta t) - \mathbf{x}(t)}{\Delta t}$$

Choose a very small discretization timestep Δt, then we have

$$\frac{\mathbf{x}(t + \Delta t) - \mathbf{x}(t)}{\Delta t} = A\mathbf{x}(t) + B\mathbf{u}(t).$$

We can calculate the state of the next time step from the current state

$$\mathbf{x}(t + \Delta t) = (I + A\Delta t)\mathbf{x}(t) + B\Delta t\mathbf{u}(t).$$

The DT system is then implemented on equally-spaced time steps $\{t_k\}_{k\in\mathbb{N}} \equiv \{k\Delta t\}_{k\in\mathbb{N}}$, such that $\mathbf{x}[k] \equiv \mathbf{x}(k\Delta t)$ and $\mathbf{u}[k] \equiv \mathbf{u}(k\Delta t)$. Then

$$\mathbf{x}[k+1] = A_d\mathbf{x}[k] + B_d\mathbf{u}[k], \tag{2.6}$$

where $A_d \triangleq I + A\Delta t$ and $B_d \triangleq B\Delta t$. While method (2.6) is the easiest to implement, it is also the least accurate. The two other discretization methods involving the approximation of the matrix exponential are more accurate.

Let the discretized input signal be piecewise-constant, i.e., for $t \in [k\Delta t, (k+1)\Delta t)$, $k \in \mathbb{N}$, $\mathbf{u}(t) \equiv \mathbf{u}(k\Delta t) \triangleq \mathbf{u}[k]$. Then

$$\mathbf{x}(k\Delta t) = e^{Ak\Delta t}\mathbf{x}_0 + \int_0^{k\Delta t} e^{A(k\Delta t-\tau)}B\mathbf{u}(\tau)d\tau$$

$$\begin{aligned}\mathbf{x}((k+1)\Delta t) &= e^{A(k+1)\Delta t}\mathbf{x}_0 + \int_0^{(k+1)\Delta t} e^{A((k+1)\Delta t-\tau)}B\mathbf{u}(\tau)d\tau \\ &= e^{A\Delta t}\left(e^{Ak\Delta t}\mathbf{x}_0 + \int_0^{(k+1)\Delta t} e^{A((k+1)\Delta t-\tau)}B\mathbf{u}(\tau)d\tau\right) \\ &= e^{A\Delta t}\mathbf{x}(k\Delta t) + e^{A\Delta t}\int_{k\Delta t}^{(k+1)\Delta t} e^{A((k+1)\Delta t-\tau)}B\mathbf{u}(\tau)d\tau\end{aligned}$$

Since $\mathbf{u}(t) = \mathbf{u}(k\Delta t)$ is a constant value for $t \in [k\Delta t, (k+1)\Delta t)$,

$$\begin{aligned}\mathbf{x}(k\Delta t) &= e^{A\Delta t}\mathbf{x}(k\Delta t) + \left(e^{A\Delta t}\int_{k\Delta t}^{(k+1)\Delta t} e^{A(k\Delta t-\tau)}d\tau\right)B\mathbf{u}(k\Delta t) \\ &= e^{A\Delta t}\mathbf{x}(k\Delta t) + \left(\int_0^{\Delta t} e^{As}ds\right)B\mathbf{u}(k\Delta t)\end{aligned}$$

Putting everything together

$$\mathbf{x}((k+1)\Delta t) = e^{A\Delta t}\mathbf{x}(k\Delta t) + e^{A\Delta t}\int_{k\Delta t}^{(k+1)\Delta t} e^{A(k\Delta t-\tau)}B\mathbf{u}(\tau)d\tau \tag{2.7}$$

Here we can also denote the formula like in (2.6). In this case, the $A_d \triangleq e^{A\Delta t}$ and $B_d \triangleq \left(\int_0^{\Delta t} e^{As}ds\right)B$.

The last discretization method is motivated from the fact that the integral in B_d may be difficult to compute. We can instead evaluate it using a power series.

$$\begin{aligned}B_d = \left(\int_0^{\Delta t} e^{As}ds\right)B &= \int_0^{\Delta t}\left(I + As + \frac{1}{2}A^2s^2 + \cdots\right)dsB \\ &= \left(\Delta t I + \frac{1}{2}A_\Delta t^2 + \frac{1}{3!}A^2\Delta t^3 + \cdots\right)B \\ &= A^{-1}\left(A\Delta t + \frac{1}{2}A^2\Delta t^2 + \frac{1}{3!}A^3\Delta t^3 + \cdots\right)B \\ &= A^{-1}(A_d - I)B\end{aligned}$$

One caveat is that this method of discretization is valid only for nonsingular A, as the formula above clearly depends on A^{-1}.

Remark 2.6 The system matrices in the CT measurement equation stay the same as in the DT case:

$$\mathbf{y}(k\Delta t) = C\mathbf{x}(k\Delta t) + D\mathbf{u}(k\Delta t) \Longrightarrow \mathbf{y}[t] = C\mathbf{x}[t] + D\mathbf{u}[t]$$

Chapter 3
State-Transition Matrix

Previously, we've seen the explicit solution to the LTI system expressed as

$$\mathbf{x}(t) = e^{A(t-t_0)}\mathbf{x}_0 + \int_{t_0}^{t} e^{A(t-s)} B\mathbf{u}(s)ds$$

with general initial condition $\mathbf{x}(t_0) = \mathbf{x}_0$. More generally, we often express this form as

$$\mathbf{x}(t) = \Phi(t, t_0)\mathbf{x}_0 + \int_{t_0}^{t} \Phi(t, s)B\mathbf{u}(s)ds$$

where $\Phi(t, s) \triangleq e^{A(t-s)}$ for any $s, t \in \mathbb{R}$ is called the *state transition matrix (STM)*. This specific form of the STM is for the CT LTI case, and it is evident that it is exactly equivalent to the time-indexed matrix exponential in the LTI case.

There are also corresponding STMs for the DT cases and the LTV case, which we will derive throughout this chapter.

3.1 STM for the DT LTI System

Since we've already extensively investigated the properties of the matrix exponential (which is the STM in the CT LTI case), we dedicate a short discussion to the DT case. For the uncontrolled LTI DT system $\mathbf{x}[t+1] = A\mathbf{x}[t]$, we assume this system is uncontrolled and the initial state is $\mathbf{x}_0$. The DT state-transition matrix can be computed simply via recursion

$$\mathbf{x}[1] = A\mathbf{x}_0 \implies \mathbf{x}[2] = A\mathbf{x}[1] = A^2\mathbf{x}_0 \implies \cdots \implies \mathbf{x}[k] = A\mathbf{x}[k-1] = A^k\mathbf{x}_0$$

SJ. Han, *Linear Systems*, KAIST Research Series,
https://doi.org/10.1007/978-981-96-9774-8_3

Matching the expression of the STM with the result of the recursion yields $\mathbf{x}[t] = \Phi[t]\mathbf{x}_0$ where $\Phi[t] \triangleq A^t$.

In the DT case, the solution space $\mathcal{X} \triangleq \{\mathbf{x}[t], t \in \mathbb{Z}^{\geq 0} : \mathbf{x}[t+1] = A\mathbf{x}[t], \mathbf{x}_0 \in \mathbb{R}^n\}$ still has dimension n, with basis $\{A^t v_1, \ldots, A^t v_n\}$. And for diagonalizable $A = V^{-1}\Lambda V$

$$\mathbf{x}[t] = A^t\mathbf{x}_0 = < w_1, \mathbf{x}_0 > \lambda_1^t v_1 + \cdots + \langle w_n, \mathbf{x}_0\rangle \lambda_n^t v_n$$

where $\{\lambda_1^t v_1, \ldots, \lambda_n^t v_n\}$ are the n modes. We see that much of the concepts are quite similar in the DT case as in the CT case. Once main difference is that the stability of each mode is determined by comparing the corresponding eigenvalue's magnitude with respect to unit circle:

- $|\lambda_k| > 1$ means unstable
- $|\lambda_k| = 1$ means marginally stable
- $|\lambda_k| < 1$ means stable.

For general $A = V^{-1}JV$, where J is the Jordan form of A, we get $\mathbf{x}[t] = A^t\mathbf{x}_0 = \sum_{k=1}^{r} V_k J_k^t \langle w_k, \mathbf{x}_0\rangle$, where the columns of $V_k J_k^t$ are modes corresponding to eigenvalue λ_k.

3.2 STM for CT LTV Systems

First, we remark that for scalar LTV systems $\dot{x}(t) = a(t)\mathbf{x}(t) \in \mathbb{R}$ with initial state is $\mathbf{x}(0) = x_0$, the integrating factor approach can be used with $\exp(-\int_0^t a(s)ds)$. In the general matrix-vector ODE, however, this integrating factor solution is incorrect. In this section, we will derive the full solution for LTV systems, including proofs of the existence and uniqueness of these solutions.

First, a few results from real analysis are needed.

Definition 3.45 (*Absolute Convergence*) A series $\sum_{n=0}^{\infty} a_n$ is said to *converge absolutely* if $\sum_{n=0}^{\infty} |a_n|$ is finite.

Definition 3.46 (*Uniform Convergence*) A sequence of function f_n, $f_n : \mathbb{S} \Longrightarrow \mathbb{R}$ is *uniformly convergent* on $\mathbb{S}$, with limit $f_n : \mathbb{S} \to \mathbb{R}$, if $\forall \epsilon > 0\ \exists N \in \mathbb{N}$, such that $\forall n \geq N, x \in \mathbb{S}, |f_n(x) - f(x)| < \epsilon$.

Lemma 3.3 (Weierstrass M-Test) *Let $\{f_n\}$ be a sequence of real (or complex) valued functions over some set $\mathbb{S}$. Suppose exist a sequence of numbers M_n, $M_n \geq 0\ \forall n$ such that*

1. *$\forall x \in \mathbb{S}$ and $n \in \mathbb{M}$, $|f_n(x)| \leq M_n$*
2. *$\sum_{n=1}^{\infty} M_n$ converges*

Then the series $\sum_{n=1}^{\infty} f_n(x)$ converges uniformly and absolutely on $\mathbb{S}$.

Lemma 3.4 (Gronwall-Bellman Inequality) *Let $\phi(t)$, $\psi(t)$ be continuous functions defined $\forall t \geq t_0$ with $\psi(t) \geq 0 \forall t \geq t_0$, and let $\alpha \geq 0$ be some constant. Then*

$$\phi(t) \leq \alpha + \int_{t_0}^{t} \psi(s)\phi(t)ds$$

and $\forall t \geq t_0$

$$\phi(t) \leq \alpha e^{\int_{t_0}^{t} \psi(s)ds}$$

We begin by proving existence of solutions because, depending on the properties of $A(t)$, the solutions $\mathbf{x}(t)$ may or may not exist. Start by constructing the solution space

$$\mathcal{X} \triangleq \{\mathbf{x}(t), t \geq 0 : \dot{\mathbf{x}}(t) = A(t)\mathbf{x}(t), \mathbf{x}_0 \in \mathbb{R}^n\},$$

where the dimension of $\mathcal{X}$ is n. Our basic assumption is that $\forall t \in \mathbb{R}^{\geq 0}$ A(t) is continuous and defined.

Construct a sequence of approximate solutions $\{\mathbf{x}_k(t)\}_{k=0}^{\infty}$ on interval $[t_0, t_0 + T]$ for some chosen arbitrary time $T > 0$. By using *Picard iterations*, we can get

$$\mathbf{x}_k(t) = \mathbf{x}_0 + \int_{t_0}^{t} A(t_1)\mathbf{x}_0 dt_1 + \int_{t_0}^{t} A(t_1) \int_{t_0}^{t_1} A(t_2)\mathbf{x}_0 dt_2 dt_1 + \cdots + \text{ term with } k \text{ integrals}$$

In fact, by factoring out the constant initial state $\mathbf{x}_0$, we can obtain the LTV STM:

$$\mathbf{x}_k(t) = \underbrace{\left(I + \int_{t_0}^{t} A(t_1)dt_1 + \int_{t_0}^{t} A(t_1) \int_{t_0}^{t_1} A(t_2)dt_2 dt_1 + \cdots + \text{ term with } k \text{ integrals}\right)}_{\triangleq \Phi(t,t_0)} \mathbf{x}_0 \tag{3.1}$$

This infinite-sum integral expression of $\Phi(t, t_0)$ is typically called the *Peano-Baker series*. One might notice it resembles an integral version of the Taylor series expansion.

Define

$$\alpha \triangleq \max_{t \in [t_0, t_0+T]} \|A(t)\|, \quad \beta \triangleq \int_{t_0}^{t_0+T} \|A(s)\mathbf{x}_0\| ds, \text{ where } ||A(t)|| \triangleq \max_{||x||=1} ||Ax||$$

Here, $\alpha, \beta < \infty$ since $A(t)$ is continuous and we are considering a finite time interval $[t_0, t_0 + T]$. By considering the successive differences between the approximate solutions, we have

$$\|\mathbf{x}_1(t) - \mathbf{x}_0\| = \left\| \int_{t_0}^{t} A(s)\mathbf{x}_0 ds \right\| \leq \int_{t_0}^{t} \|A(s)\mathbf{x}_0\| ds \leq \beta$$

$$\|\mathbf{x}_2(t) - \mathbf{x}_1(t)\| = \left\| \int_{t_0}^{t} A(s)\mathbf{x}_1(s)ds - \int_{t_0}^{t} A(s)\mathbf{x}_0 ds \right\|$$

$$\begin{aligned} &\leq \int_{t_0}^{t} \|A(s)\| \|\mathbf{x}_1(s) - \mathbf{x}_0\| ds \leq \alpha\beta(t - t_0) \\ \|\mathbf{x}_3(t) - \mathbf{x}_2(t)\| &\leq \int_{t_0}^{t} \|A(s)\| \|\mathbf{x}_2(s) - \mathbf{x}_1(s)\| ds \\ &\leq \int_{t_0}^{t} \alpha^2\beta(s - t_0)ds = \beta\frac{\alpha^2(t - t_0)^2}{2} \end{aligned}$$

In general,

$$\|\mathbf{x}_{k+1}(t) - \mathbf{x}_k(t)\| \leq \int_{t_0}^{t} \|A(s)\| \|\mathbf{x}_k(s) - \mathbf{x}_{k-1}(s)\| ds \leq \beta\frac{\alpha^k(t - t_0)^k}{k!}, \quad k \in \mathbb{N} \tag{3.2}$$

Note each $\mathbf{x}_k(t)$ can be expressed as a telescoping sum:

$$\mathbf{x}_k(t) = \mathbf{x}_0 + \sum_{i=0}^{k-1} (\mathbf{x}_{i+1}(t) - \mathbf{x}_i(t)) \tag{3.3}$$

Through (3.2) and (3.3), we can get an inequality of $||\mathbf{x}_k(t)||$,

$$||\mathbf{x}_k(t)|| \leq ||\mathbf{x}_0|| + \sum_{i=0}^{k-1} ||\mathbf{x}_{i+1}(t) - \mathbf{x}_i(t)|| \leq ||\mathbf{x}_0|| + \sum_{i=0}^{k-1} \beta\frac{\alpha^i(t - t_0)^i}{i!} \tag{3.4}$$

Apply the Weierstrass M-Test (Lemma 3.3) to the above inequality with $f_k(x) = ||\mathbf{x}_{k+1}(t) - \mathbf{x}_k(t)||$ and $M_k \triangleq \beta\frac{\alpha^k(t-t_0)^k}{k!}$.

1. According to (3.2), $||\mathbf{x}_{k+1}(t) - \mathbf{x}_k(t)|| \leq \beta\frac{\alpha^k(t-t_0)^k}{k!}$, so $f_k(x) \leq M_k$.
2. Since $t \leq t_0 + T, t - t_0 \leq T$. As $k \to \infty$,

$$\textstyle\sum_{i=0}^{k-1} \beta\frac{\alpha^i(t-t_0)^i}{i!} \leq \sum_{i=0}^{k-1} \beta\frac{(\alpha T)^i}{i!} \to \beta e^{\alpha T}.$$

Thus, Lemma 3.3 tells us that $\sum_{i=0}^{\infty} ||\mathbf{x}_{i+1}(t) - \mathbf{x}_i(t)||$ converges uniformly and absolutely on $[t_0, t_0 + T]$.

Denote $\mathbf{x}(t) \triangleq \mathbf{x}_0 + \sum_{i=0}^{\infty}(\mathbf{x}_{i+1}(t) - \mathbf{x}_i(t))$, which is also the limit of the sequence $\{\mathbf{x}_k(t)\}$. Note $\mathbf{x}$(t) is continuous on $[t_0, t_0 + T]$ since each $\mathbf{x}_k(t)$ is continuous and because the convergence $\mathbf{x}_k(t) \to \mathbf{x}(t)$ is uniform. Thus, $\mathbf{x}(t)$ satisfies $\dot{\mathbf{x}}(t) = A(t)\mathbf{x}(t)$ and because both $\mathbf{x}$ and A are continuous in t, $\dot{\mathbf{x}}(t)$ is continuous too. In conclusion, the solution exists.

Now, even if a solution exists, it isn't clear whether it is unique or not. To prove uniqueness, define $\mathbf{z}(t) \triangleq \mathbf{x}_1(t) - \mathbf{x}_2(t)$ and show that it is equal to 0. Note that $\mathbf{z}(t)$ satisfies $\dot{\mathbf{z}}(t) = A(t)\mathbf{z}(t)$ with initial condition $\mathbf{z}(t_0) = 0$. Integrating both sides of the ODE to get an integral equation instead yields

$$\mathbf{z}(t) = \int_{t_0}^{t} A(s)\mathbf{z}(s)ds \Longrightarrow ||\mathbf{z}(t)|| \leq \int_{t_0}^{t} ||A(s)|| \cdot ||\mathbf{z}(s)||ds$$

Next, apply Gronwall-Bellman (Lemma 3.4) to the above equation with $\alpha = 0$, $\phi(t) = ||\mathbf{z}(t)||$, $\psi(t) = ||A(s)||$. Then we get $||\mathbf{z}(t)|| \leq 0$. Since the vector norms are positive-definite, if $||\mathbf{z}(t)|| \leq 0$, then $||\mathbf{z}(t)|| = 0$, implying $\mathbf{z}(t) = 0$. Therefore, $\mathbf{x}_1(t) = \mathbf{x}_2(t)$. Thus, uniqueness of solutions to LTV systems is also proved.

3.3 Fundamental Matrix for CT LTV Systems

Now, how is the solution space $\mathcal{X}$ constructed? Let us define $\psi_k(t) \in \mathbb{R}^n$ to be the solution of $\dot{x} = A(t)x$ with $\mathbf{x}_0 = e_k$, where e_k is the k-th standard basis vector of $\mathbb{R}^n$:

$$e_k = \begin{pmatrix} 0 \\ \vdots \\ 1 \\ \vdots \\ 0 \end{pmatrix}$$

Here, the 1 is in the kth position. Any solution to the system can then be expressed as a linear combination of these basis solutions:

$$\mathbf{x}(t) = \mathbf{x}_1(0)\psi_1(t) + \cdots + \mathbf{x}_n(0)\psi_n(t)$$

This motivates the following definition.

Definition 3.47 (*Fundamental Matrix*) The *fundamental matrix* $\Psi(t)$ is defined as the matrix

$$\Psi(t) \equiv \begin{bmatrix}\psi_1(t) \ \dots \ \psi_n(t)\end{bmatrix} \in \mathbb{R}^{n\times n}, \quad \forall t \geq 0,$$

which satisfies the following equation

$$\dot{\Psi}(t) = A(t)\Psi(t), \quad \Psi(0) = I.$$

Remark 3.7 Note that $\Psi(t)$ is non-singular for all $t \geq 0$ because the state $\mathbf{x}(t)$ is uniquely determined from $\mathbf{x}_0$:

$$\mathbf{x}(t) = \Psi(t)\mathbf{x}_0 \iff \Psi^{-1}(t)\mathbf{x}(t) = \mathbf{x}_0.$$

Therefore, we can use the LTV state-transition matrix Φ, as defined above, and also it can be defined by the fundamental matrix:

$$\Phi(t, \tau) = \Psi(t)\Psi(\tau)^{-1}.$$

Furthermore, we have that $\mathbf{x}(t_2) = \Phi(t_2, t_1)\mathbf{x}(t_1)$ for all $0 \leq t_1 \leq t_2$.

3.4 Fundamental Matrix for DT LTV Systems

For DT LTV systems, the fundamental matrix is derived recursively as follows:

$$\mathbf{x}[t+1] = A_t\mathbf{x}[t] \Longrightarrow \mathbf{x}[t] = A_{t-1}A_{t-2}\cdots A_0\mathbf{x}_0,$$

which is denoted by $\Psi_t = A_{t-1}A_{t-2}\ldots A_0$, the DT fundamental matrix.

Remark 3.8 For DT LTV systems, Ψ_t may be singular. Consequently, time reversal is generally not possible unless every matrix A_s is nonsingular $\forall s \in \{1\cdots t\}$.

3.5 Uniqueness of STM

The fundamental matrix is not unique. One can choose any basis of $\mathbb{R}^n$ (collection of n linearly independent vectors) $\mathcal{B} \triangleq \{b_1, \ldots, b_n\} \subseteq \mathbb{R}^n$. Then any $\Psi(t)$ is the solution to the $n \times n$ matrix ODE:

$$\begin{cases} \frac{d}{dt}X(t) = A(t)X(t) \\ X(t_0) = \begin{bmatrix} b_1 & \cdots & b_n \end{bmatrix} \end{cases} \quad \text{where } X(t) \in \mathbb{R}^{n\times n} \tag{3.5}$$

We mentioned $\Psi(t)$ is non-singular $\forall t \geq t_0$ since $\mathbf{x}(t)$ is uniquely determined from $\mathbf{x}_0$:

$$\mathbf{x}(t) = \Psi(t)\mathbf{x}_0 \iff \Psi^{-1}(t)\mathbf{x}(t) = \mathbf{x}_0$$

The STM can be defined from any fundamental matrix

$$\Phi(t,s) = \Psi(t)\Psi(s)^{-1}, \; t_0 \leq s \leq t$$

The STM is also the *unique* solution of

$$\frac{d}{dt}\Phi(t,t_0) = A(t)\Phi(t,t_0), \quad \Phi(t_0,t_0) = I \tag{3.6}$$

However, you may ask that how the STM $\Phi(t,s)$ is unique even if the fundamental matrix $\Psi(t)$ is not unique. There are two ways to prove this.

1. Use existence and uniqueness argument on matrix ODE (3.6).
2. Let $\Psi_1(t)$, $\Psi_2(t)$ be two different fundamental matrices (i.e., solutions to (3.5) with different initial conditions $\Psi_1(t_0)$, $\Psi_2(t_0)$). Show that $\Phi_1(t,s) = \Phi_2(t,s)\ \forall s, t$.

Recall the change of coordinates from linear algebra. To change the coordinate from one basis $\mathcal{B}_1 \triangleq \{b_1, \ldots, b_n\}$ to another $\mathcal{B}_2 \triangleq \{v_1, \ldots, v_n\}$:

$$\begin{bmatrix} v_1 & \cdots & v_n \end{bmatrix} = \begin{bmatrix} b_1 & \cdots & b_n \end{bmatrix} C$$

for some coordinate transformation $C \in \mathbb{R}^{n \times n}$ is invertible since $B_1 \xrightarrow{C} B_2$ means $B_2 \xrightarrow{C^{-1}} B_1$.

Thus, there exists the invertible coordinate transformation $C \in \mathbb{R}^{n \times n}$ s.t.

$$\Psi_2(t_0) = \Psi_1(t_0)C$$

By assumption, $\frac{d}{dt}\Psi_1(t) = A(t)\Psi_1(t)\ \forall t \geq t_0$.
By system linearity, $\frac{d}{dt}\Psi_1(t)C = A(t)\Psi_1(t)C$

$$\frac{d}{dt}\Psi_2(t) = \frac{d}{dt}\Psi_1(t)C = A(t)\Psi_1(t)C = A(t)\Psi_2(t)$$

Thus, this C applies for all time t: $\Psi_2(t) = \Psi_1(t)C\ \forall t \geq t_0$.
The STM $\Phi_1(t, s)$ corresponding to $\Psi_1(t)$ is given by

$$\Phi_1(t, s) = \Psi_1(t)\Psi_1(s)^{-1}$$

$$\Longrightarrow \Phi_2(t, s) = \Psi_2(t)\Psi_2(s)^{-1} = \Psi_1(t)CC^{-1}\Psi_1(s)^{-1} = \Phi_1(t, s)$$

$$\therefore \Phi_1(t, s) = \Phi_2(t, s)\ \forall s, t$$

which means that the STM $\Phi(t, s)$ is always unique.

The STM has the following properties

1. $\Phi(t, t) = I$
2. $\Phi^{-1}(t, s) = [\Psi(t)\Psi(s)^{-1}]^{-1} = \Psi(s)\Psi(t)^{-1} = \Phi(s, t)$
3. $\Phi(t, t_0) = \Phi(t, \tau)\Phi(\tau, t_0),\ \forall \tau \in [t_0, t]$,

3.6 Special Topic: Periodic LTV Systems

In Sect. 1.1, we discussed equivalence transformations and realizations for linear *time-invariant* (LTI) systems. Here, we will focus on the analogous case for linear *time-varying* (LTV) systems.

$$\begin{aligned}\dot{\mathbf{x}}(t) &= A(t)\mathbf{x}(t) + B(t)\mathbf{u}(t)\\ \mathbf{y}(t) &= C(t)\mathbf{x}(t) + D(t)\mathbf{u}(t)\end{aligned}$$

Instead of a constant coordinate transformation $P \in \mathbb{R}^n$, we now have a time-varying matrix $P(t) \in \mathbb{R}^{n \times n}$ which is continuously differentiable ($\mathcal{C}^1$) and nonsingular for all t. Consider a new state representation $\tilde{\mathbf{x}}(t) \triangleq P(t)\mathbf{x}(t)$. Then

$$\dot{\tilde{\mathbf{x}}}(t) = \dot{P}(t)\mathbf{x}(t) + P(t)\dot{\mathbf{x}}(t)$$

$$= \dot{P}(t)\mathbf{x}(t) + P(t)A(t)\mathbf{x}(t) + P(t)B(t)\mathbf{u}(t)$$
$$= \underbrace{\left(\dot{P}(t) + P(t)A(t)\right) P^{-1}(t)}_{\triangleq \tilde{A}(t)} \tilde{\mathbf{x}}(t) + \underbrace{P(t)B(t)}_{\triangleq \tilde{B}(t)} \mathbf{u}(t)$$

and $\mathbf{y}(t) = C(t)P^{-1}(t)\tilde{\mathbf{x}}(t) + D(t)\mathbf{u}(t)$.

Definition 3.48 (*Equivalence: LTV Case*) Given $P(t) \in \mathbb{R}^{n \times n}$ which is $\mathcal{C}^1$ and nonsingular for all t, the LTV system $\{A(t), B(t), C(t), D(t)\}$ is *(algebraically) equivalent* to $\{\tilde{A}(t), \tilde{B}(t), \tilde{C}(t), \tilde{D}(t)\}$, where

$$\tilde{A}(t) \triangleq \left(\dot{P}(t) + P(t)A(t)\right) P^{-1}(t), \quad \tilde{B}(t) \triangleq P(t)B(t), \quad \tilde{C}(t) \triangleq C(t)P^{-1}(t), \quad \tilde{D}(t) \triangleq D(t)$$

$P(t)$ is then called an *equivalence transformation.*

Note that if we apply the equivalence transformation $P(t)$ and then $P^{-1}(t)$,[1] we return back to the original system; this is because $\left(\frac{dP^{-1}}{dt} + P^{-1}(t)\tilde{A}(t)\right) P(t) = \frac{dP^{-1}}{dt} P(t) + P^{-1}(t)\dot{P}(t) + A(t) = A(t)$.

Definition 3.49 (*Lyapunov Transformation*) A transformation $P(t)$ is a *Lyapunov transformation* if $P \in \mathcal{C}^1$, nonsingular, and both $P(t)$ and $P^{-1}(t)$ are bounded for all t. Under a Lyapunov transformation $P(t)$, $\{A(t), B(t), C(t), D(t)\}$ and $\{\tilde{A}(t), \tilde{B}(t), \tilde{C}(t), \tilde{D}(t)\}$ are said to be *Lyapunov equivalent.*

3.6.1 Fundamental Matrices Under Equivalence Transformations

Let's briefly go back to the uncontrolled system, where $B(t) \equiv 0$. Recall that any fundamental matrix $\Psi(t)$ of $\dot{\mathbf{x}}(t) = A(t)\mathbf{x}(t)$ satisfies the matrix ODE $\dot{\Psi}(t) = A(t)\Psi(t)$, $\Psi(0) = \Psi_0$.

Given an equivalence transformation $P(t)$ and a fundamental matrix $\Psi(t)$ of the LTV system $\dot{\mathbf{x}}(t) = A(t)\mathbf{x}(t)$, we can prove that $\tilde{\Psi}(t) \triangleq P(t)\Psi(t)$ is a fundamental matrix of the equivalent system $\dot{\mathbf{x}}(t) = \tilde{A}(t)\mathbf{x}(t)$, where $\tilde{A}(t)$ is given as in Definition 3.48. Note that the main property to prove in order to show that it's a fundamental matrix is that it must satisfy the matrix ODE $\dot{\tilde{\Psi}}(t) = \tilde{A}(t)\tilde{\Psi}(t)$.

First, since P and Ψ are both nonsingular for all t, note that $\tilde{\Psi}$ is also nonsingular for all t. Now let's verify the ODE equation:

$$\dot{\tilde{\Psi}}(t) = \frac{d}{dt}(P(t)\Psi(t))$$
$$= \dot{P}(t)\Psi(t) + P(t)\dot{\Psi}(t) + (\dot{P}(t) + P(t)A(t))\Psi(t)$$

[1] Note that if $P(t)$ is an equivalence transformation, $P^{-1}(t)$ is also $\mathcal{C}^1$.

$$= \underbrace{(\dot{P} + PA)P^{-1}(t)}_{\tilde{A}(t)} \tilde{\Psi}(t)$$

Now we prove the following property, which suggests the equivalence between a LTV system and a LTI system.

Theorem 3.13 *There exists an equivalence transformation $P(t)$ which transforms $A(t)$ to any given constant matrix $A_0 \in \mathbb{C}^{n\times n}$.*[2]

Proof Our proof will be divided into several steps. First, we construct an appropriate $P(t)$. Let $\Psi(t)$ be a fundamental matrix for the original LTV system $\dot{\mathbf{x}}(t) = A(t)\mathbf{x}(t)$. Note that the fundamental matrix for an LTI system $\dot{\mathbf{x}}(t) = A_0\mathbf{x}(t)$ has form $\tilde{\Psi}(t) = e^{A_0 t}\Psi_0$, and we choose initial condition $\Psi_0 = I$ for simplicity. We can rewrite $\tilde{\Psi}(t) = P(t)\Psi(t)$ as $P(t) = \tilde{\Psi}(t)\Psi^{-1}(t)$. We choose $P(t) = e^{A_0 t}\Psi^{-1}(t)$.

Obtain the derivative of the inverse $(d/dt)(\Psi^{-1}(t))$ by invoking the identity $\Psi^{-1}\Psi = I$ and taking the derivatives of both sides. We get

$$(d/dt)(\Psi^{-1}(t)) = -\Psi^{-1}(t)A(t)$$

By rote calculation and applying Definition 3.48.:

$$\begin{aligned}
\tilde{A}(t) &= (P(t)A(t) + \dot{P}(t))P^{-1}(t) \\
&= \left(e^{A_0 t}\Psi^{-1}(t)A(t) + A_0 e^{A_0 t}\Psi^{-1}(t) + e^{A_0 t}\frac{d}{dt}(\Psi^{-1}(t))\right)\Psi(t)e^{-A_0 t} \\
&= \cdots = A_0
\end{aligned}$$

Hence, this coordinate transformation $P(t)$ indeed changes LTV system $\dot{\mathbf{x}}(t) = A(t)\mathbf{x}(t)$ to LTI system $\dot{\tilde{\mathbf{x}}}(t) = A_0\tilde{\mathbf{x}}(t)$. ■

3.6.2 *Introductory Floquet Theory*

Floquet theory (attributed to Gaston Floquet, 1883) is a branch of ODE theory which studies *periodic*, linear differential equations:

$$\dot{\mathbf{x}}(t) = A(t)\mathbf{x}(t), \text{ where } \exists\, T > 0 \text{ s.t. } A(t) = A(t+T)$$

We won't go into the full details of Floquet theory in this book, but it is still an interesting topic to know.

Remark 3.9 If Ψ is a fundamental matrix of the LTV system with periodic $A(t)$, it turns out that $\Psi(t+T)$ is also a fundamental matrix. To prove this, simply show that it satisfies the matrix ODE

[2] If $A_0 \in \mathbb{R}^{n\times n}$, then $P(t)$ can be chosen to be also real-valued.

$$\dot{\Psi}(t+T) = A(t+T)\Psi(t+T) = A(t)\Psi(t+T)$$

To show that there exists a periodic equivalence transformation $P(t)$ such that $\Psi(t) = P^{-1}(t)e^{A_0 t}$, we will simply use the following two facts without proof (although, they are not too hard to show; see footnotes).

- **Fact 1**[3]: For periodic LTV $\dot{\mathbf{x}} = A(t)\mathbf{x}$, there exists a nonsingular, constant matrix $Q \in \mathbb{R}^{n\times n}$ such that $\Psi(t+T) = \Psi(t)Q$ for all t. This Q is often called the *monodromy matrix*.
- **Fact 2**[4]: For monodromy matrix $Q \in \mathbb{R}^{n\times n}$ corresponding to periodic LTV $\dot{\mathbf{x}} = A(t)\mathbf{x}$, there exists a constant matrix $A_0 \in \mathbb{C}^{n\times n}$ such that $e^{A_0 T} = Q$.

To get an explicit form of Q in terms of the fundamental matrix, we can set $t = 0$ and take advantage of the fact that Q is constant.

$$\Psi(T) = \Psi(0)Q \Longrightarrow Q = \Psi^{-1}(0)\Psi(T)$$

Now apply the construction of $P(t)$ from the proof of Theorem 3.13. Define $P(t) \triangleq e^{A_0 t}\Psi^{-1}(t)$. Note that

$$P(t+T) = e^{A_0(t+T)}\Psi^{-1}(t+T) = e^{A_0 t}e^{A_0 T} \cdot e^{-A_0 T}\Psi^{-1}(t) = P(t)$$

Thus, a period of $P(t)$ is T, the same as the period of the original LTV system.

The decomposition of the fundamental matrix $\Psi(t) = P^{-1}(t)e^{A_0 t}$ tells us that the solution trajectories $\mathbf{x}(t)$ of periodic LTV systems $\dot{\mathbf{x}} = A(t)\mathbf{x}$ can be decomposed into a periodic part and an exponential part.

[3] First, define $Q = \Psi^{-1}(0)\Psi(T)$. Note that $t \mapsto \Psi(t)Q$ is a fundamental matrix, and coincides with $t \mapsto \Psi(t+T)$ at $t = 0$. By the uniqueness theorem for ODEs, we are done.

[4] This is a special case of: If $A \in \mathbb{C}^{n\times n}$ is invertible, there exists a matrix $\log(A) \in \mathbb{C}^{n\times n}$ such that $e^{\log(A)} = A$. Proof sketch: Enough to show this for a Jordan block of the form $J = \lambda I + N$ where $\lambda \neq 0$. A suggestive formula: $\log(J) = \log(\lambda) + \log(I + N/\lambda) = \log(\lambda)I + \sum_{n\geq 1}(-1)^{n+1}(N/\lambda)^n/n$. This equation is not a formal proof, but a nice mnemonic motivated by the Taylor series of $\log(1+x)$. You should check that $e^{\log(J)} = J$.

Chapter 4
Problems and Exercises

Mathematical Review

Problem 1: Functions. Consider the vector-valued mapping $f : \mathbb{R}^3 \to \mathbb{R}^3$

$$f(\mathbf{x}) = A\mathbf{x}, \quad A \triangleq \begin{bmatrix} 1 & 2 & 0 \\ 0 & 3 & 0 \\ 0 & 0 & 4 \end{bmatrix}$$

Is f a well-defined function? Is it surjective? Injective? Justify your answers.

Problem 2: Vector Spaces. Identify whether the following objects are vector spaces or not.

(a) real continuous scalar-valued functions $\mathcal{C}^0 \triangleq \{f : \mathbb{R} \to \mathbb{R}\}$.
(b) the set of points $\mathbf{x} \in \mathbb{R}^3$ satisfying $x_1^2 + x_2^2 + x_3^2 = 1$.
(c) the set of solutions to $\dot{y}(t) + ay(t) = 0$, where $y \in \mathcal{C}^1, a \in \mathbb{R}$.
(d) $\mathcal{V} \triangleq (\mathbb{R}^n, +, \cdot)$ over field $\mathbb{F} = \mathbb{C}$ and where $+$ is the usual addition and for $c \in \mathbb{C}, v \in \mathcal{V}, cv = |c|v$

Problem 3: Subspaces. Let $\mathcal{W}_1, \mathcal{W}_2$ be two subspaces of some vector space $\mathcal{V}$ on field $\mathbb{F}$. Identify whether the following operations are also subspaces or not.

(a) the intersection $\mathcal{W}_1 \cap \mathcal{W}_2$.
(b) the union $\mathcal{W}_1 \cup \mathcal{W}_2$.
(c) the Minkowski sum of two subspaces $\mathcal{W}_1 \oplus \mathcal{W}_2 \triangleq \{\mathbf{w}_1 + \mathbf{w}_2 | \mathbf{w}_1 \in \mathcal{W}_1, \mathbf{w}_2 \in \mathcal{W}_2\}$

Problem 4: Sequence Subspaces. Consider a set of sequences $\mathcal{W} \triangleq \{f_k\}_{k=0}^{\infty}$ satisfying $f_k = f_{k-1} + f_{k-2}$, starting from arbitrary real numbers f_0 and f_1. Is $\mathcal{W}$ a subspace of the vector space $\mathcal{V} \triangleq (V, \mathbb{R}, +, \cdot)$, where V is the set of all real-numbered sequences?

SJ. Han, *Linear Systems*, KAIST Research Series,
https://doi.org/10.1007/978-981-96-9774-8_4

Problem 5: Eigenvalues and Eigenvectors. Let $A \in \mathbb{R}^n$ be a matrix, and let p be the polynomial function (1.18) defined in Sect. 1.4.4. Show that if λ is an eigenvalue of A with corresponding eigenvector $\mathbf{v}$, then $p(\lambda)$ is an eigenvalue of A with the same eigenvector $\mathbf{v}$.

Problem 6: Linear Independence. Let V be the set of 2-tuples whose entries are complex-valued rational functions. Let $\mathbf{v}_1, \mathbf{v}_2 \in V$ be defined as

$$\mathbf{v}_1 = \begin{bmatrix} \dfrac{1}{s-1} \\ \dfrac{1}{s+3} \end{bmatrix}, \quad \mathbf{v}_2 = \begin{bmatrix} \dfrac{s+2}{(s-1)(s+3)} \\ \dfrac{1}{s-1} \end{bmatrix}$$

Are $\mathbf{v}_1$ and $\mathbf{v}_2$ linearly independent over the field of rational functions? What about over the field of real numbers?

Problem 7: Basis. Let $\mathcal{W}$ be a subspace of $(\mathbb{R}^5, \mathbb{R}, +, \cdot)$, defined by

$$\mathcal{W} \triangleq \left\{[x_1, x_2, \ldots, x_5]^\top | x_1 = 3x_3, x_2 = 5x_5\right\}$$

Compute a basis for $\mathcal{W}$.

Problem 8: Jordan Forms.

(a) If $A \in \mathbb{R}^{n\times n}$ with characteristic polynomial $\chi(\lambda) = (x-\lambda_1)^{n_1} \cdots (x-\lambda_r)^{n_r}$ for some $r \in \mathbb{N}$, what is the trace of A?
(b) How many possible Jordan forms are there for a 6×6 complex matrix with characteristic polynomial $(x+2)^3(x-1)^2(x+1)$?
(c) Let $A \in \mathbb{R}^{n\times n}$ have eigenvalue λ and define

$$V_\lambda \triangleq \{\mathbf{v} \in \mathbb{R}^n : (A-\lambda I)^k \mathbf{v} = 0 \text{ for some } k$$

Show there always exists an $k \in \mathbb{N}$ such that $V_\lambda = \text{Ker}(A-\lambda I)^k$.
(d) Show that, for the value of $k \in \mathbb{N}$ which satisfies part c), $\text{Im}(A-\lambda I)^k$ and $\text{Ker}(A-\lambda I)^k$ are subspaces of $\mathbb{R}^n$ that are invariant under A. Then use the Range-Nullspace decomposition theorem to show

$$\mathbb{R}^n = \text{Im}(A-\lambda I)^k \oplus \text{Ker}(A-\lambda I)^k$$

is a decomposition of $\mathbb{R}^n$, where $\oplus$ denotes the direct sum.
(e) How would you generalize part d to show that

$$\mathbb{R}^n = \text{Ker}(A-\lambda_1 I)^{n_1} \oplus \text{Ker}(A-\lambda_2 I)^{n_2} \oplus \cdots \oplus \text{Ker}(A-\lambda_r I)^{n_r}$$

A few sentences of explanation will suffice; no formal proof is necessary.

Problem 9: Norms. Consider the inner product space $\mathcal{V}$ and two vectors $\mathbf{x}$ and $\mathbf{y}$ in $\mathcal{V}$. Show, using the properties of the inner product, the following relationship (called the *parallelogram law*).

$$\|\mathbf{x}+\mathbf{y}\|^2 + \|\mathbf{x}-\mathbf{y}\|^2 = 2\,\|\mathbf{x}\|^2 + 2\,\|\mathbf{y}\|^2$$

where $\|\cdot\|$ is the norm induced by the inner product on $\mathcal{V}$.

System Models and Classification

Problem 10. Rewrite the nth-order scalar linear ODE

$$y^{(2n)}(t) + \alpha_{2n-2}(t)y^{(2n-2)}(t) + \cdots + \alpha_0(t)y(t) = \beta_0(t)u(t) + \beta_n(t)u^{(n)}(t)$$

into state-space model form. What are the dimensions of the matrices A, B, C, and D?

Problem 11: Pendulum and Double-Pendulum. Find state ODEs to model the pendulum and double-pendulum systems shown in Fig. 4.1.

Use any small-angle approximations you can to rewrite your ODEs as state-space models.

Problem 12: State versus Output Feedback. Consider a dynamical system described by

$$\dot{\mathbf{x}} = \begin{bmatrix} 0 & 1 \\ 8 & -4 \end{bmatrix}\mathbf{x} + \begin{bmatrix} 2 \\ 1 \end{bmatrix}u, \quad y = \begin{bmatrix} 1 & 2 \end{bmatrix}\mathbf{x}$$

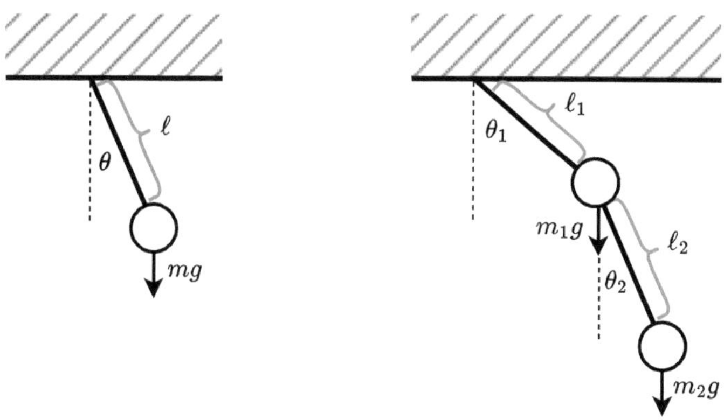

Fig. 4.1 [Left] The single pendulum. [Right] The double pendulum

For each of the cases below, derive a state-space representation of the resulting closed-loop system and determine the characteristic equation of the resulting closed loop A_{cl} matrix.

(a) $u = -\begin{bmatrix} f_1 & f_2 \end{bmatrix} \mathbf{x}$
(b) $u = -ky$

Problem 13: [1] Show that if all eigenvalues of $A \in \mathbb{R}^{n \times n}$ are distinct, then $(sI - A)^{-1}$ can be expressed as

$$\sum_{i=1}^{n} \frac{1}{s - \lambda_i} \mathbf{q}_i \mathbf{p}_i$$

where $\mathbf{q}_i, \mathbf{p}_i \in \mathbb{R}^n$ are the right and left eigenvectors, respectively, of A associated with λ_i.

Problem 14: Classification I. For each of the following systems, determine whether or not it is linear, time-invariant, or causal.

$$y[n] = e^{x[n]}, \quad y(t) = x\left(\frac{t}{4}\right), \quad y(t) = 5x(t)\sin(\omega(t+3))$$

$$y[n] = \begin{cases} +1 & \text{if } x[n] \geq 0 \\ -1 & \text{if } x[n] < 0 \end{cases}, \quad y(t) = \int_{t-1}^{t+2} x(s)ds, \quad y[n] = 3(x[n+1]u[n] - x[n]) + 10$$

Problem 15: Classification II. Classify the following systems. Clearly justify your answers.

(a) Are the following systems linear or nonlinear?

$$y[n] = \delta[n], \qquad \mathbf{y}(t) = A\mathbf{x}(t) \in \mathbb{R}^n,\ A \in \mathbb{R}^{n \times n}, \qquad y[n] = x[-n]$$

(b) Are the following systems TI or TV?

$$y(t) = x(2t), \qquad y[n] = \sin(x[n]), \qquad y(t) = \int_{-\infty}^{t} x(s)ds$$

(c) Are the following systems causal or noncausal?

$$y[n] = 3x[n] - x^2[n], \qquad y[n] = x[-n],$$

$$y[n] = x[n]\cos[\omega(n+1)],\ \omega \neq 0, \qquad y(t) = \frac{dx(t)}{dt} = x'(t)$$

(d) Are the following systems memoryless or not?

$$y[n] = 3x[n] - x^2[n], \qquad y[n] = x[-n], \qquad y(t) = x\left(\frac{t}{5}\right),$$

$$y[n] = 3, \qquad y(t) = t$$

(e) Check the invertibility of the following systems. If the system is invertible, what is its inverse? If it is not invertible, give an example of two different input signals x_1 and x_2 which produce the same output y.

$$y(t) = \alpha x(t), \alpha \neq 0, \qquad y[n] = x^2[n], \qquad y(t) = \int_{-\infty}^{0} x(s)ds$$

Problem 16: Classification III. Suppose we are given the following DT state update equation $x[k+1] = ax[k]$ with initial condition $x[0] = x_0$, and measurement output map $y[k] = x[k]$. Here, $x[0], x[1], \ldots, a \in \mathbb{R}$ are scalars. Is this a linear system? Why or why not? If so, derive the state transition function.

Problem 17: Classification IV. Suppose that the output of a system can be represented

$$y(t) = \int_{-\infty}^{t} e^{-(t-\tau)} u(\tau) d\tau$$

Show that the system is time-invariant. You may choose the input space $\mathcal{U}$ to be the set of bounded, piecewise-continuous real-valued functions on $(-\infty, \infty)$.

Problem 18: Orbital Satellite [2, 3]. Consider the system comprised of a satellite orbiting around the Earth. If we model both the Earth and the satellite as particles, the normalized equations of motion simplify to 2D, with Lagrangian

$$\mathcal{L} \triangleq \frac{1}{2}\dot{r}^2 + \frac{1}{2}r^2\dot{\theta}^2 - \frac{k}{r}$$

where (r, θ) is the polar coordinate representation of satellite relative to the surface of the Earth.

(a) Derive the equations of motion from the Lagrangian $\mathcal{L}$. You should get two equations for $\ddot{\theta}$ and $\ddot{r}$, with inputs u_1 and u_2 for the tangential and radial forces due to the thrusters.
(b) Note that the reference orbit with $u_1 = u_2 = 0$ is circular with $r(t) = p$ and $\theta(t) = \omega t$, for some constants p and ω. Linearize the equations of motion you got from part a) around this orbit. How many state variables are there?

Problem 19: Inverted Double Pendulum on a Cart. Consider the inverted *double-pendulum* on a cart system shown in Fig. 4.2. For simplicity, we will assume both pendulums have the same mass m, length l (with center of mass located at length $l/2$), and moment of inertia I. The mass of the cart is M and a force $u(t)$ is applied to it to control the double-pendulum. We assume there is no friction between the surface and the cart's wheels.

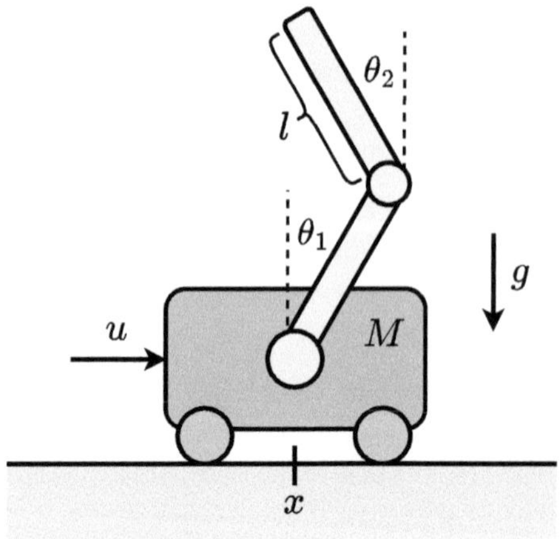

Fig. 4.2 The inverted double-pendulum on a cart

The equations of motion for this system are given by

$$(M+2m)\ddot{x}+\frac{3}{2}ml\ddot{\theta}_1\cos\theta_1+\frac{1}{2}ml\ddot{\theta}_2\cos\theta_2$$
$$-\frac{3}{2}ml\dot{\theta}_1^2\sin\theta_1-\frac{1}{2}ml\dot{\theta}_2^2\sin\theta_2=u \tag{4.1a}$$
$$\left(I+\frac{5}{4}ml^2\right)\ddot{\theta}_1+\frac{3}{2}ml\ddot{x}\cos\theta_1+\frac{1}{2}ml^2\ddot{\theta}_2\cos(\theta_1-\theta_2)$$
$$+\frac{1}{2}ml^2\dot{\theta}_2^2\sin(\theta_1-\theta_2)-\frac{3}{2}mgl\sin\theta_1=0 \tag{4.1b}$$
$$\frac{1}{2}ml\ddot{x}\cos(\theta_2)+\frac{1}{2}ml^2\ddot{\theta}_1\cos(\theta_1-\theta_2)+\left(I+\frac{1}{4}ml^2\right)\ddot{\theta}_2^2$$
$$-\frac{1}{2}ml\dot{\theta}_1^2\sin(\theta_1-\theta_2)-\frac{1}{2}mgl\sin\theta_2=0 \tag{4.1c}$$

Linearize this system around the stationary upright position $\theta_1=\dot{\theta}_1=\theta_2=\dot{\theta}_2=0$. Represent your system in state-space form with $\mathbf{x}(t)\triangleq[x,\dot{x},\theta_1,\dot{\theta}_1,\theta_2,\dot{\theta}_2]^\top$. (*Hint*: Use the small angle approximations $\sin\theta\approx\theta$ and $\cos\theta\approx 1$.)

State and Fundamental Transition Matrices

Problem 20: Matrix Exponential Properties [1]. Show that if λ is an eigenvalue of $A\in\mathbb{R}^{n\times n}$ with eigenvector $\mathbf{x}\in\mathbb{R}^n$, then $f(\lambda)$ is an eigenvalue of $f(A)$ with the same eigenvector $\mathbf{x}$.

Problem 21: Matrix Exponential Properties [1]. Show that functions of the same matrix commute, i.e., $f(A)g(A)=g(A)f(A)$. Note: consequently, we have

$Ae^{At} = e^{At}A.$

Problem 22: Matrix Exponential Properties [1]. Suppose $A \in \mathbb{R}^{n \times n}$ is a matrix where all eigenvalues are distinct. Let $\mathbf{q}_i$ be a right eigenvector of A associated with eigenvalue λ_i, i.e., $A\mathbf{q}_i = \lambda_i \mathbf{q}_i$. Define

$$Q \triangleq \begin{bmatrix} \mathbf{q}_1 \; \mathbf{q}_2 \cdots \mathbf{q}_n \end{bmatrix}, \quad P \equiv \begin{bmatrix} \mathbf{p}_1 \\ \mathbf{p}_2 \\ \vdots \\ \mathbf{p}_n \end{bmatrix} \triangleq Q^{-1}$$

Show that $\mathbf{p}_i$ is a left eigenvector of A associated with the same λ_i, i.e., $\mathbf{p}_i A = \lambda_i \mathbf{p}_i$.

Problem 23: [4] For the LTV system $\dot{\mathbf{x}}(t) = A(t)\mathbf{x}(t)$, $\mathbf{x}(t_0) = \mathbf{x}_0$, show that

$$\|\mathbf{x}(t)\| \leq \|\mathbf{x}_0\| \exp\left(\int_{t_0}^{t} \|A(s)\| \, ds\right), \quad \forall\, t \geq t_0$$

Problem 24. Given

$$A \triangleq \begin{bmatrix} -1 & 3 & -1 \\ -3 & 6 & -1 \\ -3 & 3 & 1 \end{bmatrix}$$

determine $\sin(e^A)$.

Problem 25. Suppose $A \in \mathbb{R}^{n \times n}$ is such that $\det(A) = 0$. Is $\det(e^A) = 0$?

Problem 26: Existence and Uniqueness of Solutions in the Linear Case [3]. Let $A(t)$ and $B(t)$ be respectively $n \times n$ and $n \times m$ matrices whose elements are real-valued piecewise-continuous functions on $\mathbb{R}^+$. Let u be a piecewise-continuous function from $\mathbb{R}^+$ to $\mathbb{R}^m$. Show that for any fixed such u, the differential equation

$$\dot{\mathbf{x}}(t) = A(t)\mathbf{x}(t) + B(t)\mathbf{u}(t)$$

satisfy the conditions of the fundamental theorem.

Problem 27: Perturbed Nonlinear Systems. Suppose that a physical system obeys the differential equation

$$\dot{\mathbf{x}} = f(t, \mathbf{x}), \quad \mathbf{x}(t_0) = \mathbf{x}_0, \quad \forall\, t \geq t_0$$

where f (not necessarily a linear function) obeys the conditions of the fundamental theorem. Suppose that as a result of some perturbation, the equation is transformed to

$$\dot{\mathbf{z}} = f(t, \mathbf{z}) + g(t), \quad \mathbf{z}(t_0) = \mathbf{x}_0 + \delta\mathbf{x}_0, \quad \forall\, t \geq t_0$$

Given that for $t \in [t_0, t_0 + T]$, $\|f(t)\| \leq \epsilon_1$, and $\|\delta\mathbf{x}_0\| \leq \epsilon_0$, find a bound on $\|\mathbf{z}(t) - \mathbf{x}(t)\|$ that is valid on $[t_0, t_0 + T]$.

Problem 28. Consider the uncontrolled LTI system of the form

$$\dot{\mathbf{x}}(t) = \begin{bmatrix} a & 2 \\ -2 & -1 \end{bmatrix} \mathbf{x}(t)$$

where $a \in \mathbb{R}$.

(a) Express the trace, determinant, characteristic polynomial, and eigenvalues in terms of a.
(b) For specific values of $a \in \{-6, -2, 1, 2, 3, 4, 5\}$, classify the stability of the system modes (stable, marginally stable, unstable).
(c) For each value of a in part (b), plot the solution trajectories

i. by computing the matrix exponential and directly plotting $\mathbf{x}(t)$.
ii. by discretizing the system with $\Delta t = 0.01$ and simulating the difference equation.

For each plot i. and ii., make two types of figures. The first figure is a plot of $x_1(t)$ and $x_2(t)$ versus time; the second figure is a plot of $x_1(t)$ versus $x_2(t)$. Organize all your plots in a 7×2 collection of subplots, with labeled legends where applicable. Make sure to choose various different initial conditions. (The second type of figure is called a *phase portrait*, and we will be seeing more on this when discussing stability in Part II.)

Problem 29: Numerical Integration. Suppose that a conservative physical system is modeled by the linear differential equation $\dot{\mathbf{x}} = A\mathbf{x}$, with $A \in \mathbb{R}^{n \times n}$. Further suppose it is normalized so that along any trajectory, $\|\mathbf{x}(t)\|^2$ is constant for all time.

(a) What can you say about the eigenvalues of A?
(b) Because continuous-time differential equations cannot be simulated on a computer, we commonly use one of the following three numerical methods to discretize and solve it:

- *Forward Euler.* $\boldsymbol{\xi}_{k+1} = (I + hA)\boldsymbol{\xi}_k$, $\boldsymbol{\xi}_0 = \mathbf{x}(0)$.
- *Backward Euler.* $\boldsymbol{\xi}_{k+1} = (I - hA)^{-1}\boldsymbol{\xi}_k$, $\boldsymbol{\xi}_0 = \mathbf{x}(0)$.
- *Forward and Backward Euler.* $\boldsymbol{\xi}_{k+1} = (I + \frac{h}{2}A)(I - \frac{h}{2}A)^{-1}\boldsymbol{\xi}_k$, $\boldsymbol{\xi}_0 = \mathbf{x}(0)$.

where h is the stepsize of the numerical method. Which of these three methods is appropriate to solve our differential equation, especially with $h < 2 \min_i |\lambda_i(A)|$?

Problem 30: Lipschitz [2, 3]. Consider the following two systems of ODEs.

$$(A) \triangleq \begin{cases} \dot{x}_1 = -x_1 + 2e^t \cos(x_1 - x_2) \\ \dot{x}_2 = -x_2 + 5e^t \sin(x_1 - x_2) \end{cases}, \qquad (B) \triangleq \begin{cases} \dot{x}_1 = -x_1 + 3x_1 x_2 \\ \dot{x}_2 = -x_2 \end{cases}$$

(a) Which one of these systems satisfy a global Lipschitz condition?
(b) For system (B), your classmate claims that the solution trajectories are uniquely defined for all possible initial conditions. Furthermore, he claims that they all tend to zero regardless of initial condition. Is your classmate correct?

Problem 31: Lipschitz. Consider the pendulum on the left subfigure of Fig. 4.1.

(a) Rederive the (nonlinear) equation of motion for this pendulum, now assuming there is friction with coefficient k and constant input torque T.
(b) Representing the system as $\dot{\mathbf{x}} = f(\mathbf{x})$ for appropriately-defined state $\mathbf{x}$, determine whether f is locally or globally Lipschitz. If it is locally Lipschitz, find the radius of the ball $\mathcal{B}_r = \{\mathbf{x} : \|\mathbf{x}(t)\| \leq r\}$ for which it is.

Problem 32: Discretization. Use the three methods discussed in Chap. 2 to discretize the following continuous-time system with $\Delta t = 0.1$. You may use MATLAB if you'd like.

$$\dot{\mathbf{x}}(t) = \begin{bmatrix} -2 & 1 \\ 0 & 2 \end{bmatrix} \mathbf{x}(t) + \begin{bmatrix} 1 \\ 0 \end{bmatrix} u(t),$$

Problem 33: Equivalence Transformation Proofs. Let the constant matrix $P \in \mathbb{R}^{n\times n}$ be the equivalence transformation (i.e., $\tilde{\mathbf{x}} = P\mathbf{x}$) which changes LTI system $\dot{\mathbf{x}}(t) = A\mathbf{x}(t) + B\mathbf{u}(t)$ to LTI system $\dot{\tilde{\mathbf{x}}}(t) = \tilde{A}\tilde{\mathbf{x}}(t) + \tilde{B}\mathbf{u}(t)$.

(a) Prove that $(I - \tilde{A})^{-1} = P(I - A)^{-1}P^{-1}$.
(b) Use *induction* to prove that $\tilde{A}^k = PA^kP^{-1}$ for all $k \in \mathbb{Z}_{\geq 0}$.

Problem 34. Compute the fundamental and state transition matrices of the following systems:

$$\text{(a) } \dot{\mathbf{x}}(t) = \begin{bmatrix} 0 & 1 \\ 0 & t \end{bmatrix} \mathbf{x}(t) \quad \text{(b) } \dot{\mathbf{x}}(t) = \begin{bmatrix} -1 & e^{2t} \\ 0 & -1 \end{bmatrix} \mathbf{x}(t) \quad \text{(c) } \dot{\mathbf{x}}(t) = \begin{bmatrix} \sin t & \cos t & 1 \\ 0 & \sin t & \cos t \\ 0 & 0 & \sin t \end{bmatrix} \mathbf{x}(t)$$

Problem 35: [3] Compute the fundamental and state transition matrices when the A matrix of a CT linear system is given as follows:

$$\text{(a) } A(t) = \begin{bmatrix} -1 & 0 \\ 2 & -3 \end{bmatrix} \quad \text{(b) } \begin{bmatrix} -2t & 0 \\ 1 & -1 \end{bmatrix} \quad \text{(c) } \begin{bmatrix} 0 & \omega(t) \\ -\omega(t) & 0 \end{bmatrix}$$

Hint: for part (c), define $\Omega(t) \triangleq \int_0^t \omega(s)ds$ and consider the matrix

$$\begin{bmatrix} \cos\Omega(t) & \sin\Omega(t) \\ -\sin\Omega(t) & \cos\Omega(t) \end{bmatrix}$$

Problem 36. Consider the general, uncontrolled linear system

$$\mathbb{R}^n \ni \dot{\mathbf{x}}(t) = A(t)\mathbf{x}(t), \quad \mathbf{x}(0) = \mathbf{x}_0 \tag{4.2}$$

with continuous $A(t)$ and state-transition matrix $\Phi(t, \tau)$, where $n \in \mathbb{N}$ is the state dimension.

(a) Consider also the linear system

$$\dot{\mathbf{z}}(t) = A(t)\mathbf{z}(t) + \mathbf{x}(t), \quad \mathbf{z}(0) = \mathbf{z}_0$$

Express both $\mathbf{x}(t)$ and $\mathbf{z}(t)$ as a function of $\mathbf{x}_0$, $\mathbf{z}_0$, and Φ. (There should be no integrals in your solutions.)

(b) Now, suppose $n = 2$ and (4.2) is an LTI system with $A(t) \equiv A$ given by

$$A = \begin{bmatrix} 0 & 1 \\ -10 & -7 \end{bmatrix}$$

Compute its state transition matrix Φ. Use change-of-coordinates to compute its characteristic modes. Classify the stability type of each mode.

(c) Repeat part (b) when A is instead given by

$$A = \begin{bmatrix} 3 & 2 \\ -2 & -1 \end{bmatrix}$$

(d) Now, suppose $n = 2$ and (4.2) is LTV with $A(t)$ given by

$$A(t) = \begin{bmatrix} -\alpha & \cos(\omega t) \\ \sin(\omega t) & -\beta \end{bmatrix}$$

where $\alpha, \beta, \omega > 0$ are constants. Prove that (4.2) is uniformly asymptotically stable if $2\alpha\beta > \sqrt{\alpha^2 + \beta^2}$.

Problem 37: Solving Matrix ODEs [4]. Consider the $n \times n$ matrix ODE $\dot{X}(t) = X(t)A(t)$, with initial condition $X(t_0) = X_0$. Express the solution in terms of an appropriate transition matrix. Use this to determine a closed-form expression of the solution to

$$\dot{X}(t) = A_1(t)X(t) + X(t)A_2^\top(t) + Q(t), \quad X(t_0) = X_0$$

where $A(t)$, $A_1(t)$, $A_2(t)$ are not all necessarily the same matrix. We will see a lot of equations that look like this when we discuss *Lyapunov equations* in Part II. (*Hint*: Take the transpose of the original matrix ODE. Use Leibniz's Rule.)

Problem 38: Extras on Discretization. A process $\{\mathbf{x}(t), t \geq 0\}$ (or $\{\mathbf{x}_t\}_{t \in \mathbb{N}}$ in discrete-time) is said to be *time-reversible* if the dynamics of the process remain well-defined when time runs backwards (i.e., there is a one-to-one mapping from $\mathbf{x}(t)$ to $\mathbf{x}(s)$ for any $s \leq t$). In continuous-time linear systems, we've seen this hold true for any $A(t)$ which is piecewise continuous, but for discrete-time systems, this is only true if each $\{A_s, s = 1, \ldots, t\}$ is nonsingular.

(a) Can you give an example of a continuous-time linear system which is nonreversible? If no such example exists, explain why (rigorous proof not required).

(b) Consider the following discretized system:

$$\dot{\mathbf{x}}(t) = A\mathbf{x}(t) + B\mathbf{u}(t) \Longrightarrow \mathbf{x}_{t+1} = \underbrace{e^{A\Delta t}}_{\triangleq A_d} \mathbf{x}_t + \underbrace{\left(\int_0^{\Delta t} e^{As} ds\right) B}_{\triangleq B_d} \mathbf{u}_t$$

This is sometimes known as a *sampled-data system* with sampling period $\Delta t \in \mathbb{R}^+$, i.e., a system where continuous-time dynamics are controlled with a piecewise-continuous input signal $u(t)$ (e.g., a digital clock). Is this system time-reversible? Why or why not?

(c) Consider the class of DT systems called *finite memory systems*, whose trajectories look like $x_t = \Phi_t \mathbf{x}_0$ with $A_\tau = 0$ after some time $\tau \in \mathbb{N}$ (i.e., the initial state influences the system evolution for only up to τ, and its free motion is zero after τ). Is this system time-reversible? Why or why not?

References

1. Chi-Tsong Chen. *Linear System Theory and Design*. Saunders College Publishing, 1984.
2. Shankar Sastry. *Lecture notes for EE221A: Linear Systems*. 2013.
3. Claire Tomlin. *Lecture notes for EE221A: Linear Systems*. 2017.
4. Wilson J. Rugh. *Linear System Theory*. 2nd ed. Prentice Hall, 1996.

Part II
Linear Stability

Chapter 5
Input-Output Stability

An important prerequisite question that must be addressed for any dynamical system is *stability*. how does the system behave when there is no input signal $\mathbf{u}(t)$ or $\mathbf{w}(t)$ given to it (i.e., "open-loop stability")? If we give it a bounded input signal $\mathbf{u}(t)$ such that $\|\mathbf{u}(t)\| \leq \bar{u}$ for some $\bar{u}$, does it also yield a bounded output signal $\mathbf{y}(t)$? What is the behavior of all signals inside the system, i.e., not just the input $\mathbf{u}(t)$ and output $\mathbf{y}(t)$, but the state $\mathbf{x}(t)$ as well?

The main idea of stability is to ensure that every bounded or finite-variance input signal (and possibly finite-variance disturbance) results in an output which does not "blow up" to infinity. We can easily imagine unstable systems in real-world scenarios, such as a bipedal robot falling down while walking, or an autonomous car which veers off the road. Consequently, one may also realize that every type of control problem (to be studied in detail in Part III) is essentially concerned with making the system stable in some sense. In disturbance-rejection problems, such as power outage mitigation in a power system network, the goal is to ensure that the output signal's deviation away from some nominal value gradually goes to zero. Reference-tracking problems like robotic path-planning address the disturbance-rejection problem for a time-varying reference signal rather than a fixed value.

There are many different notions of stability for dynamical systems (and not necessarily systems which are linear). The most appropriate notion of stability to use for analysis depends greatly on the application of interest. In this chapter, we investigate two common notions linear stability– *bounded-input, bounded-output (BIBO)* stability and *internal* stability. The discussion of internal stability, in particular, leads naturally to *Lyapunov stability*, which is one of the most popular types of internal stability due to its applicability to more general nonlinear autonomous systems.

SJ. Han, *Linear Systems*, KAIST Research Series,
https://doi.org/10.1007/978-981-96-9774-8_5

5.1 BIBO Stability for LTI Systems

Definition 5.50 (*Bounded*) A signal $\mathbf{u}(t)$ is *bounded* if $\exists \bar{u} \in \mathbb{R}^+$ s.t. $\|\mathbf{u}(t)\| \leq \bar{u}\ \forall\ t \in \mathbb{R}$.

Definition 5.51 (*BIBO Stability*) Given a bounded input signal $\mathbf{u}(t)$, system $\mathcal{H}$ is *bounded-input, bounded-output (BIBO)* stable if every bounded $\mathbf{u}(t)$ yields bounded output $\mathbf{y}(t)$,

$$\text{i.e., } \exists \bar{y} \in \mathbb{R}^+ \text{ s.t. } \|\mathbf{y}(t)\| = \|\mathcal{H}\{\mathbf{u}\}(t)\| \leq \bar{y} \forall t \in \mathbb{R}$$

While the above definition can be applied to general systems, we focus on specifically the linear case.

Lemma 5.5 *A linear system is BIBO stable if for $x_0 = 0$, $\exists M < \infty$ s.t.*

$$\sup_{t\geq 0} \|\mathbf{y}(t)\| \leq M \cdot \sup_{t\geq 0} \|\mathbf{u}(t)\|$$

Recall the impulse response $h(t)$, $t \geq 0$ (in scalar notation) is the output of system $\mathcal{H}$ when $u(t) = \delta(t)$, $x_0 = 0$. The impulse response can be used to compute the output given any general input $u(t)$:

$$y(t) = (h * u)(t) \Longrightarrow y(s) = H(s)u(s) \Longrightarrow H(s) = \frac{y(s)}{u(s)}$$

Theorem 5.14 *A CT SISO LTI system is BIBO stable if and only if impulse response $h(t)$ is absolutely integrable:*

$$\int_0^\infty |h(t)| dt < \infty$$

Proof (*Sufficiency*) First we show that if $h(t)$ is absolutely integrable, then every bounded input excites a bounded output. Let $u(t)$ be an arbitrary input with $|u(t)| \leq u_m \leq \infty$ for all $t \geq 0$. Then,

$$\begin{aligned} |y(t)| = \left| \int_0^t h(\tau)u(t-\tau)d\tau \right| &\leq \int_0^t |h(\tau)||u(t-\tau)|d\tau \\ &\leq u_m \int_0^\infty |h(\tau)|d\tau \leq u_m M \end{aligned}$$

Thus the output is bounded.

(Necessity.) Next we show intuitively that if $h(t)$ is not absolutely integrable, then the system is not BIBO stable. If $h(t)$ is not absolutely integrable, then there exists a t_1 such that

$$\int_0^{t_1} |h(\tau)| d\tau = \infty.$$

Let us choose

$$u(t_1 - \tau) = \begin{cases} 1 & \text{if } h(\tau) \geq 0 \\ -1 & \text{if } h(\tau) < 0 \end{cases}$$

Clearly u is bounded. However, the output excited by this input equals

$$y(t_1) = \int_0^{t_1} h(\tau)u(t_1 - \tau)d\tau = \int_0^{t_1} |h(\tau)|d\tau = \infty$$

which is not bounded. Thus the system is not BIBO stable. The proof is concluded by contradiction, showing that a non-absolutely integrable impulse response leads to BIBO instability. ■

For continuous-time (CT) MIMO LTI systems, we can use norm instead of absolute value, and for discrete-time (DT) systems, the sum is used instead of the integral.

Corollary 5.1 *A CT MIMO LTI system is BIBO stable if and only if impulse response $h(t)$ is absolutely integrable:*

$$\int_0^{\infty} \|h(t)\|dt < \infty$$

Corollary 5.2 *A DT MIMO LTI system is BIBO stable if and only if impulse response $h(t)$ is absolutely summable:*

$$\sum_{t=0}^{\infty} \|h(t)\| < \infty$$

Example 5.6 Consider the SISO system represented by the differential equation

$$y''(t) + 5y'(t) + 6y(t) = u(t)$$

By applying the Laplace transform, we obtain

$$y(s)(s^2 + 5s + 6) = u(s) \Longrightarrow H(s) = \frac{1}{s^2 + 5s + 6}$$

We can factorize the denominator as

$$s^2 + 5s + 6 = (s + 2)(s + 3)$$

Using partial fraction decomposition, we find

$$\frac{1}{s^2 + 5s + 6} = \frac{A}{s + 2} + \frac{B}{s + 3} \text{ for some } A, B$$

Solving for A and B, we get $A = -\frac{1}{4}$, $B = \frac{1}{4}$. Therefore,

$$H(s) = \frac{-\frac{1}{4}}{s+2} + \frac{\frac{1}{4}}{s+3}$$

which gives us

$$h(t) = -\frac{1}{4}e^{-2t} + \frac{1}{4}e^{-3t}$$

The integral of $|h(t)|$ is finite:

$$\int_0^\infty |h(t)|dt = \frac{1}{4}\left(\int_0^\infty e^{-2t}dt + \int_0^\infty e^{-3t}dt\right) = \frac{1}{8} + \frac{1}{12} < \infty$$

Therefore, the system is BIBO stable. □

Example 5.7 Consider first, the system

$$y_t = 2u_t + 3u_{t-1} \Longrightarrow h_t = 2\delta_t + 3\delta_{t-1}$$

This system is BIBO stable since

$$\sum_{t=0}^{\infty} |h_t| = 5 < \infty$$

Now consider instead the system

$$y_t + 2y_{t-1} = 3u_t$$

Applying the Z-transform, we get

$$zy(z) + 2y(z) = 3zu(z) \Longrightarrow H(z) = \frac{3z}{z+2}$$

From the table of Z-transforms,

$$\mathcal{Z}\{\alpha^t\} = \frac{z}{z-\alpha}, \ t \in \mathbb{Z} \geq 0$$

Therefore,

$$h_t = 3(-2)^t$$

and

$$\sum_{t=0}^{\infty} |h_t| \Longrightarrow \infty$$

Thus, the system is not BIBO stable. □

5.2 BIBO Stability for LTV Systems

BIBO stability can also be extended to LTV systems. Recall the zero-state response of CT LTV systems:

$$y(t) = \int_{t_0}^{t} C(t)\Phi(t, s)B(s)u(s)ds + D(t)u(t)$$

Then impulse response is

$$h(t) = \int_{t_0}^{t} C(t)\Phi(t, s)B(s)\delta(s)ds + D(t)\delta(t)$$

$$\Longrightarrow \|h(t)\| \leq \int_{t_0}^{t} \|C(t)\Phi(t, s)B(s)\|ds + \|D(t)\|$$

by the triangle inequality.

A sufficient condition for LTV BIBO stability is

$$\sup_{t\geq 0} \int_{0}^{t} \|h(t, \tau)\|d\tau < \infty \quad \text{and} \quad \sup_{t\geq 0} \|D(t)\| < \infty$$

where $h(t, \tau) \triangleq C(t)\Phi(t, \tau)B(\tau)$ is the impulse response with time-shifted impulse function $u(t) = \delta(t - \tau)$.

Chapter 6
Internal Stability

The previous chapter defined a notion of stability which is dependent only on the boundedness of the input and output signals. Here, we consider a stronger notion of stability which also investigates the boundedness of the internal state signal $\mathbf{x}(t)$.

The tests that we will use for internal stability can be applied more generally to nonlinear autonomous systems of the form $\dot{\mathbf{x}}(t) = f(\mathbf{x}(t))$, where $\mathbf{x} \in \mathbb{R}^n$, $f : \mathbb{R}^n \to \mathbb{R}^n$. In order to have existence and uniqueness of solutions $\mathbf{x}(t)$ to the system, we must impose that f is (locally) Lipschitz continuous over a domain $D \subseteq \mathbb{R}^n$: there exists $L > 0$ such that $\|f(\mathbf{x}) - f(\mathbf{y})\| \leq L \|\mathbf{x} - \mathbf{y}\|$ for all $\mathbf{x}, \mathbf{y} \in D$. We will first present the discussion beyond the scope of linear systems, then treat the linear case as a special instance.

6.1 Linearization

In order to justify the treatment of the nonlinear case before discussing the linear case, we first describe a technique used to relate the two: *linearization*.

To make the system easier to analyze, simplifying these systems around a local neighborhood of $\mathbf{x}$ is often necessary.

Definition 6.1 (*Equilibrium Point*) For autonomous systems $\dot{\mathbf{x}}(t) = f(\mathbf{x}(t))$, $\mathbf{x}^* \in \mathbb{R}^n$ is an *equilibrium point (fixed point)* if $f(\mathbf{x}^*) = 0$.

Note: It is possible for a system to have multiple equilibrium points.

Figure 6.1 are special types of graphs called *phase portraits*. They describe the geometrical evolution of state trajectories on the phase plane $(x_1, \ldots, x_n)$. Conventionally, the coordinate plane of $\mathbb{R}^2$ is often shifted so that the equilibrium point lies at the origin. While phase portraits can be drawn for higher-dimensional systems, they are often most useful in the analysis of 2D or 3D systems.

SJ. Han, *Linear Systems*, KAIST Research Series,
https://doi.org/10.1007/978-981-96-9774-8_6

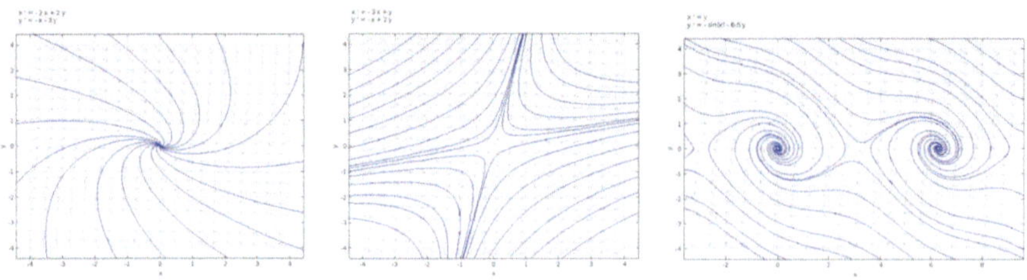

Fig. 6.1 Sample phase portraits

To linearize a nonlinear system, we select a neighborhood around an equilibrium point. The Taylor expansion for a function $f(\mathbf{x})$ near an equilibrium point $\mathbf{x}^*$, considering a small deviation $\Delta\mathbf{x}$, is given by:

$$\text{For } \mathbf{x} \text{ near } \mathbf{x}^* + \Delta\mathbf{x},\ \dot{\mathbf{x}} = f(\mathbf{x}) = f(\mathbf{x}^* + \Delta\mathbf{x}) = f(\mathbf{x}^*) + \nabla f(\mathbf{x})\Big|_{x=\mathbf{x}^*} \cdot \Delta\mathbf{x} + \text{h.o.t.}$$

where $f(\mathbf{x})$ is a vector function, "h.o.t" refers to higher-order terms, and $\nabla f(\mathbf{x})$ is the *Jacobian matrix*:

$$f(\mathbf{x}) = \begin{bmatrix} f_1(\mathbf{x}) \\ \vdots \\ f_n(\mathbf{x}) \end{bmatrix}, \quad \nabla f(\mathbf{x}) \equiv \begin{bmatrix} \frac{\partial f_1}{\partial x_1} & \cdots & \frac{\partial f_1}{\partial x_n} \\ \vdots & \ddots & \vdots \\ \frac{\partial f_n}{\partial x_1} & \cdots & \frac{\partial f_n}{\partial x_n} \end{bmatrix}$$

Since the equilibrium point is typically set to 0 by convention, the above can be rewritten as $\mathbf{x}(t) = \mathbf{x}^* + \Delta\mathbf{x}(t) = \Delta\mathbf{x}(t)$. Therefore, the linearized dynamics of a nonlinear system, around equilibrium point $\mathbf{x}^* = 0$, are

$$\Delta\dot{\mathbf{x}} = \left(\nabla f(\mathbf{x})\Big|_{x=\mathbf{x}^*}\right)\Delta\mathbf{x}$$

where $\nabla f(\mathbf{x}^*)$ is analogous to A in $\dot{\mathbf{x}} = A\mathbf{x}$ form.

Example 6.1 (*Van der Pol Oscillator*) Consider the Van der Pol oscillator, a nonlinear system characterized by the equation:

$$\ddot{\mathbf{x}} - \mu(1 - x^2)\dot{\mathbf{x}} + \mathbf{x} = 0$$

where $\mathbf{x}$ is its position, $\mu > 0$ is a parameter that controls the nonlinearity and damping of the oscillator. To analyze the system's behavior near an equilibrium point, we first convert it into a system of first-order differential equations. Letting $x_1 \triangleq x$ and $x_2 \triangleq \dot{x}$, we obtain:

$$\begin{cases} \dot{x}_1 = x_2, \\ \dot{x}_2 = \mu(1 - x_1^2)x_2 - x_1. \end{cases}$$

To linearize this system around the equilibrium point $(x_1, x_2) = (0, 0)$, we compute the Jacobian matrix with respect to x_1 and x_2 at the origin to get $A = \begin{bmatrix} 0 & 1 \\ -1 & \mu \end{bmatrix}$. Thus, the linearized system near the equilibrium point (0, 0) is:

$$\begin{bmatrix} \dot{x}_1 \\ \dot{x}_2 \end{bmatrix} = \begin{bmatrix} 0 & 1 \\ -1 & \mu \end{bmatrix} \begin{bmatrix} x_1 \\ x_2 \end{bmatrix}$$

It is important to remember that this linearization represents the system's dynamics only near the equilibrium point. Nonlinear effects, especially those introduced by the term $\mu(1 - x_1^2)x_2$, lead to complex behaviors such as limit cycles that are not captured by this linear model. As the parameter μ increases, these nonlinear phenomena become more pronounced, and the utility of the linear approximation diminishes away from the equilibrium. □

6.2 Determining Stability via Eigenvalues

We are now ready to present the three main types of internal stability notions for general autonomous nonlinear systems.

Definition 6.2 (*Main Types of Stability*) A CT nonlinear autonomous system $\dot{\mathbf{x}}(t) = f(\mathbf{x}(t))$ with equilibrium point $\mathbf{x}^* \in \mathbb{R}^n$

1. is *stable* if for all $R > 0$ there exists $r > 0$ such that if $\|\mathbf{x}_0 - \mathbf{x}^*\| < r$, then $\|\mathbf{x}(t) - \mathbf{x}^*\| < R$ for all $t \geq t_0$. The size of R relative to r does not matter. An example of a stable trajectory is $\mathbf{x}(t) = \sin t$ because $|\mathbf{x}(t)| \leq 1$ for all t.
2. is *asymptotically stable* if it is stable and there exists $r > 0$ such that for all $\|\mathbf{x}_0 - \mathbf{x}^*\| < r$, we have $\lim_{t\to\infty} \mathbf{x}(t) = 0$. An example of an asymptotically stable trajectory is $x(t) = \frac{1}{t+1}$ since for $t \geq 0$, $|x(t)| \leq 1$ and $\lim_{t\to\infty} x(t) = 0$.
3. is *exponentially stable* if there exists $M, \alpha > 0$ such that $\|\mathbf{x}(t) - \mathbf{x}^*\| \leq M \|\mathbf{x}_0 - \mathbf{x}^*\|^{\alpha(t-t_0)}$ for all $t \geq t_0$. An example of an exponentially stable trajectory is $x(t) = e^{-t}$; $M = \alpha = 1$ in this case.

Definition 6.3 (*Unstable*) CT nonlinear autonomous system $\dot{\mathbf{x}}(t) = f(\mathbf{x}(t))$ with equilibrium point $\mathbf{x}^* = 0$ is *unstable* if there exists an initial condition $\mathbf{x}_0 \in \mathbb{R}^n$ s.t. $\lim_{t\to\infty} \|\mathbf{x}(t)\| = \infty$.

Definition 6.4 (*Marginal Stability*) CT nonlinear autonomous system $\dot{\mathbf{x}}(t) = f(\mathbf{x}(t))$ with equilibrium point $\mathbf{x}^* = 0$ is *marginally stable* if:

1. It is *stable* (in the sense of Definition 6.2).
2. It is not *asymptotically stable* (in the sense of Definition 6.2), meaning that there exists at least one solution $\mathbf{x}(t)$ such that:

$$\lim_{t\to\infty} \|\mathbf{x}(t) - \mathbf{x}^*\| \neq 0.$$

In other words, while trajectories starting near $\mathbf{x}^*$ remain close to the equilibrium point, they do not necessarily converge to $\mathbf{x}^*$ as $t \to \infty$.

Remark 6.1 Exponential stability implies asymptotic stability, but the converse does not hold for general nonlinear systems. In the linear case, however, both directions hold because the state trajectories of linear systems naturally follow an exponential path.

Checking stability using Definition 6.2 are often difficult to verify directly. For linear systems, there are a collection of equivalent conditions that may be used.

Theorem 6.1 *For CT LTI system $\mathcal{H}$, the following are equivalent:*

1. *$\mathcal{H}$ is asymptotically stable.*
2. *$\mathcal{H}$ is exponentially stable.*
3. *A is Hurwitz, i.e. all eigenvalues of A are in the open left half complex plane $\mathbb{C}^-$.*

Moreover, $\mathcal{H}$ being unstable means one of the following cases holds:

1. *A has eigenvalues on the open right half complex plane $\mathbb{C}^+$.*
2. *A has at least one repeating eigenvalue of the $j\omega$ axis and marginally stable if A has some eigenvalues on the $j\omega$ axis which are simple.*

There are also analogous tests for the DT linear systems, i.e., $\mathbf{x}[t+1] = A_t\mathbf{x}[t]$.

Corollary 6.1 *For DT LTI system $\mathcal{H}$, the following are equivalent:*

1. *$\mathcal{H}$ is asymptotically stable.*
2. *$\mathcal{H}$ is exponentially stable.*
3. *A is Schur, i.e. all eigenvalues of A are in the open unit disk $\{|z| < 1\} \subset \mathbb{C}$.*

Moreover, $\mathcal{H}$ is unstable if A has eigenvalues outside the closed unit disk $\{|z| > 1\}$. And $\mathcal{H}$ is marginally stable if A has some eigenvalues on the unit circle which are all simple.

6.3 Classifying Types of Equilibria

For CT LTI systems with $n = 2$ (i.e. $\mathbf{x}(t) \in \mathbb{R}^2$), there are 4 types of equilibria depending on the eigenvalues of the A matrix.

Type	Condition
Node	$\lambda_1, \lambda_2 \in \mathbb{R} \setminus \{0\}$ and $\text{sgn}(\lambda_1)=\text{sgn}(\lambda_2)$
Saddle	$\lambda_1, \lambda_2 \in \mathbb{R} \setminus \{0\}$ and $\text{sgn}(\lambda_1) \neq \text{sgn}(\lambda_2)$
Focus	$\lambda_1, \lambda_2 \in \mathbb{C}$ and $\text{Re}(\lambda_1) = \text{Re}(\lambda_2) \neq 0$
Center	$\lambda_1, \lambda_2 \in \mathbb{C}$ and $\text{Re}(\lambda_1) = \text{Re}(\lambda_2) = 0$

As we've seen in the previous section, the exact sign of the eigenvalues determines the stability, and the two characteristics could be combined together to describe the equilibrium point more precisely. For example, if $\lambda_1, \lambda_2 \in \mathbb{R}\backslash\{0\}$ and $\text{sgn}(\lambda_1) = \text{sgn}(\lambda_2) < 0$, then it is a stable node.

Example 6.2 Consider the system given by

$$\dot{\mathbf{x}} = f(\mathbf{x}) = \begin{bmatrix} x_1^2 + 4x_1 - 12 \\ x_1 - 2x_2 \end{bmatrix} \in \mathbb{R}^2$$

Then its equilibria are $\mathbf{x}_1^* = (2, 1), \mathbf{x}_2^* = (-6, -3)$. After linearization at each of these equilibria:

$$\nabla_{\mathbf{x}} f(\mathbf{x}) = \begin{bmatrix} 2x_1 + 4 & 0 \\ -1 & 2 \end{bmatrix} \Longrightarrow \begin{cases} \nabla_{\mathbf{x}} f(\mathbf{x})|_{\mathbf{x}=\mathbf{x}_1^*} = \begin{bmatrix} 8 & 0 \\ -1 & 2 \end{bmatrix} & \lambda = 8, 2 \Longrightarrow \text{unstable node} \\ \nabla_{\mathbf{x}} f(\mathbf{x})|_{\mathbf{x}=\mathbf{x}_2^*} = \begin{bmatrix} -8 & 0 \\ -1 & 2 \end{bmatrix} & \lambda = -8, 2 \Longrightarrow \text{saddle} \end{cases}$$

□

Now, how do the *phase portraits* of each type of equilibria look like? Let $\mathbf{v}_1, \mathbf{v}_2$ be the two eigenvectors corresponding to eigenvalues λ_1, λ_2.

6.3.1 Node Equilibria

- **Case 1**: $\lambda_1, \lambda_2 \in \mathbb{R}$ $(\lambda_1 \neq \lambda_2)$, assume $|\lambda_1| < |\lambda_2|$.
  Denote the initial state $\mathbf{x}_0 = c_1\mathbf{v}_1 + c_2\mathbf{v}_2$, where the $\mathbf{v}_1, \mathbf{v}_2 \in \mathbb{R}^2$ are eigenvectors of A. Then the state at time t is generally written as

$$\mathbf{x}(t) = c_1 e^{\lambda_1 t}\mathbf{v}_1 + c_2 e^{\lambda_2 t}\mathbf{v}_2.$$

  (We can do this because the eigenvectors always from a basis of $\mathbb{R}^2$). Therefore,

$$x_1(t) = c_1 e^{\lambda_1 t} v_{11} + c_2 e^{\lambda_2 t} v_{21}$$
$$x_2(t) = c_1 e^{\lambda_1 t} v_{12} + c_2 e^{\lambda_2 t} v_{22}$$

  Then we have

$$\frac{dx_2}{dx_1} = \frac{c_1\lambda_1 e^{\lambda_1 t} v_{12} + c_2\lambda_2 e^{\lambda_2 t} v_{22}}{c_1\lambda_1 e^{\lambda_1 t} v_{11} + c_2\lambda_2 e^{\lambda_2 t} v_{21}} = \frac{c_1\lambda_1 v_{12} + c_2\lambda_2 e^{(\lambda_2-\lambda_1)t} v_{22}}{c_1\lambda_1 v_{11} + c_2\lambda_2 e^{(\lambda_2-\lambda_1)t} v_{21}} \tag{6.1}$$

  1. Subcase (sgn(λ_1) = sgn(λ_2) < 0).
     In Eq. 6.1, if $c_1 \neq 0$, as $t \Longrightarrow \infty$ $\frac{dx_2}{dx_1} = \frac{v_{12}}{v_{11}}$, and when $c_1 = 0$, $\frac{dx_2}{dx_1} = \frac{v_{22}}{v_{21}}$. The phase portrait is shown in Fig. 6.2.
  2. Subcase (sgn(λ_1) = sgn(λ_2) > 0 ($\lambda_1 < \lambda_2$)).
     In this situation, the equilibrium point is an unstable ordinary node. In Fig. 6.3, the direction gets reversed compare with sgn(λ_1) = sgn(λ_2) < 0.

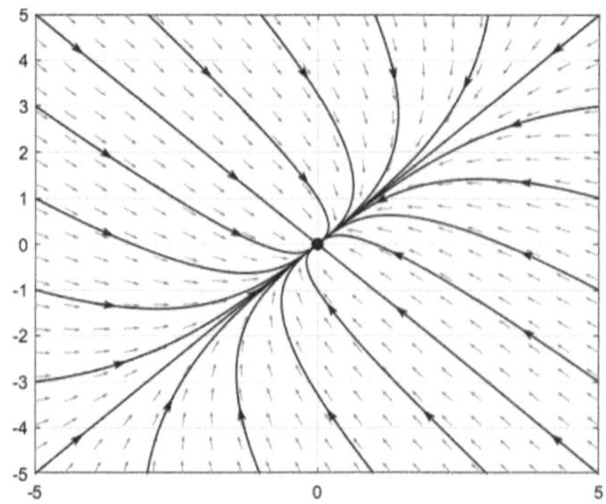

Fig. 6.2 Phase portrait for stable ordinary node

- **Case 2**: $\lambda_1, \lambda_2 \in \mathbb{R}$ $(\lambda_1 = \lambda_2 = \lambda)$

1. Subcase $A = V\Lambda V^{-1}$ where $\Lambda = \begin{bmatrix} \lambda & 0 \\ 0 & \lambda \end{bmatrix}$.

$$\dot{x}_1(t) = \lambda x_1(t)$$
$$\dot{x}_2(t) = \lambda x_2(t)$$

We can get that $x_1(t) = c_1 e^{\lambda t}$, and $x_2(t) = c_2 e^{\lambda t}$, where c_1, c_2 are the initial state. $\frac{dx_2}{dx_1} = \frac{c_2}{c_1}$. The phase portraits are shown in Figs. 6.4 and 6.5.

2. Subcase $A = V\Lambda V^{-1}$ where $\Lambda = \begin{bmatrix} \lambda & 1 \\ 0 & \lambda \end{bmatrix}$.

$$\dot{x}_1(t) = \lambda x_1(t) + x_2(t)$$
$$\dot{x}_2(t) = \lambda x_2(t)$$

In this case, shown in Figs. 6.6 and 6.7, matrix A only has one eigenvector v_1 such that $Av_1 - \lambda v_1 = 0$. The other eigenvector v_2 is the generalized eigenvector generated from v_1. If $x_0 = c_1 v_1 + c_2 v_2$, then $x(t) = e^{Jt} x_0$ after change of coordinates

$$\begin{bmatrix} x_1 \\ x_2 \end{bmatrix} = \begin{bmatrix} e^{\lambda t} & te^{\lambda t} \\ 0 & e^{\lambda t} \end{bmatrix} \begin{bmatrix} c_1 v_{11} + c_2 v_{21} \\ c_1 v_{12} + c_2 v_{22} \end{bmatrix}.$$

The node is called *degenerate* node or *singular* node (Fig. 6.4).

Fig. 6.3 Phase portrait for unstable ordinary node

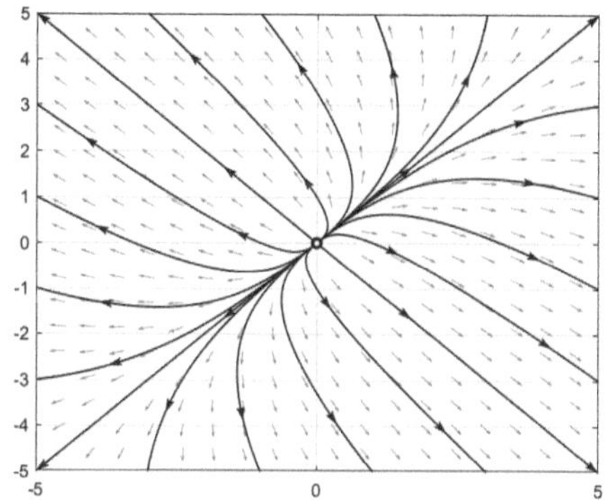

Fig. 6.4 Phase portrait for stable dicritical node

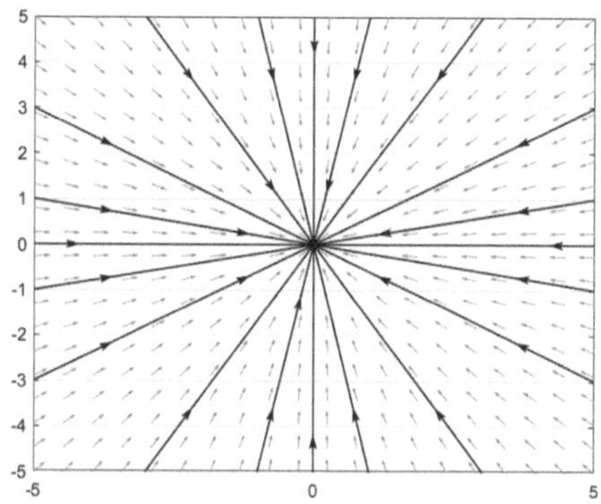

Fig. 6.5 Phase portrait for unstable dicritical node

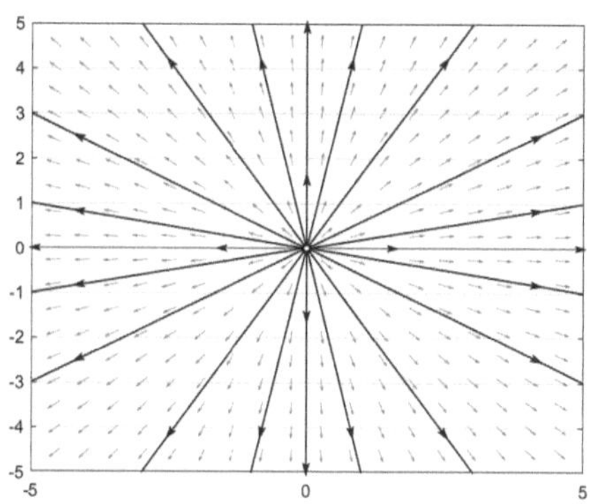

6.3.2 Saddle Equilibria

Saddle equilibria occur when $\lambda_1, \lambda_2 \in \mathbb{R}$, $\text{sgn}(\lambda_1) \neq \text{sgn}(\lambda_2) < 0$. The eigenvectors become separatrices and trajectories are hyperbolic. Like shown in Figs. 6.8 and 6.9, the direction of the trajectory is dependent on the sign of λ_1 and λ_2. All saddle equilibria have opposite sign eigenvalues, indicating instability. Eigenvectors become separatrices and trajectories are hyperbolic.

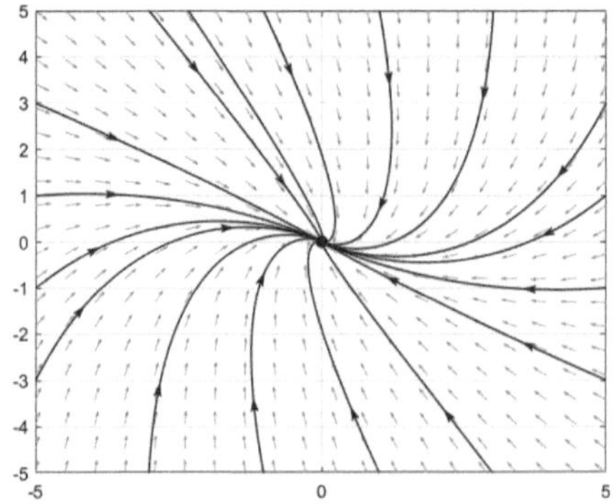

Fig. 6.6 Phase portrait for stable degenerate node

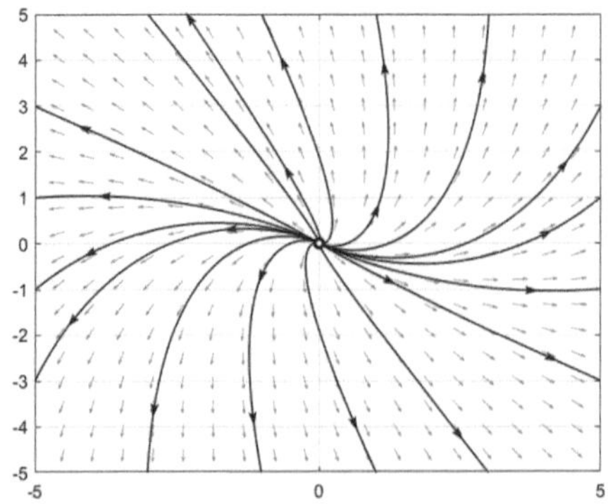

Fig. 6.7 Phase portrait for unstable degenerate node

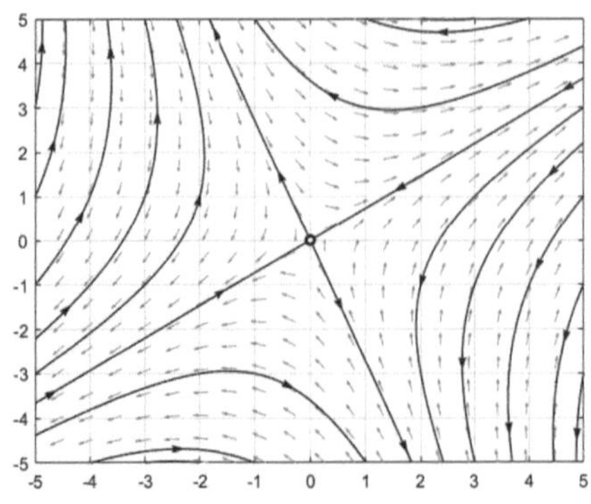

Fig. 6.8 Phase portrait for saddle when $\lambda_1 < 0$

6.3.3 Focus Equilibria

Focus equilibria occur when $\lambda_1, \lambda_2 \in \mathbb{C}$ s.t. $\text{Re}(\lambda_1) \neq 0$, $\text{Re}(\lambda_2) \neq 0$. Recall that conjugate pairs are always both eigenvalues/eigenvectors, i.e., we can write $\lambda_1 = \sigma + j\omega$, $\lambda_2 = \sigma - j\omega$ where $\sigma, \omega \in \mathbb{R}$, and $\mathbf{v}_1 = \mathbf{u} + j\mathbf{w}$, $\mathbf{v}_2 = \mathbf{u} - j\mathbf{w}$, where $\mathbf{u}, \mathbf{w} \in \mathbb{R}^2$. Now consider

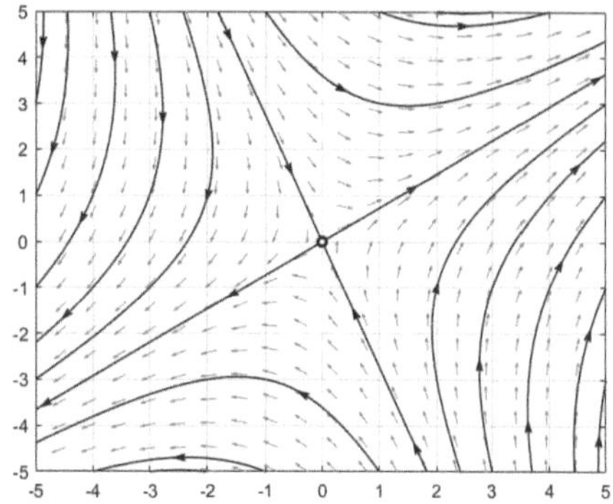

Fig. 6.9 Phase portrait for saddle when $\lambda_1 > 0$

$$
\begin{aligned}
e^{\lambda_1 t}\mathbf{v}_1 = e^{(\sigma+j\omega)t}(\mathbf{u}+j\mathbf{w}) &= e^{\sigma t}(\cos\omega t + j\sin\omega t)(\mathbf{u}+j\mathbf{w}) \\
&= e^{\sigma t}(\mathbf{u}\cos\omega t - \mathbf{w}\sin\omega t) + je^{\alpha t}(\mathbf{u}\sin\omega t + \mathbf{w}\cos\omega t).
\end{aligned}
$$

Likewise,

$$
e^{\lambda_2 t}\mathbf{v}_2 = e^{\sigma t}(\mathbf{u}\cos\omega t - \mathbf{w}\sin\omega t) - je^{\alpha t}(\mathbf{u}\sin\omega t + \mathbf{w}\cos\omega t).
$$

Thus general solutions $\mathbf{x}(t)$ can be expressed as a linear combo of the real part and imaginary part:

$$
\mathbf{x}_0 = c_1\mathbf{v}_1 + c_2\mathbf{v}_2
$$

Here, $c_1, c_2 \in \mathbb{C}$ in order for $\mathbf{x}_0 \in \mathbb{R}^2$, since $\mathbf{v}_1, \mathbf{v}_2 \in \mathbb{C}^2$. Introduce $r_1, r_2 \in \mathbb{R}$ so that

$$
\mathbf{x}(t) = r_1 e^{\alpha t}(\mathbf{u}\cos\omega t - \mathbf{w}\sin\omega t) + r_2 e^{\alpha t}(\mathbf{u}\sin\omega t + \mathbf{w}\cos\omega t).
$$

Thus, $\mathbf{x}(t)$ can also be written in terms of the basis vectors $\mathbf{u}, \mathbf{w}$

$$
\mathbf{x}(t) \equiv \alpha(t)\mathbf{u} + \beta(t)\mathbf{w}.
$$

Choose $r_1 = c\cos\theta t, r_2 = c\sin\theta t$ for some auxiliary angle θ. Then $\alpha(t) = ce^{\sigma t}\sin(\omega t + \theta)$ and $\beta(t) = ce^{\sigma t}\cos(\omega t + \theta)$. The phase portraits are shown in Figs. 6.10 and 6.11 The phase trajectories are spirals that depend on sgn(σ).

6.3.4 Center Equilibria

Center equilibria occur when $\lambda_1, \lambda_2 \in \mathbb{C}$ s.t. $\text{Re}(\lambda_1) = \text{Re}(\lambda_2) = 0$. The possible phase portraits for center equilibria are shown in Fig. 6.12. All center equilibria are marginally stable. Phase trajectories become nonintersecting ellipses. We leave it as an exercise to derive conditions for phase portrait curve orientation.

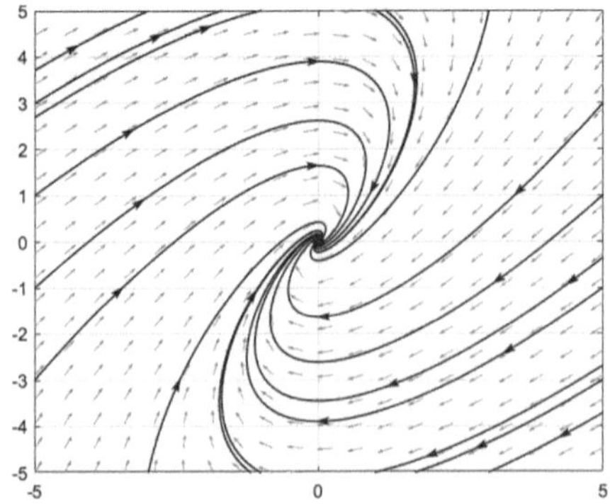

Fig. 6.10 Phase portrait for stable focus ($\text{sgn}(\sigma) < 0$)

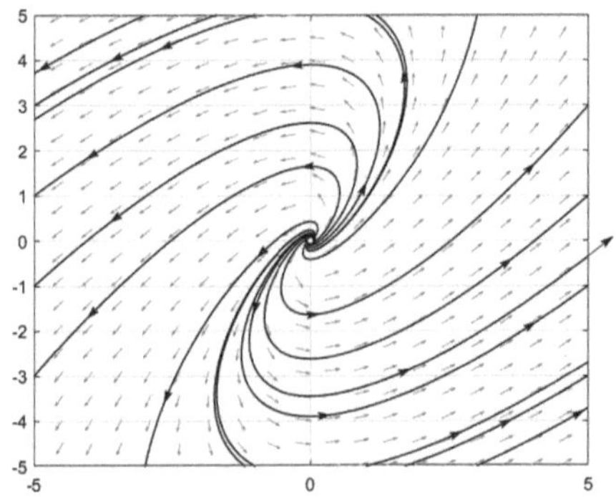

Fig. 6.11 Phase portrait for unstable focus ($\text{sgn}(\sigma) > 0$)

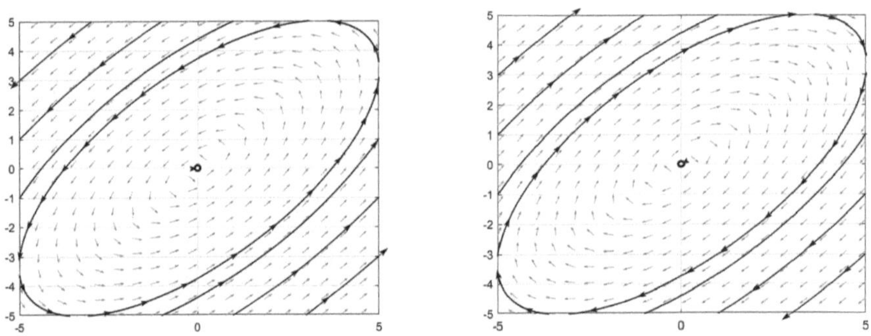

Fig. 6.12 Phase portrait for center

Example 6.3 Draw the phase portrait of the system $\dot{\mathbf{x}} = \begin{bmatrix} -3 & 4 \\ -2 & 3 \end{bmatrix} \mathbf{x}$.

Solution 2: Let's denote $A = \begin{bmatrix} -3 & 4 \\ -2 & 3 \end{bmatrix}$. First calculate the eigenvalues and eigenvectors of A. Through $\det |\lambda I - A| = 0$, we can get $\lambda_1 = 1$ and $\lambda_2 = -1$. Meanwhile, the eigenvector $\mathbf{v}_1 = [1\ 1]^\top$, $\mathbf{v}_2 = [2\ 1]^\top$, respectively.

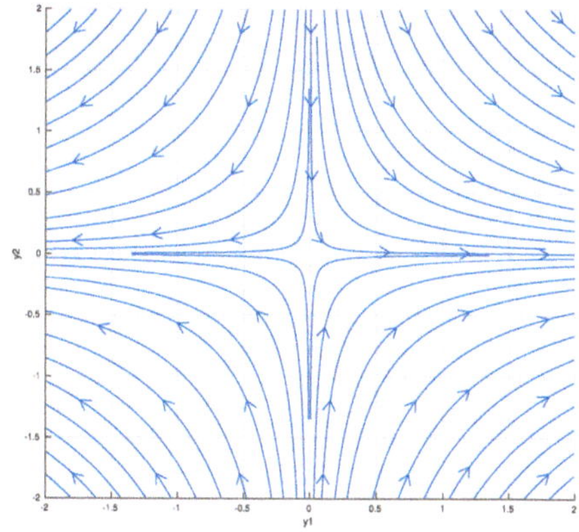

Fig. 6.13 Phase portrait under $\mathbf{y}$ frame

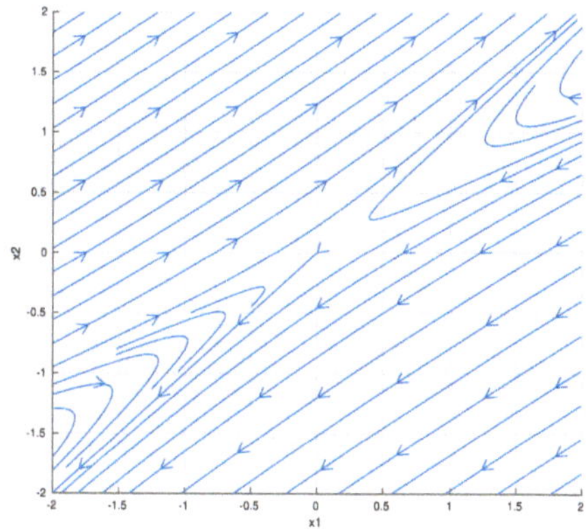

Fig. 6.14 Phase portrait under $\mathbf{x}$ frame

Then let denote $P = \begin{bmatrix} 1 & 2 \\ 1 & 1 \end{bmatrix}$, $\Lambda = \begin{bmatrix} 1 & 0 \\ 0 & -1 \end{bmatrix}$ and $\mathbf{x} = P\mathbf{y}$. Therefore $\dot{\mathbf{y}} = \Lambda \mathbf{y}$. For this system, it is easy to see that $\dot{y}_1 = y_1$ and $\dot{y}_2 = -y_2$, implies $y_1(t) = c_1 e^t$ and $y_2(t) = c_2 e^{-t}$. The phase portrait is shown in Fig. 6.13. Through the coordinate changes $\mathbf{x} = P\mathbf{y}$. The $[1\ 0]^\top$ in the $\mathbf{y}$ frame is $[1\ 1]^\top$ in the $\mathbf{x}$ frame, and also the $[0\ 1]^\top$ in the $\mathbf{y}$ frame is $[2\ 1]^\top$ in the $\mathbf{x}$ frame. All these are shown in Fig. 6.14. $\square$

Chapter 7
Lyapunov Stability

The evolution of the fundamental concepts of system and trajectory stabilities went through a long history, with many fruitful advances and developments, until the celebrated Ph.D. thesis of A. M. Lyapunov, The General Problem of the Stability of Motion, published in 1892 [1]. This monograph is so fundamental that its ideas and techniques are virtually leading all kinds of basic research and applications regarding stabilities of dynamical systems today. In fact, not only dynamical behavior analysis in modern physics, but also controllers design in engineering systems depend on the principles of Lyapunov's stability theory.

Lyapunov's theory is very powerful, but there are cases where it fails to demonstrate stability in specific situations. We will also examine another powerful theory, Lasalle's Invariance principle [2], which can be applied in such situations.

Lyapunov stability is a very common method of checking internal stability in modern control theory and often used to verify stability for autonomous linear or nonlinear systems [3]. It describes properties of the system's equilibrium point. We can make future states $x(t),\ \forall t > t_0$ arbitrarily close to the equilibrium by taking the initial condition to be close enough.

7.1 A Motivating Example

Let's consider the pendulum example we've seen before. In this case, we add a damping term related to the dissipation of the pendulum's energy.

$$mL\ddot{\theta}(t) = -mg\sin\theta(t) - kL\dot{\theta}(t)\ \ ,\ \ k \geq 0$$

The state-space equations becomes

SJ. Han, *Linear Systems*, KAIST Research Series,
https://doi.org/10.1007/978-981-96-9774-8_7

$$\begin{bmatrix} \dot{x_1} \\ \dot{x_2} \end{bmatrix} = \begin{bmatrix} x_2 \\ -\frac{g}{L}\sin x_1 - \frac{k}{m}x_2 \end{bmatrix} \begin{bmatrix} x_1 \\ x_2 \end{bmatrix}$$

where $x_1 = \theta$, $x_2 = \dot{\theta}$. Then, the system's equilibria can be found by $\dot{x} = \begin{bmatrix} x_1 \\ x_2 \end{bmatrix} = 0$. This tells us $x_2 = 0$ and $x_1 = n\pi$ for every $n \in \mathbb{Z}$.

Physically, the system has only two equilibria. One is pendulum hanging down, and the other is hanging up. A natural way to investigate stability of equilibria is to see how much energy it dissipates over time.

- Case : $k = 0$

$$E(x) = KE(x) + PE(x) = \frac{1}{2}x_2^2 + \int_0^{x_1} \frac{g}{L}\sin z\, dz = \frac{1}{2}x_2^2 + \frac{g}{L}(1 - \cos(x_1))$$

Applying the Jacobian,

$$\dot{E}(x) = \frac{\partial E}{\partial x_1}\dot{x}_1 + \frac{\partial E}{\partial x_2}\dot{x}_2 = \frac{g}{L}\sin x_1 \cdot x_2 + x_2 \cdot \left(-\frac{g}{L}\sin x_1\right) = 0$$

It means that energy remains constant over time and there is no energy dissipation.
- Case : $k > 0$

$$\dot{E}(x) = \frac{g}{L}\sin x_1(x_2) + x_2\left(-\frac{g}{L}\sin x_1 - \frac{k}{m}x_2\right) = -\frac{k}{m}x_2^2 \leq 0 \quad \forall x_2 \in \mathbb{R}.$$

This means that energy is decreasing, which adheres to the intuition of stability (e.g., dissipative energy).

Through the pendulum example, the concept of Lyapunov stability (stability in the sense of Lyapunov) takes this similar "energy" perspective of the system and intuition for the Lyapunov function. The Lyapunov function $V : \mathbb{R}^n \to \mathbb{R}$ is a chosen energy function such that when it is decreasing over time, it tells us the stability of the system's equilibrium.

There are two common methods associated with the Lyapunov approach. But before we begin, let's make some important function definition.

7.2 Lyapunov-Sense Stability

Stability *in the sense of Lyapunov* is often characterized with respect to an equilibrium point (see Definition 6.1). However, autonomous systems don't always converge to an equilibrium point.

7.2.1 Additional Definitions

Definition 7.56 (*Limit Cycle*) Let $\mathcal{O} := \{\mathbf{x}(t)|\mathbf{x}(t) = \mathbf{x}(t+T)\}$ be a periodic orbit with period $T > 0$. Then $\mathcal{O}$ is defined to be a *limit cycle* of the system $\dot{\mathbf{x}} = f(\mathbf{x})$ if for all $\mathbf{y} \in \mathcal{O}$, there exists a sequence of times $\{t_n\} \subseteq \mathbb{R}^+$ such that $\lim_{n\to\infty} t_n = +\infty$ and $\lim_{n\to\infty} \mathbf{x}(t_n) = \mathbf{y}$. Intuitive illustrations of the two concepts are in Fig. 7.1.

As before, we will define most of our following concepts with respect to an equilibrium point, though they can be easily extended to limit cycles as well. This type of stability is commonly referred to as *Lyapunov-sense stability*.

Lyapunov stability is a type of internal stability that describes the properties of the system's equilibrium point(or limit cycle). We can make future states $\mathbf{x}(t), t > t_0$ arbitrarily close to the equilibrium by taking the initial condition to be "close enough". Often used to verify stability for autonomous nonlinear systems.

Definition 7.57 (*Lyapunov-Sense Stability*) Let system $\dot{\mathbf{x}}(t) = f(\mathbf{x}(t)) \in \mathbb{R}^n$ have equilibrium point $\mathbf{x}^* \in \mathbb{R}^n$. Then $\mathbf{x}^*$ is

- *stable* if $\forall \epsilon > 0, \exists \delta > 0$ such that if $||\mathbf{x}_0 - \mathbf{x}^*|| \leq \delta$ then $||\mathbf{x}(t) - \mathbf{x}^*|| \leq \epsilon \; \forall t \geq t_0$.
- *(locally) asymptotically stable* if $\forall \epsilon > 0$, $\exists \delta > 0$ such that $||\mathbf{x}_0 - \mathbf{x}^*|| \leq \delta$ then $||\mathbf{x}(t) - \mathbf{x}^*|| \leq \epsilon \; \forall t \geq t_0$ and $\lim_{t\to\infty} \mathbf{x}(t) = \mathbf{x}^*$.
- *(locally) exponentially stable* if $\exists \epsilon, M, \alpha > 0$ such that $||\mathbf{x}_0 - \mathbf{x}^*|| \leq \delta$ then $||\mathbf{x}(t) - \mathbf{x}^*|| \leq Me^{-\alpha t}||\mathbf{x}(t) - \mathbf{x}^*|| \;\; \forall t \geq t_0$.

Remark 7.11 *Global* asymptotic/exponential stability occurs when you can start the initial condition $\mathbf{x}_0$ anywhere in $\mathbb{R}^n$ and the trajectories still converge.

Like the three main types of stability in Definition 6.2, the conditions presented in Definition 7.57 are difficult to verify directly. For LTI systems, we can characterize

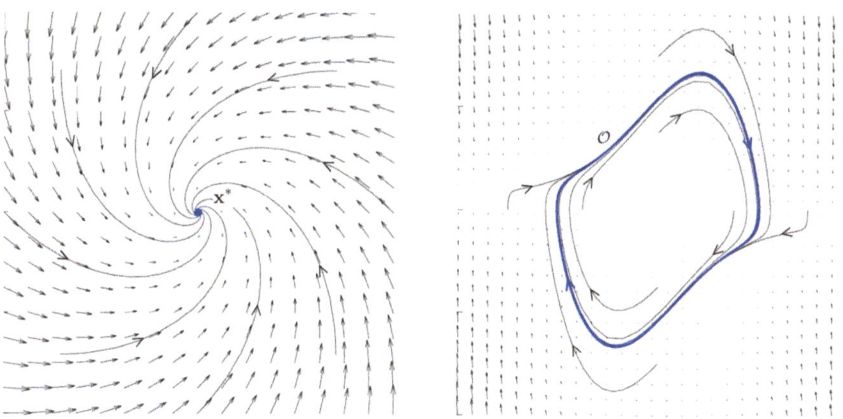

Fig. 7.1 Trajectory converging towards an equilibrium point(left) and a limit cycle (right)

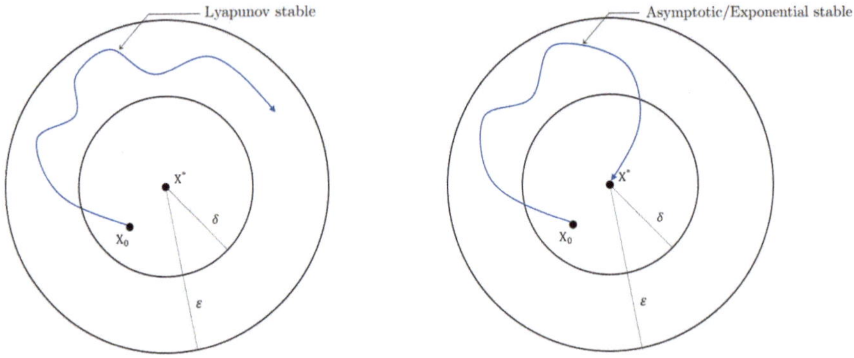

Fig. 7.2 Trajectory for a stable equilibrium point (left) versus an asymptotic or exponentially stable equilibrium point (right). Essentially, stability requires the trajectory to remain within a bounded error ball of the equilibrium point, while asymptotic or exponential stability requires convergence towards the equilibrium point

stability by computing the eigenvalues. For more general systems and LTV systems, we instead have two main *Lyapunov methods* to check whether a system is stable and of what type (Fig. 7.2).

7.2.2 Preliminary Mathematical Reviews

Before we dive into a thorough discussion of Lyapunov stability methods, we first review a few mathematical preliminaries.

Definition 7.58 (*Symmetric and Skew-Symmetric Matrices*) Let $A \in \mathbb{R}^{n\times n}$ be a square matrix.

1. A is *symmetric* if $A^\top = A$. Some common properties of symmetric matrices are:
   - all eigenvalues are real
   - all eigenvectors are orthogonal: $A = V\Lambda V^{-1} = V\Lambda V^\top$
2. A is skew-symmetric if $A^\top = -A$. Some common properties of skew-symmetric matrices are:
   - all eigenvalues are purely imaginary
   - diagonalized by unitary matrix $V^{-1} = V^*$

Definition 7.59 (*Quadratic Form*) A *quadratic form* corresponding to symmetric Q is

$$f(\mathbf{x}) = \mathbf{x}^\top Q\mathbf{x},\ \ \mathbf{x} \in \mathbb{R}^n.$$

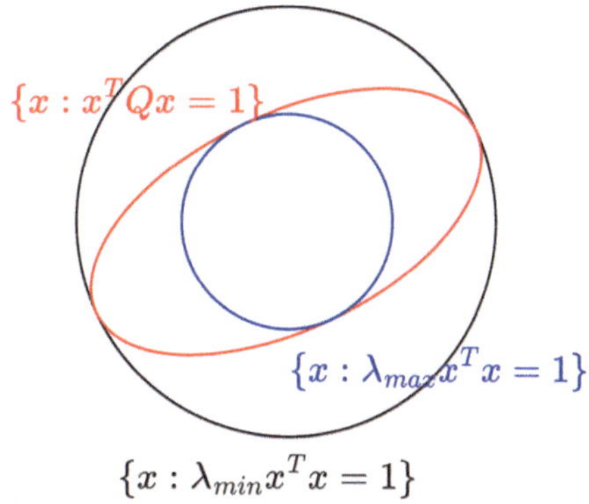

Fig. 7.3 Quadratic form in 2D (ellipsoid)

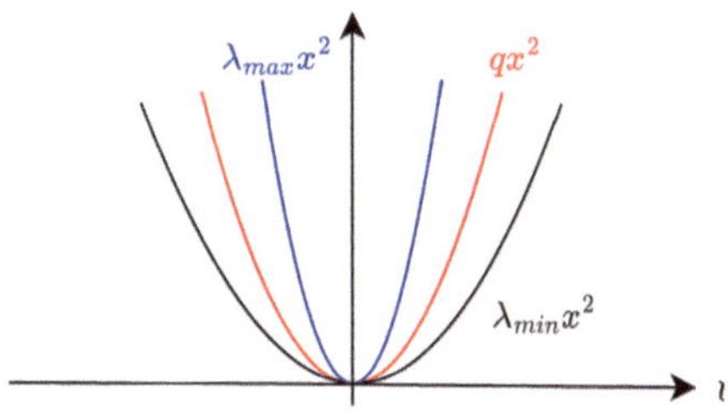

Fig. 7.4 Quadratic form in 1D (function)

The presentation of the quadratic form is unique:

$$\mathbf{x}^\top Q\mathbf{x} = \mathbf{x}^\top \tilde{Q}\mathbf{x}$$

so the $Q = \tilde{Q}$. When Q is not symmetric, we can still derive a corresponding quadratic form with a symmetric matrix:

$$f(\mathbf{x}) = \mathbf{x}^\top Q\mathbf{x} = \frac{1}{2}\mathbf{x}^\top Q\mathbf{x} + \frac{1}{2}\mathbf{x}^\top Q^\top \mathbf{x} = \mathbf{x}^\top \left[\frac{1}{2}(Q + Q^\top)\right]\mathbf{x}$$

let $\tilde{Q} = \frac{1}{2}(Q + Q^\top)$, so $\tilde{Q}$ is symmetric.

Let $\lambda_{\min} < \cdots < \lambda_{\max}$ be the sorted eigenvalues of Q. Then $\forall \mathbf{x} \in \mathbb{R}^n$ $\lambda_{\min}||x||^2 \leq \mathbf{x}^\top Q\mathbf{x} \leq \lambda_{\max}||x||^2$. This means quadratic forms can be bounded (Figs. 7.3 and 7.4).

Definition 7.60 (*Definite and Semidefinite*) Symmetric matrix $Q \in \mathbb{R}^{n\times n}$ is

- *positive semidefinite* ($Q \succcurlyeq 0$) if $\mathbf{x}^\top Q\mathbf{x} \geq 0 \; \forall \mathbf{x} \in \mathbb{R}^n$
- *positive definite* ($Q \succ 0$) if $\mathbf{x}^\top Q\mathbf{x} > 0 \; \forall \mathbf{x} \in \mathbb{R}^n$
- *negative semidefinite* ($Q \preccurlyeq 0$) if $\mathbf{x}^\top Q\mathbf{x} \leq 0 \; \forall \mathbf{x} \in \mathbb{R}^n$
- *positive definite* ($Q \prec 0$) if $\mathbf{x}^\top Q\mathbf{x} < 0 \; \forall \mathbf{x} \in \mathbb{R}^n$

Here, $Q \succcurlyeq 0$ means all eigenvalues of Q are not less than 0, and also means $\exists P \in \mathbb{R}^{n\times m}$ such that $Q = PP^\top$ (since $\mathbf{x}^\top Q\mathbf{x} = \mathbf{x}^\top PP^\top \mathbf{x} = \mathbf{y}^\top \mathbf{y} = ||\mathbf{y}||^2 \geq 0$).

$Q \succ 0$ means all eigenvalues of Q are bigger than 0, and also means $\exists$ onto $P \in \mathbb{R}^{n\times m}$ such that $Q = PP^\top$.

We introduce the notation $\mathcal{B}_\epsilon = \mathcal{B}(\mathbf{x}^*, \epsilon)$, representing the collection of all points situated within a hyperball centered at the equilibrium point $\mathbf{x}^*$ with a radius of ϵ, denoted as, i.e. $\mathcal{B}_\epsilon = \{\mathbf{x} \in \mathbb{R}^n \,|\, \|\mathbf{x} - \mathbf{x}^*\| \leq \epsilon\}$. The property is

- *local*, if it holds for all $\mathbf{x}$ within a hyperball $\mathcal{B}_\epsilon$
- *global*, if it holds for all $\mathbf{x} \in \mathbb{R}^n$
- *uniform*, if it holds for all $t_0 \geq 0$.

Definition 7.61 (*Positive Definite Functions (pdf)*) A function $V : \mathbb{R}^n \to \mathbb{R}$ is called *positive definite* for some continuous, strictly-increasing $\alpha : \mathbb{R}^{\geq 0} \to \mathbb{R}$ if and only if,

$$V(\mathbf{0}) = 0, \quad V(\mathbf{x}) \geq \alpha(\|\mathbf{x}\|) \text{ and } \lim_{\|\mathbf{x}\| \Longrightarrow \infty} \alpha(\|\mathbf{x}\|) = \infty \tag{7.1}$$

(i.e. α is *radially unbounded*). In short

$$\begin{aligned} &V(\mathbf{x}) > 0 \text{ for } \mathbf{x} \neq \mathbf{0} \text{ and} \\ &V(\mathbf{x}) = 0 \text{ for } \mathbf{x} = \mathbf{0}. \end{aligned} \tag{7.2}$$

Definition 7.62 (*Positive Semidefinite Functions (psdf)*) A function $V : \mathbb{R}^n \to \mathbb{R}$ is called *positive definite* if and only if it takes nonnegative values, with the exception of yielding zero when $\mathbf{x} = 0$. In short

$$\begin{aligned} &V(\mathbf{x}) \geq 0 \text{ for } \mathbf{x} \neq \mathbf{0} \text{ and} \\ &V(\mathbf{x}) = 0 \text{ for } \mathbf{x} = \mathbf{0}. \end{aligned} \tag{7.3}$$

Likewise, a function $V(\mathbf{x})$ is called *negative definite*, if $-V(\mathbf{x})$ is positive definite and is called *negative semidefinite*, if $-V(\mathbf{x})$ is positive semidefinite.

Definition 7.63 (*Decrescent Functions*) A continuous function $V : \mathbb{R}^n \to \mathbb{R}$ is *decrescent* if for some $\epsilon > 0$ and some continuous, strictly-increasing $\beta : \mathbb{R}^{\geq 0} \to \mathbb{R}$,

$$V(\mathbf{x}) \leq \beta(\|\mathbf{x}\|) \;\; \forall \mathbf{x} \in \mathcal{B}_\epsilon \subseteq D \tag{7.4}$$

7.3 Lyapunov Indirect Method

The first method is called the *Lyapunov Indirect Method*, which relies on the linearization of a system. Define

$$A \triangleq \left.\frac{\partial f}{\partial \mathbf{x}}\right|_{\mathbf{x}=\mathbf{x}^*}$$

to be the linearization of the nonlinear system $\dot{\mathbf{x}} = f(\mathbf{x})$ about $\mathbf{x} = \mathbf{x}^*$. Note that $f(\mathbf{x})$ must be twice continuously differentiable.

1. If $\mathrm{Re}(\lambda_i(A)) < 0$ for all i then $\mathbf{x}^*$ is locally asymptotically stable.
2. If $\mathrm{Re}(\lambda_i(A)) > 0$ for some i then $\mathbf{x}^*$ is unstable.
3. If $\mathrm{Re}(\lambda_i(A)) = 0$ for some i then it cannot be determined whether $\mathbf{x}^*$ is stable or not.

If the linearized system is asymptotically stable (in the sense of Lyapunov), then $\mathbf{x}^* = \mathbf{0}$ is a locally asymptotically stable equilibrium point of the original system. The following conclusions can be drawn for the equilibria $\mathbf{x}^*$ of the system:

1. If all eigenvalues of the system matrix A have a negative real part, then the equilibrium $\mathbf{x}^*$ of the system is locally asymptotically stable in the sense of Lyapunov.
2. If an eigenvalue of the system matrix A has a real part, then the equilibrium $\mathbf{x}^*$ of the system is locally unstable.
3. When at least one eigenvalue of the system matrix A has a real part equal to zero and none have a positive real part, no clear conclusions can be drawn about the equilibrium $\mathbf{x}^*$ of the system. The equilibrium $\mathbf{x}^*$can be locally asymptotically stable, locally stable in the sense of Lyapunov, or locally unstable.

The motivation for this type of test is derived from linear systems. Consider a scalar linear system $\dot{x} = ax$. The solution is $x(t) = x_0 e^{at}$. If $a < 0$, then $\lim_{t\to\infty} x(t) = 0$, which implies local asymptotic stability according to the definition above. If $a > 0$, then the trajectory blows up and is clearly unstable.

7.4 Lyapunov Direct Method

A more refined test of stability is the *Lyapunov Direct Method.* We first shift the dynamics of the system such that its equilibrium point is at the origin: $\dot{\mathbf{x}} = f(\mathbf{x} - \mathbf{x}^*)$. Then we construct a continuously differentiable, real-valued, positive-definite function $V(\mathbf{x})$ over some domain D that contains $\mathbf{x}^* \equiv 0$. This function is called the *Lyapunov function*, and can be thought of as a potential energy-like function such that a decrease in its value implies a transition to a stabler, lower-energy state.

1. If $\dot{V}(\mathbf{x}) = \frac{\partial V}{\partial \mathbf{x}}\dot{\mathbf{x}} = \frac{\partial V}{\partial \mathbf{x}} f(\mathbf{x})$ is nonpositive for all $x \in D$, then the system is stable.
2. If $\dot{V}(\mathbf{x}) < 0$ with strict inequality for all $\mathbf{x} \in D$, then the system is locally asymptotically stable.
3. If $V(\mathbf{x})$ is radially unbounded, i.e., $\lim_{\|\mathbf{x}\|\to\infty} V(\mathbf{x}) = \infty$, then the system is globally asymptotically stable.

The choice of V is not unique. In each case, the existence of a single V that satisfies the condition is all that is needed to show the corresponding form of stability. Some popular forms are $V(\mathbf{x}) = \mathbf{x}^\top P\mathbf{x}$ for positive definite P or $V(\mathbf{x}) = \sum_i c_i x_i^2$ for $c_i > 0$. Furthermore, the Direct method gives no insight into the instability of a system. In fact, to prove instability, one could take the converse of the first test, which means one would need to show that $\dot{V}(\mathbf{x}) > 0$ for all possible Lyapunov functions V. Clearly this is an infeasible task.

Example 7.11 Show that the following system is stable at its equilibrium points.

$$\begin{bmatrix} \dot{x}_1 \\ \dot{x}_2 \end{bmatrix} = \begin{bmatrix} x_2 - x_1^3 \\ -x_1 - x_2^3 \end{bmatrix}$$

Note that $(0, 0)$ is the only equilibrium of this system. Further, choose Lyapunov function $V(\mathbf{x}) = x_1^2 + x_2^2$. Note that it satisfies $\mathcal{C}^1$, $V(0) = 0$, $V(\mathbf{x}) > 0$ for all $\mathbf{x} \neq 0$, and it is decrescent with $\beta(\|\mathbf{x}\|) = 2\|\mathbf{x}\|^2 = 2(x_1^2 + x_2^2)$. $V(\mathbf{x})$ is also radially unbounded:

$$\dot{V}(\mathbf{x}) = 2x_1\dot{x}_1 + 2x_2\dot{x}_2 = 2x_1(x_2 - x_1^3) + 2x_2(-x_1 - x_2^3) = -2x_1^4 - 2x_2^4$$

$$\implies \dot{V}(\mathbf{x}) < 0 \quad \forall \mathbf{x} \neq 0$$

Altogether, the system is globally asymptotically stable. Note that the choice of V is not unique, and the stability proof of the above system also works with $V(\mathbf{x}) = 2(x_1^2 + x_2^2)$. □

7.5 Exponential Stability

The theorem presented so far makes statements about the stability of the equilibria but does not provide any information about the type of convergence. Accordingly, we define exponential stability for systems whose solutions decay exponentially.

Definition 7.64 (*Exponential Stability*) Suppose we are given the autonomous nonlinear system

$$\dot{\mathbf{x}}(t) = f(\mathbf{x}(t)) \tag{7.5}$$

with $f : D \to \mathbb{R}^n$ locally Lipschitz over a domain $D \subseteq \mathbb{R}^n$, and the equilibrium point $\mathbf{x}^* = \mathbf{0} \in D$. Then $\mathbf{x}^*$ is *exponentially stable* if there exist constants $\alpha, M, c > 0$ such that for all $\|\mathbf{x}(t_0)\| < c$,

$$\|\mathbf{x}(t)\| \leq M\|\mathbf{x}(t_0)\|e^{-\alpha(t-t_0)}. \tag{7.6}$$

If (7.6) holds for all $\mathbf{x}(t_0) \in \mathbb{R}^n$, then the equilibrium point is globally exponentially stable.

Theorem 7.16 (Exponential Stability) *Let $\boldsymbol{x}^* = \boldsymbol{0}$ be an equilibrium of* (7.5) *and $V : \mathbb{R}^n \to \mathbb{R}$ be a continuously differentiable function. If*

$$\begin{aligned} c_1 \|\boldsymbol{x}\|^2 \leq V(\boldsymbol{x}) &\leq c_2 \|\boldsymbol{x}\|^2, \\ \dot{V}(\boldsymbol{x}) &\leq -c_3 \|\boldsymbol{x}\|^2, \\ \left\| \frac{\partial V(\boldsymbol{x})}{\partial \boldsymbol{x}} \right\| &\leq c_4 \|\boldsymbol{x}\| \end{aligned} \tag{7.7}$$

holds locally and $\forall t \geq 0$ for some constants $c_1, c_2, c_3, c_4 > 0$, then $\boldsymbol{x}^* = 0$ is exponentially stable. If (7.7) *holds globally (i.e., for all $\boldsymbol{x} \in \mathbb{R}^n$), then $\boldsymbol{x}^* = \boldsymbol{0}$ is globally exponentially stable.*

Now the question may arise, how is (7.7) related to (7.6)? Therefore, we first transform the first term and insert it into the second term. This results in

$$\dot{V}(\mathbf{x}) \leq -c_3 \|\mathbf{x}\|^2 \leq -\frac{c_3}{c_2} V(\mathbf{x}). \tag{7.8}$$

The next step is to treat the new term like a scalar differential inequality:

$$\dot{z}(t) \leq -\frac{c_3}{c_2} z(t) \tag{7.9}$$

with initial condition $z(0) = V(\mathbf{x}(0))$. By applying the Gronwall-Bellman inequality and considering that all terms are nonnegative, we can obtain

$$\|\mathbf{x}(t)\| \leq \sqrt{\frac{c_2}{c_1}} \|\mathbf{x}(0)\| e^{-\frac{c_3}{2c_2} t}. \tag{7.10}$$

Therefore, we can choose $M \leq \sqrt{\frac{c_2}{c_1}}$ and $\alpha \geq \frac{c_3}{2c_2}$.

For linear time-invariant (LTI) systems, we can also prove exponential stability using $V(\mathbf{x}) = \mathbf{x}^\top P \mathbf{x}$ for $P \succ 0$ as the Lyapunov function.

7.6 Lyapunov Stability for LTI Systems

For LTI systems where $\dot{\mathbf{x}}(t) = Ax(t)$, we can simply choose a quadratic form to be our Lyapunov function:

$$V(\mathbf{x}) = \mathbf{x}^\top P \mathbf{x}, \quad P \succeq 0 \text{ is a } n \times n \text{ symmetric matrix.}$$

Because $P \succ 0.$, V is a positive-definite function. We can verify that V is $\mathcal{C}^1$, and satisfies $V(0) = 0$, decrescence with $\beta(\|\mathbf{x}\|) = \lambda_{\max}(P)\|\mathbf{x}\|^2$. Moreover, $V(\mathbf{x}) \geq \lambda_{\max}(P)\|\mathbf{x}\|^2$ is radially-unbounded. so, asymptotic stability is global.

$$\dot{V}(\mathbf{x}) = \mathbf{x}^\top P\dot{\mathbf{x}} + \dot{\mathbf{x}}^\top P\mathbf{x} = \mathbf{x}^\top A^\top P\mathbf{x} + \mathbf{x}^\top PA\mathbf{x} = \mathbf{x}^\top(A^\top P + PA)\mathbf{x}$$

This results gives us useful information for Lyapunov stability for LTI systems.

Theorem 7.17 *Such a LTI system is asymptotically stable if*

$$A^\top P + PA < 0 \tag{7.11}$$

Equivalently, there must exist $n \times n$ *symmetric matrix* $Q \succ 0$ *such that*

$$A^\top P + PA = -Q \tag{7.12}$$

where (7.12) is called the Lyapunov equation.

The MATLAB commands `lyap` for CT, `dlyap` for DT. Standard guesses for Q are I_n or qI_n for some $q \in \mathbb{R}^+$.

Theorem 7.18 *Suppose we are given autonomous nonlinear system* $\dot{\mathbf{x}}(t) = f(\mathbf{x}(t))$ *with* $f : D \to \mathbb{R}^n$ *locally Lipschitz over domain* $D \subseteq \mathbb{R}^n$, $0 \in D$, *and equilibrium point* $\mathbf{x}^* = 0$. *Then* $\mathbf{x}^* = 0$ *is locally exponentially stable iff* $\exists \epsilon > 0$ *and Lyapunov function* $V : \mathbb{R}^n \to \mathbb{R}$ *and* $\alpha_1, \alpha_2, \alpha_3, \alpha_4 > 0$ *s.t.*

$$\begin{cases} \alpha_1\|\mathbf{x}\|^2 \le V(\mathbf{x}) \le \alpha_2\|\mathbf{x}\|^2 \\ \dot{V}(\mathbf{x}) \le -\alpha_3\|\mathbf{x}\|^2 \\ \|\nabla V(\mathbf{x})\| \le \alpha_4\|\mathbf{x}\| \end{cases} \quad \forall \mathbf{x} \in B_\epsilon \subseteq D \tag{7.13}$$

How is Theorem 7.18 related to the original condition for exponential stability from Definition 7.57? That is, $\|\mathbf{x}(t)\| \le Me^{-\alpha t}\|\mathbf{x}_0\|$? Let's combine the first two conditions in (7.13):

$$\begin{aligned} -\alpha_1\|\mathbf{x}\|^2 &\ge -V(\mathbf{x}) \ge -\alpha_2\|\mathbf{x}\|^2 \\ \dot{V}(\mathbf{x}) &\le -\alpha_3\|\mathbf{x}\|^2 \le -\frac{\alpha_3}{\alpha_2}V(\mathbf{x}) \end{aligned}$$

Treat this like a scalar differential inequality $\dot{z}(t) \le -\frac{\alpha_3}{\alpha_2}z(t)$ with initial condition $z(0) = V(\mathbf{x}(0))$. By Gronwall-Bellman, $z(t) \le z(0)e^{-\frac{\alpha_3}{\alpha_2}t}$, therefore,

$$V(\mathbf{x}) \le V(\mathbf{x}_0)e^{-\frac{\alpha_3}{\alpha_2}t} \implies \|\mathbf{x}\|^2 \le \frac{\alpha_2}{\alpha_1}\|\mathbf{x}_0\|^2 e^{-\frac{\alpha_3}{\alpha_2}t} \implies \|\mathbf{x}\| \le \sqrt{\frac{\alpha_2}{\alpha_1}}\|\mathbf{x}_0\|e^{-\frac{\alpha_3}{2\alpha_2}t}$$

since all terms are non-negative, we can choose $M \le \sqrt{\frac{\alpha_2}{\alpha_1}}$, $\alpha \ge \frac{\alpha_3}{2\alpha_2}$.

For LTI systems, we can also prove exponential stability using the same Lyapunov function as before: $V(\mathbf{x}) = \mathbf{x}^\top P\mathbf{x}$, $P \succ 0$.

Theorem 7.19 *A continuous LTI system $\dot{\boldsymbol{x}} = A\boldsymbol{x}$ is asymptotically (exponentially) stable if and only if $\forall Q \succ 0,\ Q = Q^\top \in \mathbb{R}^{n\times n}$, there exists a unique $P \succ 0,\ P = P^\top \in \mathbb{R}^{n\times n}$ such that the Lyapunov equation (7.12) holds.*

Proof *(Sufficiency.)* Suppose $A^\top P + PA = -Q$ (the Lyapunov equation holds)

$$\dot{V}(\mathbf{x}) = \mathbf{x}^\top(A^\top P + PA)\mathbf{x} = -\mathbf{x}^\top Q\mathbf{x}$$

$$\Longrightarrow -\dot{V}(\mathbf{x}) = \mathbf{x}^\top Q\mathbf{x} \geq \lambda_{\min}(Q)\|\mathbf{x}\|^2 \text{ and } V(\mathbf{x}) = \mathbf{x}^\top P\mathbf{x} \leq \lambda_{\max}(P)\|\mathbf{x}\|^2$$

Therefore,

$$-\dot{V}(\mathbf{x}) \geq \frac{\lambda_{\min}(Q)}{\lambda_{\max}(P)}\|\mathbf{x}\|^2 \geq \alpha \text{ for } \mathbf{x} \neq 0, \alpha = \frac{\lambda_{\min}(Q)}{\lambda_{\max}(P)}$$

$$\Longrightarrow \dot{V}(\mathbf{x}) \leq -\alpha V(\mathbf{x})$$

Treat this like a scalar differential inequality $\dot{z}(t) \leq -\alpha z(t)$ with initial condition $z(0) = V(\mathbf{x}_0)$ as before. By Gronwall-Bellman,

$$z(t) \leq z(0)e^{-\alpha t} \Longrightarrow V(\mathbf{x}(t)) \leq V(\mathbf{x}_0)e^{-\alpha t}$$

$$\Longrightarrow V(\mathbf{x}(t)) \Longrightarrow 0 \text{ exponentially as } t \to \infty$$

Since $V(\mathbf{x}(t)) \geq \lambda_{\min}(P)\|\mathbf{x}(t)\|^2$, we also have $\lambda_{\min}(P)\|\mathbf{x}(t)\|^2 \Longrightarrow 0$ as $t \to \infty$. Since $\lambda_{\min}(P) > 0$, this means,

$$\|\mathbf{x}(t)\| \Longrightarrow 0 \text{ exponentially as } t \to \infty \text{ too.}$$

(Necessity.) Suppose $\dot{\mathbf{x}} = A\mathbf{x}$ is asymptotically stable. This means all eigenvalues $\lambda_i(A)$ have strictly negative real parts. This further implies that A is nonsingular, and we can get the unique solution $\mathbf{x}(t) = e^{At}\mathbf{x}_0$ given $\mathbf{x}(0) = \mathbf{x}_0$.

Let $Q \succ 0$ be some $n \times n$ matrix. Then $e^{A^\top t}Qe^{At} \succ 0$ for all $t \geq 0$. Consider

$$P \triangleq \int_0^\infty e^{A^\top t}Qe^{At}dt. \tag{7.14}$$

Note that such a P exists, is unique, is positive definite, and satisfies $\|P\|_2 < \infty$. Then,

$$
\begin{aligned}
A^\top P + PA &= \int_0^\infty \left(A^\top e^{A^\top t} Q e^{At} + e^{A^\top t} Q e^{At} A\right) dt \\
&= \int_0^\infty \frac{d}{dt}\left\{e^{A^\top t} Q e^{At}\right\} dt \\
&= \lim_{t\to\infty} e^{A^\top t} Q e^{At} - Q = -Q.
\end{aligned}
$$

Therefore, $A^\top P + PA = -Q$. ■

7.7 Invariant Set Theorem

LaSalle's Invariant Set Theorem can help us conclude asymptotic stability even though the Lyapunov function we've constructed has negative semidefinite derivative so that it only lets us conclude stability. Suppose $\Omega_c := \{\mathbf{x}|V(\mathbf{x}) \le c\}$ is bounded and $\dot{V}(\mathbf{x}) \le 0$ on Ω_c. Let D be the largest invariant set in $\{\mathbf{x}|\dot{V}(\mathbf{x}) = 0\}$; that is, every trajectory $\mathbf{x}(t)$ that enters D at some time τ never leaves D for all $t \ge \tau$. Then all $\mathbf{x}(t)$ such that $\mathbf{x}_0 \in \Omega_c$ tends to D as $t \to \infty$.

There is a similar version of LaSalle's principle for limit cycles, called the *Poicare-Bendixson Theorem*. Informally stated, let $\mathcal{C}$ be a compact (forward) invariant set, i.e., once a trajectory enters $\mathcal{C}$, it never leaves it. If $\mathcal{C}$ contains no equilibrium points, a limit cycle that must exist in it. It must clearly be the case that $\dot{V}(\mathbf{x}) = 0$ for all $\mathbf{x}$ on the limit cycle. Thus, all trajectories which begin in $\mathcal{C}$ ultimately converge to the limit cycle in $\mathcal{C}$.

This section introduces *LaSalle's invariance principle*, which has two main applications. On the one hand, it is an extension of the stability analysis of Lyapunov. In some cases, with this principle, we can examine stability without $V(\mathbf{x})$ being (locally) positive definite. Secondly, it can be used to show that trajectories which start in a particular area, converge to an invariant set. An invariant set is defined as a set, where trajectories can no longer leave once they have entered.

Definition 7.65 (*Invariant Set*) A set $S \subset \mathbb{R}^n$ is said to be *(positively) invariant*, also called *(forward) invariant*, for a $\dot{\mathbf{x}} = \mathbf{f}(\mathbf{x})$ if $\forall \mathbf{x}_0 \in S$ and $t_0 \ge 0$, $\mathbf{x}(t) \in S\ \forall t \ge t_0$. That is, every trajectory of $\dot{\mathbf{x}} = \mathbf{f}(\mathbf{x})$ which enters set S at time τ never leaves $S\ \forall t \ge \tau$.

Theorem 7.20 *Let* $V : \mathbb{R}^n \to \mathbb{R}$ *be locally positive definite, and let*

$$\Omega_c \triangleq \{\boldsymbol{x} \in \mathbb{R}^n | V(\boldsymbol{x}) \le c,\ \dot{V}(\boldsymbol{x}) \le 0\}, \tag{7.15}$$

where $c > 0$, *be a compact, invariant set with respect to* $\dot{\boldsymbol{x}} = \boldsymbol{f}(\boldsymbol{x})$. *Further define*

$$M \triangleq \{\boldsymbol{x} \in \mathbb{R}^n | \dot{V}(\boldsymbol{x}) = 0\} \tag{7.16}$$

Let D be the largest invariant set in $M \cap \Omega_c$. Then all $\mathbf{x}(t)$ such that $\mathbf{x}_0 \in \Omega_c$ converges towards D as $t \to \infty$. And if $D = \{0\}$, then 0 is (locally) asymptotically stable (Fig. 7.5).

Remark 7.12 Sets of the form $\{\mathbf{x} \in \mathbb{R}^n : V(\mathbf{x}) = c\}$, $c \in \mathbb{R}$ are often called the level set of V (at level c), while sets $\{\mathbf{x} \in \mathbb{R}^n : V(\mathbf{x}) \leq c\}$ are often called sublevel set of V (at level c).
(If V is continuous, both level sets and sublevel sets of V are closed.)

Example 7.12 Recall the pendulum-with-friction example described in Sect. 7.1.

$$\dot{x}_1 = x_2, \; \dot{x}_2 = -\frac{g}{L}\sin x - 1 - \frac{k}{m}x_2$$

Choose energy function to be the Lyapunov function

$$V(\mathbf{x}) = \frac{1}{2}x_2^2 + \frac{g}{L}(1 - \cos x_1)$$

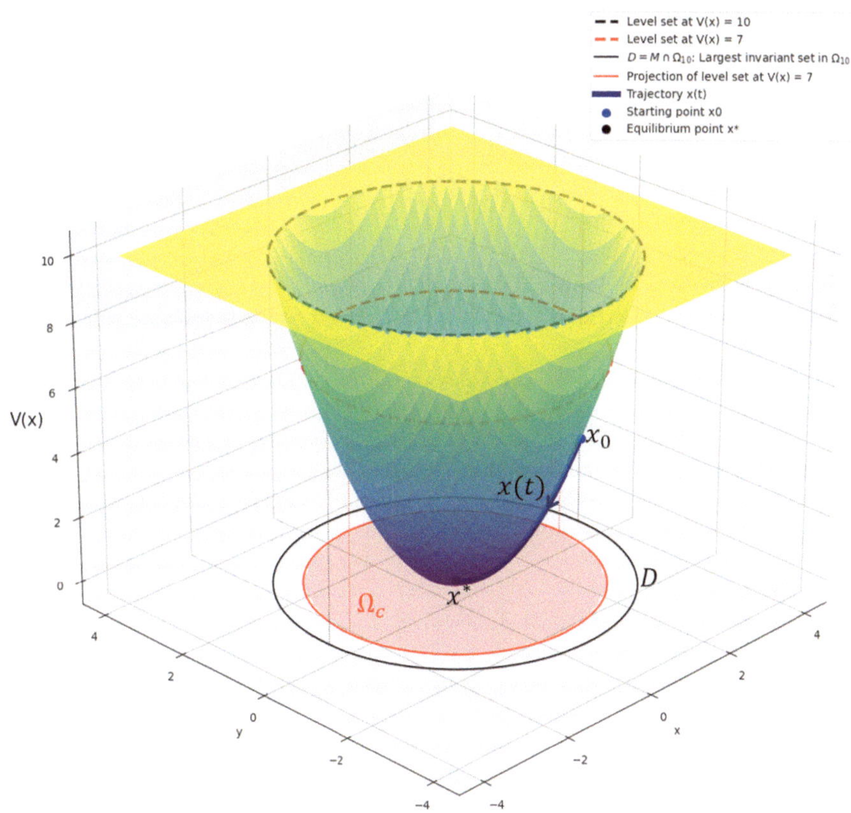

Fig. 7.5 A Lyapunov function portrayed over a domain D and an invariant set Ω_c

We saw that $\dot{V}(\mathbf{x}) = -\frac{k}{m}x_2^2 \le 0 \Longrightarrow$ negative semidefinite. The Lyapunov direct method only tells us that $(0, 0)$ is a stable equilibrium. We will try to arrive at a stronger conclusion with LaSalle's principle.

Note that in order for $\dot{V}(\mathbf{x}) = 0$, all $\mathbf{x}(t)$ must be such that $x_2(t) = 0\ \forall t$.

$$M = \{\mathbf{x} \in \mathbb{R}^n | \dot{V}(\mathbf{x}) = 0\} = \{\mathbf{x} \in \mathbb{R}^n | x_2 = 0\}$$

Choose Ω_c corresponding to $x_1 \in (-\pi, \pi), x_2 = 0$:

$$\Omega_c \triangleq \left\{\mathbf{x} \in \mathbb{R}^n | V(\mathbf{x}) < \frac{g}{L}(1 - \cos\pi) = \frac{2g}{L}\right\} \tag{7.17}$$

For any $\mathbf{x} \in \Omega_c$, the system can maintain $\dot{V}(\mathbf{x}) = 0$ only when $x_1 = 0$. This means that $\{0\}$ is the largest invariant set in $M \cap \Omega_c$ and so $(0, 0)$ is an asymptotically stable equilibrium point. □

Example 7.13 Consider the system

$$\dot{\mathbf{x}} = f(\mathbf{x}) = \begin{bmatrix} x_2 \\ x_1 - x_1^2 x_2 \end{bmatrix}$$

Construct Lyapunov function $V(x) = x_1^2 + x_2^2$. Clearly this is positive definite. Then $\dot{V}(x) = 2x_1\dot{x}_1 + 2x_2\dot{x}_2 = -2x_1^2x_2^2$. But this is only negative semidefinite.

Consider the set $D = \left\{x \in \mathbb{R}^n | \dot{V}(x) = 0\right\} = \{x \in \mathbb{R}^n | x_1 = 0 \text{ or } x_2 = 0\}$. Note that:

$$\begin{aligned} &x_1(t) \equiv 0 \Longrightarrow \dot{x}_1(t) \equiv 0 \Longrightarrow x_2 \equiv 0 \\ &x_2(t) \equiv 0 \Longrightarrow x_2(t) \equiv 0 \text{ and } \dot{x}_2(t) \equiv 0 \Longrightarrow x_1(t) \equiv 0 \text{ since } x_2 \equiv 0 \end{aligned}$$

So the only invariant set that the system converges to is $D = \{x \in \mathbb{R}^n | x_1 \equiv x_2 \equiv 0\}$. Therefore, $\mathbf{x}^* = 0$ is asymptotically stable. □

Example 7.14 Consider the following system

$$\begin{aligned} \dot{x}_1 &= -4x_2 + x_1(1 - \frac{1}{4}x_1^2 - x_2^2) \\ \dot{x}_2 &= x_1 + x_2(1 - \frac{1}{4}x_1^2 - x_2^2) \end{aligned}$$

Again use Lyapunov function $V(x) = \frac{1}{2}\left(\frac{1}{4}x_1^2 + x_2^2\right)$. Then

$$\dot{V}(x) = \frac{1}{4}x_1\dot{x}_1 + x_2\dot{x}_2 = \underbrace{\left(\frac{1}{4}x_1^2 + x_2^2\right)}_{\text{always } \ge 0}\left(1 - \frac{1}{4}x_1^2 - x_2^2\right)$$

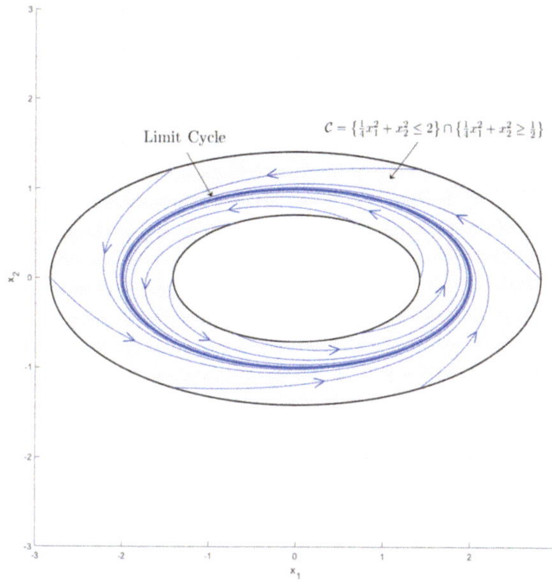

Fig. 7.6 Trajectories of the system when initial conditions are on the ellipse $\left\{\frac{1}{4}x_1^2 + x_2^2 = 2\right\}$ or on the ellipse $\left\{\frac{1}{4}x_1^2 + x_2^2 = \frac{1}{2}\right\}$. Indeed, convergence is towards the ellipse $\left\{\frac{1}{4}x_1^2 + x_2^2 = 1\right\}$

Consider the annulus $C = \left\{\frac{1}{4}x_1^2 + x_2^2 \leq 2\right\} \cap \left\{\frac{1}{4}x_1^2 + x_2^2 \geq \frac{1}{2}\right\}$. Note that on the outer boundary, $\dot{V}(x) \leq 0$, whereas on the inner boundary, $\dot{V}(x) \geq 0$, so all trajectories that cross either boundary go inside the annulus and towards the invariant set $\dot{V}(x) = 0$, given by the ellipse $\left\{\frac{1}{4}x_1^2 + x_2^2 = 1\right\}$. Thus C is a forward Limit Cycle invariant set that contains no equilibrium points (the only other equilibrium point that the system has is $(x_1, x_2) = 0$). All trajectories that begin in C must therefore converge to the limit cycle $\left\{\frac{1}{4}x_1^2 + x_2^2 = 1\right\}$.

Figure 7.6 illustrates the behavior of the system for some sample initial conditions. □

References

1. Alexandr Mikhailovich Liapunov. The general problem of the stability of motion. In: 1892.
2. Joseph P. LaSalle and Solomon Lefschetz. Stability by Liapunovs direct method with applications. Vol. 4. Academic Press New York, 1973.
3. Hassan K Khalil. Nonlinear Systems. Prentice Hall, New York, NY, 2002.

Chapter 8
Uniform Stability

Most stability *in the sense of Lyapunov (i.s.L.)* definitions can be extended straightforwardly to the nonautuonomous case: $\dot{\mathbf{x}}(t) = f(t, \mathbf{x}(t)), \dot{\mathbf{x}}(t) = A(t)\mathbf{x}(t), \mathbf{x}(t_0) = \mathbf{x}_0$. Assume f satisfies the standard conditions for existence and uniqueness of solutions.

Definition 8.66 (*Stability for Nonautonomous Systems*) Nonautonomous system $\dot{\mathbf{x}}(t) = f(t, \mathbf{x})$ is *stable* around $\mathbf{x}^*$ at t_0 if $\forall\epsilon > 0\ \exists\delta(t_0) > 0$ s.t.

$$\|\mathbf{x}_0 - \mathbf{x}^*\| < \delta(t_0, \epsilon) \implies \|\mathbf{x}(t) - \mathbf{x}^*\| < \epsilon \quad \forall t \geq t_0.$$

Remark 8.13 Compared to the autonomous case $\dot{\mathbf{x}} = f(\mathbf{x})$, which had $\delta \equiv \delta(\epsilon)$ depend only on ϵ, our new $\delta \equiv \delta(t_0, \epsilon)$ also depends on time.

Definition 8.67 (*Uniform Stability*) $\dot{\mathbf{x}}(t) = f(t, \mathbf{x})$ is *uniformly stable* around $\mathbf{x}^*$ if it is stable i.s.L. with $\delta \equiv \delta(t_0)$. That is, we can choose δ independently of time t_0.

Definition 8.68 (*Uniform Asymptotic Stability*) Nonautnomous system $\dot{\mathbf{x}}(t) = f(t, \mathbf{x})$ is asymptotically stable at t_0 around $\mathbf{x}^*$ if it is stable i.s.L. and $\exists\delta(t_0) > 0$ s.t.

$$\|\mathbf{x}_0 - \mathbf{x}^*\| < \delta(t_0) \implies \lim_{t\to\infty} \|\mathbf{x}(t) - \mathbf{x}^*\| = 0.$$

In addition, it is *uniformly asymptotically stable (UAS)* if it is uniformly stable and $\exists\delta > 0$ s.t.

$$\|\mathbf{x}_0 - \mathbf{x}^*\| < \delta \implies \lim_{t\to\infty} \|\mathbf{x}(t) - \mathbf{x}^*\| = 0.$$

Note that in the definition of UAS, δ is independent of time.

Definition 8.69 (*Uniform Exponential Stability*) Nonautonomous system $\dot{\mathbf{x}}(t) = f(t, \mathbf{x})$ is *uniformly exponentially stable (UES)* at t_0 around $\mathbf{x}^*$ if $\exists\delta(t_0) > 0$, $M > 0$, $\alpha > 0$ s.t.

$$\|\mathbf{x}_0 - \mathbf{x}^*\| < \delta(t_0) \implies \|\mathbf{x}(t) - \mathbf{x}^*\| \leq M\|\mathbf{x}_0 - \mathbf{x}^*\|e^{-\alpha t} \quad \forall t \geq t_0.$$

SJ. Han, *Linear Systems*, KAIST Research Series,
https://doi.org/10.1007/978-981-96-9774-8_8

Remark 8.14 UES implies UAS.

Example 8.15 Consider the LTV system,

$$\dot{x}(t) = (5t \sin t - 2t)x(t),\ x(t_0) = x_0$$

check whether this system is stable and uniformly stable.

Using the separation of variables,

$$\frac{dx}{x} = (5t \sin t - 2t)dt$$

$$\begin{aligned}\ln x(t) - \ln x_0 =& \int_{t_0}^{t} (5r \sin r - 2r)dr\\ =& -5(t \cos t - \sin t - t^2 + 5(t_0 \cos t_0 - \sin t_0) + t_0^2 \equiv g(t, t_0)\end{aligned}$$

Rearranging variables, our solution is

$$x(t) = x_0 e^{g(t,t_0)}. \tag{8.1}$$

For fixed t_0, $g(t, t_0)$ will eventually be dominated by $-t^2$ term, and $e^{g(t,t_0)} \to 0$. This implies that there exists constant $c(t_0)$ s.t. $|x(t)| < c(t_0)|x(t_0)|$ for all $t \geq t_0$.

If we choose $\delta(t_0) = \frac{\epsilon}{c(t_0)}$, then

$$|x(t_0)| < \delta \implies |x(t)| < c(t_0)|x(t_0)| < c(t_0)\frac{\epsilon}{c(t_0)} = \epsilon.$$

$\therefore$ This system is stable i.s.L.

Considering the sequence $\{t_0^{(n)}\}_{n\in\mathbb{N}}$, $t_0^{(n)} \triangleq 2\pi n$, then

$$\frac{x(t)}{x_0} = \frac{x(t_0^{(n)} + \pi)}{x(t_0^{(n)})} = \frac{x\left((2n+1)\pi\right)}{x(2\pi n)} = e^{g((2n+1)\pi, 2\pi n)} = Me^{\alpha n}$$

for $M = e^{\pi(5-\pi)} > 0$ and $\alpha = 4\pi(5 - \pi) > 0$, where the third equality comes from (8.1).

If $n \to \infty$, then $Me^{\alpha n} \to \infty$. Therefore, $|\frac{x(t)}{x_0}| < |c(t_0)| \to \infty$. Since the dependence on t_0 is needed,

Since the dependence on t_0 is needed, this system is not uniformly stable. □

8.1 Lyapunov Direct Method for Nonautonomous Systems

Theorem 8.22 *Let $\dot{\mathbf{x}}(t) = f(t, \mathbf{x})$ have equilibrium $\mathbf{x}^* = 0$ with the usual existence and uniqueness conditions. Construct Lyapunov function $V(t, \mathbf{x})$, $V : \mathbb{R} \times \mathbb{R}^n \to \mathbb{R}$, $V \in C^{(1,2)}$. Suppose that there exits continuous positive definite functions α_1, α_2 and α_3, $\alpha_i : \mathbb{R}^n \to \mathbb{R}$ s.t.*

$$\alpha_1(\mathbf{x}) \leq V(t, \mathbf{x}) \leq \alpha_2(\mathbf{x}),$$

$$\dot{V}(t, \mathbf{x}) \leq -\alpha_3(\mathbf{x}),$$

then 0 is UAS.

Let's make some comparisons with the autonomous-system version we saw in the previous chapter:

1. $V(\mathbf{x}) > 0$ for $\mathbf{x} \neq 0$ (locally positive definite) becomes $\alpha_1(\mathbf{x}) \leq V(t, \mathbf{x}) \leq \alpha_2(\mathbf{x})$. V needs to be sandwiched in between two positive definite functions that are independent of time t.
2. $\dot{V}$ has a partial derivative with respect to t. $\dot{V}(\mathbf{x}) < 0$ for $x \neq 0$ becomes $\dot{V}(t, \mathbf{x}) \leq -\alpha_3(\mathbf{x})$. $\dot{V}(t, \mathbf{x})$ is bounded by a negative definite function that is independent of time t.

8.1.1 LTV Case

For LTV system

$$\dot{\mathbf{x}}(t) = A(t)\mathbf{x}(t), \mathbf{x}(t_0) = \mathbf{x}_0, \mathbf{x}^* = 0,$$

show that

$$V(t, \mathbf{x}) \triangleq \mathbf{x}^\top(t)P(t)\mathbf{x}(t)$$

is a Lyapunov function for the system with the following statements:

1. Suppose that there exists symmetric, positive definite, and C^1 function $P : \mathbb{R} \to \mathbb{R}^{n\times n}$ s.t.
$$c_1 I \leq P(t) \leq c_2 I \text{ for some } c_1 > 0, c_2 > 0 \quad \forall t \geq t_0. \tag{8.2}$$
2. Assume that $P(t)$ also satisfies
$$-\dot{P}(t) = A^\top(t)P(t) + P(t)A(t) + Q(t) \tag{8.3}$$
for some $Q(t)$ that is continuous, symmetric and positive definite s.t. $Q(t) \geq c_3 I$, $c_3 > 0\ \forall t \geq t_0$.

Proof By Theorem 8.22, check whether $V(t, \mathbf{x})$ satisfies

$$\alpha_1(\mathbf{x}) \leq V(t, \mathbf{x}) \leq \alpha_2(\mathbf{x}) \text{ and } \dot{V}(t, \mathbf{x}) \leq -\alpha_3(\mathbf{x})$$

with some continuous positive definite functions α_1, α_2 and α_3, $\alpha_i : \mathbb{R}^n \to \mathbb{R}$.

Using (8.2), we have

$$c_1\|\mathbf{x}\|^2 \leq \mathbf{x}^\top P(t)x \leq c_2\|\mathbf{x}\|^2.$$

Taking $\alpha_1 = c_1\|x|^2$ and $\alpha_2 = c_2\|\mathbf{x}\|^2$, it satisfies $\alpha_1(\mathbf{x}) \leq V(t, \mathbf{x}) \leq \alpha_2(\mathbf{x})$.

Furthermore, we have

$$\begin{aligned} \dot{V}(t, \mathbf{x}) =& \dot{\mathbf{x}}^\top Px + \mathbf{x}^\top \dot{P}x + \mathbf{x}^\top P\dot{\mathbf{x}} \\ =& \mathbf{x}^\top(A^\top P + \dot{P} + PA)x \\ =& -\mathbf{x}^\top Qx \\ \leq& -c_3\|\mathbf{x}\|^2, \end{aligned} \tag{8.4}$$

where the third equality comes from (8.3). Taking $\alpha_3 = c_3\|\mathbf{x}\|^2$, it satisfies $\dot{V}(t, \mathbf{x}) \leq -\alpha_3(\mathbf{x})$.

$$\therefore V(t, \mathbf{x}) = \mathbf{x}^\top(t)P(t)\mathbf{x}(t) \text{ is a Lyapunov function.}$$

■

By Lyapunov Direct Method, $\dot{\mathbf{x}}(t) = A(t)\mathbf{x}(t)$ is *UAS* if there exists symmetric, positive definite, and $\mathcal{C}^1$ function P s.t.

1. $c_1 I \leq P(t) \leq c_2 I$ for some $c_1 > 0, c_2 > 0 \quad \forall t \geq t_0$.
2. $-\dot{P}(t) = A^\top(t)P(t) + P(t)A(t) + Q(t)$ for some $Q(t)$ that is continuous, symmetric and positive definite s.t. $Q(t) \geq c_3 I, c_3 > 0 \quad \forall t \geq t_0$.

Note that 1. and 2. are only *sufficient* conditions for UAS of LTV system. Because eigenvalues are changing over time, they are difficult to use in stability proofs. Instead, we can use the STM Φ.

8.2 Comparison Functions

Before we move on to the stability proof using the STM, let's introduce some definitions for convenience.

Definition 8.70 Continuous function $\alpha : [0, a) \Longrightarrow [0, \infty)$ is *class-$\mathcal{K}$* if it is strictly increasing and $\alpha(0) = 0$.

Definition 8.71 Continuous function $\alpha : [0, \infty) \Longrightarrow [0, \infty)$ is *class-$\mathcal{K}_\infty$* if it is class-$\mathcal{K}$ and radially-unbounded ($\alpha(r) \to \infty$ as $r \to \infty$).

Definition 8.72 Continuous function $\beta : [0, a) \times [0, \infty) \Longrightarrow [0, \infty)$ is *class-$\mathcal{KL}$* if

1. $\forall$ fixed s, $\beta(\cdot, s) : [0, a) \Longrightarrow [0, \infty)$ is class-$\mathcal{K}$.
2. $\forall$ fixed r, $\beta(r, \cdot) : [0, \infty) \Longrightarrow [0, \infty)$ is decreasing s.t. $\lim_{s\to\infty} \beta(r, s) = 0$.

Comparison functions have some properties:

1. $\alpha \in \mathcal{K} \implies \alpha^{-1} \in \mathcal{K}$
2. $\alpha_1, \alpha_2 \in \mathcal{K} \implies \alpha_1 \circ \alpha_2 \in \mathcal{K}$
3. $\alpha \in \mathcal{K}_\infty \implies \alpha^{-1} \in \mathcal{K}_\infty$
4. $\alpha_1, \alpha_2 \in \mathcal{K}, \beta \in \mathcal{KL} \implies \alpha_1(\beta(\alpha_2(r), s)) \in \mathcal{KL}$

We can rewrite the conditions of uniform stability and uniform asymptotic stability in terms of class-$\mathcal{K}$ and class-$\mathcal{KL}$ functions:

1. The system is *uniformly stable* iff $\exists$ class-$\mathcal{K}$ function α and $\delta > 0$ that is independent of t_0 s.t.

$$\|\mathbf{x}(t_0) - \mathbf{x}^*\| < \delta \implies \|\mathbf{x}(t) - \mathbf{x}^*\| \le \alpha(\|\mathbf{x}(t_0) - \mathbf{x}^*\|) \quad \forall t \ge t_0.$$

2. The system is *uniformly asymptotically stable* iff $\exists$ class-$\mathcal{KL}$ function β and $\delta > 0$ that is independent of t_0 s.t.

$$\|\mathbf{x}(t_0) - \mathbf{x}^*\| < \delta \implies \|\mathbf{x}(t) - \mathbf{x}^*\| \le \beta(\|\mathbf{x}(t_0) - \mathbf{x}^*\|, t - t_0) \quad \forall t \ge t_0.$$

Note that exponential stability is achieved with $\beta(r, s) = Mre^{-\alpha s}$.

8.3 Proof of Stability Using STM

Theorem 8.23 *LTV system* $\dot{\mathbf{x}}(t) = A(t)\mathbf{x}(t)$, $\mathbf{x}(t_0) = \mathbf{x}_0$ *is UAS at* $\mathbf{x}^* = 0$ *iff* $\exists M, \alpha > 0$ *s.t.*

$$\|\Phi(t, t_0)\| \le Me^{-\alpha(t-t_0)}, \quad \forall t \ge t_0.$$

Here, Theorem 8.23 shows that UAS $\iff$ exponential stability (ES).

Proof (*Sufficiency*) Suppose $\|\Phi(t, t_0)\| \le Me^{-\alpha(t-t_0)}$ for all $t \ge t_0$. By the definition of STM, $\mathbf{x}(t) = \Phi(t, t_0)\mathbf{x}(t_0)$.

$$\begin{aligned}
\|\mathbf{x}(t)\| &\le \|\Phi(t, t_0)\mathbf{x}(t_0)\| \\
&\le \|\Phi(t, t_0)\| \|\mathbf{x}(t_0)\| \\
&\le M\|\mathbf{x}(t_0)\|e^{-\alpha(t-t_0)}
\end{aligned}$$

As $t \to \infty$, $\|\mathbf{x}(t)\| \to 0$ at a rate α which is independent of t_0. Therefore, this system is UAS.

Moreover, if $M\|\mathbf{x}(t_0)\|e^{-\alpha(t-t_0)}$ holds for any $\mathbf{x}(t_0)$, it is also global UAS.
(*Necessity*) Suppose $\dot{\mathbf{x}}(t) = A(t)\mathbf{x}(t)$ is UAS at $\mathbf{x}^* = 0$. Then, there exists class-$\mathcal{KL}$ function β s.t.

$$\|\mathbf{x}(t)\| \leq \beta(\|\mathbf{x}(t_0)\|, t - t_0) \quad \forall t \geq t_0.$$

Using the definition of induced matrix norm, we have

$$\|\Phi(t,s)\| = \max_{\|\mathbf{x}\|=1} \|\Phi(t,s)x\| \leq \max_{\|\mathbf{x}_0\|=1} \beta(\|\mathbf{x}\|, t-s) = \beta(1, t-s).$$

By Definition 8.72, $\beta(1, t-s) \to 0$ as $t \to \infty$ for fixed s. Therefore, there exists $T > 0$ s.t. $\beta(1, T) \leq e^{-1}$.

For any $t \geq t_0$, let $N \in \mathbb{N}$ be the smallest positive number s.t. $t_0 + (N-1)T \leq t \leq t_0 + NT$ and partition $[t_0, t_0 + (N-1)T]$ into $\{[t_0, t_0 + T), [t_0 + T, t_0 + 2T), \ldots, [t_0 + (N-2)T, t_0 + (N-1)T), [t_0 + (N-1)T, t)\}$. Using this partition, rewriting transition matrix leads to

$$\Phi(t, t_0) = \Phi(t, t_0 + (N-1)T) \prod_{k=1}^{N-1} \Phi(t_0 + kT, t_0 + (k-1)T)$$

$$\begin{aligned}
\|\Phi(t, t_0)\| \leq & \|\Phi(t, t_0 + (N-1)T)\| \prod_{k=1}^{N-1} \|\Phi(t_0 + kT, t_0 + (k-1)T)\| \\
\leq & \beta(1, t - t_0 - (N-1)T)\beta(1, T)^{N-1} \\
\leq & \beta(1, 0)e^{1-N} \\
= & e\beta(1, 0)e^{-N} \\
\leq & e\beta(1, 0)e^{-\frac{1}{T}(t-t_0)} \\
= & Me^{-\alpha(t-t_0)}
\end{aligned}$$

$$\therefore \|\Phi(t, t_0)\| \leq Me^{-\alpha(t-t_0)}$$

■

Example 8.16 Consider $\dot{\mathbf{x}}(t) = \begin{bmatrix} -2 & t \\ 0 & -2 \end{bmatrix} \mathbf{x}(t) = \left(\begin{bmatrix} -2 & 0 \\ 0 & -2 \end{bmatrix} + \begin{bmatrix} 0 & t \\ 0 & 0 \end{bmatrix} \right) \mathbf{x}(t)$.

Then STM is given by

$$\Phi(t, t_0) = \begin{bmatrix} e^{-2(t-t_0)} & \frac{1}{2}(t-t_0)^2 e^{-2(t-t_0)} \\ 0 & e^{-2(t-t_0)} \end{bmatrix} = \begin{bmatrix} 1 & \frac{1}{2}(t-t_0)^2 \\ 0 & 1 \end{bmatrix} e^{-2(t-t_0)}.$$

Since $\|\Phi(t, t_0)\|$ is dominated by $e^{-2(t-t_0)}$, $\|\Phi(t, t_0)\| \to 0$ as $t \to \infty$. Therefore, this system is UAS. □

8.4 Converse Theorem for LTV Systems

In this section, we will prove the system is UES by showing the Lyapunov function exists.

Theorem 8.24 *Let $\dot{\mathbf{x}}(t) = A(t)\mathbf{x}(t)$ with continuous and bounded $A(t)$ that has UES equilibrium point $\mathbf{x}^* = 0$. Assume that $Q(t)$ be continuous, bounded, symmetric and positive definite $\forall t > 0$. Then, there exists $\mathcal{C}^1$ bounded, symmetric, positive definite $P(t)$ s.t.*

$$-\dot{P}(t) = A^\top(t)P(t) + P(t)A(t) + Q(t) \tag{8.5}$$

and $V(t, \mathbf{x}) \triangleq \mathbf{x}^\top P(t)\mathbf{x}$ is a Lyapunov function for this system.

We already showed sufficiency via Lyapunov Direct Method ($\mathbf{x}^\top P\mathbf{x}$ Lyapunov function $\Longrightarrow$ UES) in Theorem 8.22.

Proof By Theorem 8.22, check whether $V(t, \mathbf{x})$ satisfies

$$\alpha_1(\mathbf{x}) \le V(t, \mathbf{x}) \le \alpha_2(\mathbf{x}) \text{ and } \dot{V}(t, \mathbf{x}) \le -\alpha_3(\mathbf{x})$$

with some continuous positive definite functions α_1, α_2 and α_3, $\alpha_i : \mathbb{R}^n \to \mathbb{R}$.

Construct $P(t) = \int_t^\infty \Phi^\top(s, t)Q(s)\Phi(s, t)ds$. Then,

$$\forall \mathbf{x} \in \mathbb{R}^n, \mathbf{x}^\top P\mathbf{x} = \int_t^\infty \mathbf{x}^\top \Phi^\top(s, t)Q(s)\Phi(s, t)\mathbf{x}ds.$$

Note that in the above equation, $\Phi(s, t)\mathbf{x}$ is the solution trajectory obtained by starting the system from state $\mathbf{x}$ at time t.

1. Show $\alpha_1(\mathbf{x}) \le V(t, \mathbf{x}) \le \alpha_2(\mathbf{x})$.
   Since we assumed that $Q(t)$ is bounded and positive definite, there exists $c_3 > 0$ and $c_4 > 0$ s.t. $c_3 I \le Q(t) \le c_4 I$ for all t. Then,

$$\begin{aligned}
\mathbf{x}^\top Px &= \int_t^\infty \mathbf{x}^\top \Phi^\top(s, t)Q(s)\Phi(s, t)\mathbf{x}ds \\
&\le \int_t^\infty c_4\|\Phi(s, t)\mathbf{x}\|^2 ds \\
&\le \int_t^\infty c_4 \left(M^2 e^{-2\alpha(s-t)}\right) \|\mathbf{x}\|^2 ds \\
&= c_4 M^2 \|\mathbf{x}\|^2 \left(-\frac{1}{2\alpha} e^{-2\alpha(s-t)}\right)\Bigg|_t^\infty
\end{aligned}$$

$$=\frac{c_4 M^2}{2\alpha}\|\mathbf{x}\|^2,$$

where the second inequality is due to Theorem 8.23.
Furthermore, using the assumption that the matrix $A(t)$ is bounded, i.e., $\|A(t)\| \leq a$, we get

$$\mathbf{x}^\top P\mathbf{x} \geq \frac{c_3}{2a}\|\mathbf{x}\|^2.$$

If we choose $\alpha_1(r) = \frac{c_3}{2a}r^2$ and $\alpha_2(r) = \frac{c_4 M^2}{2\alpha}r^2$, it satisfies $\alpha_1(\|\mathbf{x}\|) \leq V(t, \mathbf{x}) \leq \alpha_2(\|\mathbf{x}\|)$.

2. Show $\dot{V}(t, \mathbf{x}) \leq -\alpha_3(\mathbf{x})$.
First, we will prove (8.5). Using that $\partial_t \Phi(s, t) = -\Phi(s, t)A(t)$ holds for each fixed s, we have

$$\begin{aligned}
\dot{P}(t) &= \int_t^\infty \partial_t \Phi^\top(s, t)Q(s)\Phi(s, t)ds + \int_t^\infty \Phi^\top(s, t)Q(s)\partial_t\Phi(s, t)ds - Q(t) \\
&= -\int_t^\infty A^\top(t)\Phi^\top(s, t)Q(s)\Phi(s, t)ds - \int_t^\infty \Phi^\top(s, t)Q(s)\Phi(s, t)A(t)ds - Q(t) \\
&= -A^\top(t)P(t) - P(t)A(t) - Q(t).
\end{aligned}$$

If (8.5) holds, we already showed $\dot{V}(t, \mathbf{x}) \leq -\alpha_3(\mathbf{x})$ when $\alpha_3(r) = c_3 r^2$ in (8.4).

$$\therefore \mathbf{x}^\top P\mathbf{x} \text{ is a valid Lyapunov function.}$$

■

With Theorems 8.22 and 8.24, (8.5) becomes a necessary and sufficient condition to prove UES.

8.5 Lyapunov Indirect Method for Nonautonomous Systems

Consider nonautonomous system $\dot{\mathbf{x}} = f(t, \mathbf{x})$ s.t. $f(t, 0) = 0$ for all t. In addition, define Jacobian $A(t) \triangleq \nabla_{\mathbf{x}} f(t, \mathbf{x})|_{\mathbf{x}^*=0}$.

By Taylor expansion, we know the remainder

$$g(t, \mathbf{x}) \triangleq f(t, \mathbf{x}) - A(t)\mathbf{x}$$

satisfies $\lim_{t\to 0} \|g(t, \mathbf{x})\| = 0$ for each fixed t. However, in order to approach 0 uniformly, we need a stronger condition:

$$\lim_{\mathbf{x}\to 0} \sup_{t\geq 0} \frac{\|g(t,\mathbf{x})\|}{\|\mathbf{x}\|} = 0 \tag{8.6}$$

Theorem 8.25 *For nonautonomous system* $\dot{\mathbf{x}} = f(t, \mathbf{x})$ *with bounded* $A(t)$ *and* $f(t, 0) = 0$ *for all* t*, assume that* (8.6) *also holds true. If 0 is UAS for* $\dot{\mathbf{x}} = A(t)\mathbf{x}$*, then it is locally UAS for* $\dot{\mathbf{x}} = f(t, \mathbf{x})$.

Chapter 9
Problems and Exercises

Input-Output and BIBO Stability

Problem 1. For each of the following systems, (1) calculate its impulse response $h(t)$, and (2) determine whether it is BIBO stable or not. It may help to consult a table of Laplace/z transforms.

(a) the scalar LTV system $y(t) = \frac{1}{2}(1 - e^{-2t})u(t)$
(b) the CT first-order LTI system

$$\dot{y}(t) + ay(t) = \dot{u}(t) - bu(t), \quad a, b \in \mathbb{R}, a, b > 0$$

(c) the DT second-order LTI system

$$y_t + 0.9y_{t-1} + 0.2y_{t-2} = u_t, \quad t \in \mathbb{Z}^{\geq 0}$$

Problem 2. For each of the following systems, (1) sketch its phase portrait by hand, (2) characterize the type of equilibria, and (3) verify your answers to a and b by plotting the phase portrait with MATLAB.

$$\text{(a) } \dot{\mathbf{x}}(t) = \begin{bmatrix} 1 & 1 \\ 4 & -2 \end{bmatrix} \mathbf{x}(t) \qquad \text{(b) } \dot{\mathbf{x}}(t) = \begin{bmatrix} 4 & -1 \\ 2 & 1 \end{bmatrix} \mathbf{x}(t) \qquad \text{(c) } \dot{\mathbf{x}}(t) = \begin{bmatrix} 1 & -1 \\ 1 & 1 \end{bmatrix} \mathbf{x}(t)$$

(d) A simple harmonic oscillator, which models the vibrations of a mass m hanging from a spring: $m\ddot{x}(t) + kx(t) = 0$. What is the physical interpretation of the orbits in your phase portrait?

Problem 3. A very common system used to motivate the necessity of control theory is the *inverted pendulum on a cart*. The system is comprised of a pendulum attached to a cart which is only allowed to move side-to-side along a horizontal track. In the

SJ. Han, *Linear Systems*, KAIST Research Series,
https://doi.org/10.1007/978-981-96-9774-8_9

stability problem, the cart's objective is to maintain the upright position of the pole. See, for example, [Link] for a video demo. In the control problem (which we will get to late in the course), the objective of the cart is to get to its upright position by itself from any initial condition (i.e., swing the pole autonomously). See, for example, [Link] for a video demo.

(a) Use Newton's laws (and/or the Euler-Lagrange equations) to derive the equations of motions as

$$(M+m)\ddot{x}(t)+b\dot{x}(t)+mL\ddot{\theta}(t)\cos\theta(t)-mL\dot{\theta}^2(t)\sin\theta(t)=F(t)$$
$$(I+mL^2)\ddot{\theta}(t)+mgL\sin\theta(t)+mL\ddot{x}(t)\cos\theta(t)=0$$

where x is the position of the cart, θ is the angle of the pole with respect to the vertical line (down), M is the mass of the cart, L is the length of the pole, m is the mass of the bob at the end of the pole, b is the friction coefficient of the cart along the track, I is the mass moment of inertia of the pendulum, and F is the force input to the cart.

(b) How many equilibria does this system have? Choose an appropriate state-space representation, then linearize the system dynamics around each equilibrium in terms of the parameters in part (a). Characterize the type of each equilibrium.

(c) [*MATLAB Coding*] Choose values $M=3$, $m=1$, $b=0.1$, $L=0.5$, $I=0.6$, $g=9.81$ and simulate the system. Try injecting various force inputs F and various initial conditions, then plot the system trajectories over time. Draw the phase portraits of each equilibria from part (b).

Problem 4: Stiff Differential Equations [1]. In the simulation of several engineering systems, there are often "parasitic elements" which result in the differential equation becoming stiff (e.g., parasitic capacitances and inductances in electronic circuits). This phenomenon results in some state variables changing much more rapidly than others.

(a) Consider a $(n+m)$-dimensional system with $\mathbf{x}_1\in\mathbb{R}^n$ representing the "slow" variables and $\mathbf{x}_2\in\mathbb{R}^m$ representing the "fast" variables.

$$\dot{\mathbf{x}}_1=A_{11}\mathbf{x}_1+A_{12}\mathbf{x}_2$$
$$\dot{\mathbf{x}}_2=\frac{1}{\epsilon}A_{21}\mathbf{x}_1+\frac{1}{\epsilon}A_{22}\mathbf{x}_2$$

Here, A_{22} is nonsingular. Show that m eigenvalues tend to ∞ like $\sigma(A_{22})/\epsilon$ and the other n tend to $\sigma(A_{11}-A_{12}A_{22}^{-1}A_{21})$ as $\epsilon\to 0$.

(b) In circuit theory, the system

$$\dot{\mathbf{x}}_1=A_{11}\mathbf{x}_1+A_{12}\mathbf{x}_2$$
$$0=A_{21}\mathbf{x}_1+A_{22}\mathbf{x}_2$$

is often referred to as *singularly-perturbed* or a *low-frequency approximation*. In addition to parasitic (small) capacitances, electronic circuits can also be affected by coupling (large) capacitances.

$$\dot{\mathbf{x}}_1 = A_{11}\mathbf{x}_1 + A_{12}\mathbf{x}_2 + A_{13}\mathbf{x}_3$$
$$\dot{\mathbf{x}}_2 = \frac{1}{\epsilon}A_{21}\mathbf{x}_1 + \frac{1}{\epsilon}A_{22}\mathbf{x}_2 + \frac{1}{\epsilon}\mathbf{x}_3$$
$$\mathbf{x}_2 = \frac{1}{\rho}A_{21}\mathbf{x}_1 + \frac{1}{\rho}A_{22}\mathbf{x}_2 + \frac{1}{\rho}\mathbf{x}_3$$

where $\epsilon > 0$ is small, and $\rho > 0$ is large. A *mid-frequency model* takes $\epsilon = 0$ and $\rho = \infty$. A *low-frequency model* takes $\epsilon = 0$ and $\rho = \infty$ in the $\tau = t/\rho$ timescale, while a *high-frequency model* takes it in the $\tau = t/\epsilon$ timescale. What is the relationship among the eigenvalues in these three regimes?

Problem 5 [2]. Consider a simple heat exchanger where f_C and f_H are the (assumed constant) flows of cold and hot water, T_C and T_H are the temperatures in the cold and hot compartments, and V_C and V_H are the volumes of the cold and hot water, respectively. The temperatures in both compartments evolve according to

$$V_C\frac{dT_C}{dt} = f_C(T_{C0} - T_C) + \beta(T_H - T_C)\; V_H\frac{dT_H}{dt} = f_H(T_{H0} - T_H) - \beta(T_H - T_C)$$

where T_{C0} and T_{H0} are the respective initial temperatures of each compartment.

(a) Write the state and output equations for this system in state-space model form.
(b) In the absence of any input, what are y_1 and y_2?
(c) Is the system BIBO stable? Show why or why not.

Problem 6. Consider a 2D autonomous LTI system with the dynamics

$$\dot{\mathbf{x}}(t) = \begin{bmatrix} a & b \\ c & d \end{bmatrix}\mathbf{x}(t), \quad \mathbf{x}(0) = \mathbf{x}_0$$

for some values $a, b, c, d \in \mathbb{R}$. Suppose it has an equilibrium at $\mathbf{x}^*$ and two eigenvalues λ_1, λ_2.

(a) Recall that $\mathbf{x}^*$ was a *degenerate node* if $\lambda_1 = \lambda_2 \equiv \lambda$ and A was not diagonalizable. What are the conditions for determining the "direction" of each trajectory? See Fig. 9.1 for examples.

(b) Recall that $\mathbf{x}^*$ was a *focus* if $\lambda_1, \lambda_2 \in \mathbb{C}$ and their real parts were nonzero. What are the conditions for determining the orientation (clockwise or counter-clockwise) of each spiral?
(Hint: Write down the characteristic polynomial in terms of a, b, c, d.)

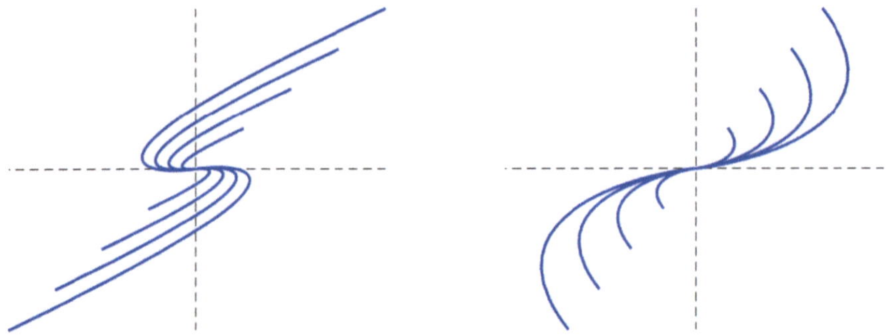

Fig. 9.1 Two types of "direction" in degenerate node trajectories

(c) Recall that $\mathbf{x}^*$ was a *center* if $\lambda_1, \lambda_2 \in \mathbb{C}$ and their real parts were zero. What are the conditions for determining the orientation (clockwise or counter-clockwise) of each orbit?

(d) `ode45` is one of the most common built-in functions of MATLAB which is used to solve nonstiff ODEs. Choose any sample values of a, b, c, d and verify your answers in parts (a)–(c). Plot a few phase portraits to justify your answers. Use `ode45` to simulate your linear systems without any of the three discretization methods we discussed in class.

(e) Do any trajectories in a phase portrait, starting from different initial conditions, intersect with each other? Why or why not? Does your answer depend on the type of equilibrium?

Problem 7 [2]. The equations of errors in an inertial navigation system can be approximated via the following system.

$$\dot{\delta x} = \delta v, \quad \dot{\delta v} = -g\delta\psi + E_A, \quad \dot{\delta\psi} = \frac{1}{R}\delta v + E_G$$

where $\delta x, \delta v, \delta\psi$ are the position error, velocity error, and tilt of the platform, respectively, R is the radius of the Earth, and E_A, E_G are the biases in the accelerometer and gyroscope, respectively.

(a) Is this system internally stable? Is it BIBO stable for any output $\mathbf{y} = C[\delta x, \delta v, \delta\psi]^\top$?

(b) Consider a constant gyro bias E_G and zero accelerometer bias $E_A = 0$. Discuss what happens to the position error as a function of time by deriving a concrete expression for it.

Problem 8. Is the SISO LTI system with transfer function $G(s) = 1/(s^2 + 4)$ BIBO stable?

Lyapunov Stability, Direct and Indirect Methods

Problem 9. For each of the following systems, use Lyapunov's methods (direct or indirect) to classify the stability of the origin.

$$\text{(a)} \begin{bmatrix}\dot{x}_1\\ \dot{x}_2\end{bmatrix} = \begin{bmatrix}x_2\\ -x_1 - \epsilon x_1^2 x_2\end{bmatrix}, \quad \text{(b)} \begin{bmatrix}\dot{x}_1\\ \dot{x}_2\end{bmatrix} = \begin{bmatrix}-x_1^3 + 2x_2^3\\ -2x_1 x_2^2\end{bmatrix}, \quad \text{(c)} \begin{bmatrix}\dot{x}_1\\ \dot{x}_2\end{bmatrix} = \begin{bmatrix}-3x_1^3 - x_2\\ x_1^5 - 2x_2^3\end{bmatrix}$$

(*Hint*: In part (a), make sure to specify some conditions on $\epsilon \in \mathbb{R}$. In part (c), try $V(\mathbf{x}) = ax_1^{2c} + bx_2^{2d}$ and choose appropriate constants a, b, c, d.)

Problem 10. Consider the following system

$$\dot{x}_1 = -4x_2 + x_1\left(1 - \frac{1}{4}x_1^2 - x_2^2\right)$$

$$\dot{x}_2 = x_1 + x_2\left(1 - \frac{1}{4}x_1^2 - x_2^2\right)$$

Are there any equilibrium points or limit cycles in this system? Use the Lyapunov direct method and LaSalle's invariant set theorem to classify the stability type of this system.

Problem 11. Suppose there exist positive definite matrices $P, Q \in \mathbb{R}^{n\times n}$ and some $\lambda > 0$ such that

$$A^\top P + PA - 4\lambda P = -Q$$

What can you say about the eigenvalues of A?

Problem 12. Consider the linear map $\mathcal{L} : \mathbb{R}^{n\times n} \to \mathbb{R}^{n\times n}$ defined by $\mathcal{L}(P) \triangleq A^\top P + PA$. Show that is $\lambda_i + \lambda_j \neq 0$ for any eigenvalues λ_i, λ_j of A, then the equation

$$A^\top P + PA = Q$$

has a unique symmetric solution for a given symmetric Q.

Asymptotic and Exponential Stability

Problem 13. Consider the following non-autonomous system

$$\begin{bmatrix}\dot{x}_1\\ \dot{x}_2\end{bmatrix} = \begin{bmatrix}-x_1^3 + \alpha(t)x_2\\ -\alpha(t)x_1 - x_2^3\end{bmatrix}$$

where $\alpha(t)$ is a continuous, bounded function of time. Determine whether the system is exponentially stable or not.

Problem 14. Given the matrix

$$A \triangleq \begin{bmatrix} -2 & 1 & 0 & 0 & 0 & 0 & 0 \\ 0 & -2 & 1 & 0 & 0 & 0 & 0 \\ 0 & 0 & -2 & 1 & 0 & 0 & 0 \\ 0 & 0 & 0 & -5 & 1 & 0 & 0 \\ 0 & 0 & 0 & 0 & -5 & 1 & 0 \\ 0 & 0 & 0 & 0 & 0 & 0 & 0 \\ 0 & 0 & 0 & 0 & 0 & 0 & 0 \end{bmatrix},$$

is the system $\dot{\mathbf{x}} = A\mathbf{x}$ exponentially stable?

Problem 15. If $A(t) = A^\top(t) \in \mathbb{R}^{n \times n}$ and the smallest eigenvalue of $A(t)$ satisfies $\lambda_{\min}(A(t)) \leq -\epsilon$ for all t, show that the state transition matrix of $A(t)$ is asymptotically stable.

Uniform Stability

Problem 16. We return to one of the previous problems seen before.

$$\begin{bmatrix} \dot{x}_1 \\ \dot{x}_2 \end{bmatrix} = \begin{bmatrix} -x_1^3 + \alpha(t)x_2 \\ -\alpha(t)x_1 - x_2^3 \end{bmatrix}$$

where $\alpha(t)$ is a continuous, bounded function of time. Use the Lyapunov function $V(x) = (1/2)(x_1^2 + x_2^2)$ to determine whether the origin is globally uniformly asymptotically stable or not.

Problem 17. In this chapter, we presented a theorem which showed that for LTV systems, uniform asymptotic stability (UAS) implies exponential stability. Now, consider the following example. Suppose we have the scalar system $x(t_0) = x_0$, $\dot{x}(t) = -(1/t)x(t)$ for $t \geq t_0 > 0$. By the separation of variables, it has the solution $x(t) = (t_0/t)x_0$, which shows that the equilibrium 0 is UAS:

- as $t \to \infty$, $x(t) \to 0$ regardless of t_0 and x_0
- for any $\epsilon > 0$, the choice of $\delta = \epsilon$ means $|x_0| < \delta$ implies $|x| = |t_0/t||x_0| < \delta = \epsilon$ since $t \geq t_0$.

However, this system is clearly not exponentially stable, since t_0/t decays at a rate that is slower than any exponential of the form $e^{-\alpha t}$, $\alpha > 0$. Then is this scalar LTV system a contradiction to the theorem? If not, can you find anything wrong with the argument presented above?
(*Hint*: There are alternative definitions of UAS; it may help to use them in your argument.)

References

1. Shankar Sastry. *Lecture notes for EE221A: Linear Systems*. 2013.
2. Claire Tomlin. *Lecture notes for EE221A: Linear Systems*. 2017.

Part III
Linear Control and Estimation

Chapter 10
Canonical Forms

So far, I and II focused mostly on *uncontrolled* systems, where there is no direct control input into the system. In this part, we will begin the study of controlled systems.

First, we begin this part of the book by discussing *canonical forms*, which are crucial in linear systems theory because they provide simplified, standardized representations of systems that make it easier to analyze their properties, such as controllability, observability, and stability. By transforming a system into a canonical form, its essential features are made more apparent, aiding in both theoretical understanding and practical applications like simulation and real-time control.

Canonical forms are most meaningful in the context of LTI systems, and so this chapter will focus exclusively on the canonical forms for LTI systems.

10.1 System Realizations

We just discussed that the characteristic modes of a system can be more easily revealed by a change of coordinates $\tilde{\mathbf{x}}(t) \triangleq V^{-1}\mathbf{x}(t)$ (see Chap. 2). In particular, we can choose V such that $A \triangleq V\tilde{A}V^{-1}$ such that $\tilde{A}$ is the Jordan form of A.

Definition 10.73 (*Algebraically Equivalent Systems*) Two LTI systems (A, B, C, D) and $(\tilde{A}, \tilde{B}, \tilde{C}, \tilde{D})$ are said to be *algebraically equivalent* if they have the same solution space $\mathcal{X} = \tilde{\mathcal{X}}$.

Theorem 10.26 *For any nonsingular matrix* $V \in \mathbb{R}^{n\times n}$, *choosing* $\tilde{A} = V^{-1}AV$, $\tilde{B} = V^{-1}B$, $\tilde{C} = CV$, *and* $\tilde{D} = D$ *yields* $(\tilde{A}, \tilde{B}, \tilde{C}, \tilde{D})$ *which is algebraically equivalent to* (A, B, C, D).

The proof of this theorem is straightforward by implementing a change of coordinates on the state $\mathbf{x}$ using the change-of-basis matrix V.

SJ. Han, *Linear Systems*, KAIST Research Series,
https://doi.org/10.1007/978-981-96-9774-8_10

Definition 10.74 (*Zero-State Equivalence*) Two transfer functions $H(s)$ and $\tilde{H}(s)$ are said to be *zero-state equivalent* if $C(sI - A)^{-1}B + D = \tilde{C}(sI - \tilde{A})^{-1}\tilde{B} + \tilde{D}$, where (A, B, C, D) and $(\tilde{A}, \tilde{B}, \tilde{C}, \tilde{D})$ are the state-space equations to $H(s)$ and $\tilde{H}(s)$, respectively.

By combining Definitions 10.74 and 10.73, we can see equivalent state-space models are always zero-state equivalent because they have the same characteristic polynomial, eigenvalues, and the same transfer function.

Lemma 10.6 *Two state-space models* (A, B, C, D) *and* $(\tilde{A}, \tilde{B}, \tilde{C}, \tilde{D})$ *are zero-state equivalent iff*

1. $D = \tilde{D}$
2. $CA^kB = \tilde{C}\tilde{A}^k\tilde{B}$

The proof follows directly from applying the power-series expansion of $(sI - A)^{-1}$ to both sides and matching the coefficients of s:

$$(sI - A)^{-1} = s^{-1}(I - s^{-1}A)^{-1} = s^{-1}\sum_{k=0}^{\infty}(s^{-1}A)^k$$

We leave the details to the interested reader.

Remark 10.15 A transfer function $H(s) \triangleq N(s)/D(s)$ is realizable iff it is a rational proper matrix. The reason is that we have a realizable $H(s) = C(sI - A)^{-1}B + D = C\frac{\text{adj}(sI-A)}{\det(sI-A)}B + D$. The denominator $\det(sI - A)$ (the determinant) is a polynomial in s with degree n, while the numerator $\text{adj}(sI - A)$ (the adjugate) has matrix entries which are polynomials of degree $\leq n - 1$. Indeed, all n^2 entries of $H(s)$ are proper transfer functions.

A realizable $H(s)$ has infinitely many realizations, but among them, there are some which are more conventional. In the following sections, we will introduce and discuss three main ones: controllable, observable, and modal canonical forms.

10.2 Three Main Canonical Forms

The three main canonical forms—controllable, observable, and modal—each emphasize distinct system properties that are valuable in control applications. The controllable canonical form explicitly shows state controllability, making it easier to design inputs that reach any desired state. The observable canonical form highlights observability, ensuring that all internal states can be inferred from output measurements. Finally, the modal canonical form organizes the system around its characteristic modes (eigenvalues), simplifying stability analysis.

In the following sections, we will explore each form in depth, examining how their structures align with various control objectives. Later, in Chap. 14, we will discuss the Kalman canonical form (also called the Kalman decomposition), which provides a system representation which reveals both the controllable and (un)observable subspaces (to be defined later).

10.3 Controllable Canonical Form

Let $H(s)$ be a proper $k \times m$ transfer matrix. Decompose $H(s)$ as

$$H(s) = H_{\text{sp}}(s) + H_\infty,$$

where

- $H_{\text{sp}}(s) = \frac{N_{\text{sp}}(s)}{D_{\text{sp}}(s)}$ is the strictly proper part of $H(s)$, with $\deg N_{\text{sp}}(s) < \deg D_{\text{sp}}(s)$,
- $H_\infty \in \mathbb{R}^{n \times m}$ is the steady-state part of $H(s)$.

Consider $H_{\text{sp}}(s) = \frac{N_{\text{sp}}(s)}{D_{\text{sp}}(s)}$. Write it in the form:

$$\frac{N_{\text{sp}}(s)}{D_{\text{sp}}(s)} = \left(\frac{1}{s^r + \alpha_1 s^{r-1} + \cdots + \alpha_{r-1}s + \alpha_r}\right)\left(N_1 s^{r-1} + N_2 s^{r-2} + \cdots + N_{r-1}s + N_r\right)$$

where

- $r \in \mathbb{N}$,
- $D_{\text{sp}}(s) \triangleq s^r + \alpha_1 s^{r-1} + \cdots + \alpha_{r-1}s + \alpha_r$ is the least common denominator of all entries in $H_{\text{sp}}(s)$ (and we require D_{sp} to be monic, i.e., the coefficient of s^r is 1),
- $N_i \in \mathbb{R}^{k \times m}$, $1 \le i \le r$, are constant matrix coefficients.

The Controllable Canonical Form (CCF) is then obtained as

$$\dot{\mathbf{x}} = \underbrace{\begin{bmatrix} -\alpha_1 I_m & -\alpha_2 I_m & \cdots & -\alpha_{r-1} I_m & -\alpha_r I_m \\ I_m & 0 & \cdots & 0 & 0 \\ \vdots & I_m & \ddots & \vdots & \vdots \\ 0 & \ddots & \ddots & 0 & 0 \\ 0 & \cdots & 0 & I_m & 0 \end{bmatrix}}_{A} \mathbf{x} + \underbrace{\begin{bmatrix} I_m \\ 0 \\ \vdots \\ 0 \end{bmatrix}}_{B} u,$$

$$\mathbf{y} = \underbrace{\begin{bmatrix} N_1 & N_2 & \cdots & N_r \end{bmatrix}}_{C} \mathbf{x} + \underbrace{H_\infty}_{D} u.$$

Remark 10.16 In the CCF, the state dimensions are $\mathbf{x}(t) \in \mathbb{R}^n$ (where $n = mr$), $u(t) \in \mathbb{R}^m$, $\mathbf{y}(t) \in \mathbb{R}^k$.

Example 10.17 (*CCF*) Let

$$H(s) = \frac{Y(s)}{U(s)} = \frac{s^3 + 5s^2 + 10s + 15}{s^3 + 3s^2 + 6s + 9} = 1 + \frac{2s^2 + 4s + 6}{s^3 + 3s^2 + 6s + 9}$$

where $n = mr = 3$. Thus, $H(s) = H_\infty + H_{sp}(s)$, with $H_\infty = 1$ and $H_{sp}(s) = \frac{2s^2+4s+6}{s^3+3s^2+6s+9}$.

Choose $x_1(t)$ such that

(1)

$$\frac{Y(s)}{X_1(s)} = s^3 + 5s^2 + 10s + 15$$

(2)

$$\frac{X_1(s)}{U(s)} = \frac{1}{s^3 + 3s^2 + 6s + 9}$$

From (1),

$$\mathbf{y}(t) = \dddot{x}_1(t) + 5\ddot{x}_1(t) + 10\dot{x}_1(t) + 15x_1(t) = \dddot{x}_1 + 5x_3 + 10x_2 + 15x_1$$

From (2),

$$u(t) = \dddot{x}_1(t) + 3\ddot{x}_1(t) + 6\dot{x}_1(t) + 9x_1(t)$$

which implies

$$\dddot{x}_1(t) = u(t) - 3\ddot{x}_1(t) - 6\dot{x}_1(t) - 9x_1(t)$$

and

$$\dot{x}_3(t) = u(t) - 3x_3(t) - 6x_2(t) - 9x_1(t)$$

Choose

$$x_2 = \dot{x}_1, \quad x_3 = \ddot{x}_1.$$

Thus,

$$\mathbf{y}(t) = u(t) + 2x_3(t) + 4x_2(t) + 6x_1(t)$$

The CCF is:

$$\begin{bmatrix} \dot{x}_3 \\ \dot{x}_2 \\ \dot{x}_1 \end{bmatrix} = \begin{bmatrix} -3 & -6 & -9 \\ 1 & 0 & 0 \\ 0 & 1 & 0 \end{bmatrix} \begin{bmatrix} x_3 \\ x_2 \\ x_1 \end{bmatrix} + \begin{bmatrix} 1 \\ 0 \\ 0 \end{bmatrix} u, \quad \mathbf{y} = \begin{bmatrix} 2 & 4 & 6 \end{bmatrix} \begin{bmatrix} x_3 \\ x_2 \\ x_1 \end{bmatrix} + u$$

Note that there are multiple equivalent CCFs (and other canonical forms in general) depending on how you order the state. For example,

$$\begin{bmatrix} x_1 \\ \vdots \\ x_n \end{bmatrix} \quad \text{versus} \quad \begin{bmatrix} x_n \\ \vdots \\ x_1 \end{bmatrix}$$

We can represent the CCF using block diagrams. The diagram for this specific example is shown below.

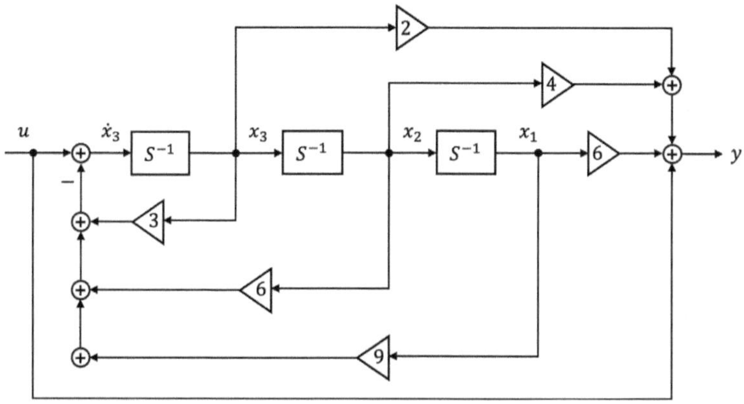

□

10.4 Observable Canonical Form

As before, decompose $H(s) = H_{\text{sp}}(s) + H_\infty$ with

$$H_{\text{sp}}(s) = \left(\frac{1}{s^r + \alpha_1 s^{r-1} + \cdots + \alpha_{r-1}s + \alpha_r}\right)\left(N_1 s^{r-1} + N_2 s^{r-2} + \cdots + N_{r-1}s + N_r\right).$$

Then Observable Canonical Form (OCF) is obtained as

$$\dot{\mathbf{x}} = \underbrace{\begin{bmatrix} -\alpha_1 I_m & I_m & 0 & \cdots & 0 \\ -\alpha_2 I_m & 0 & I_m & \ddots & \vdots \\ \vdots & \vdots & \ddots & \ddots & 0 \\ \vdots & 0 & \cdots & 0 & I_m \\ -\alpha_r I_m & 0 & \cdots & 0 & 0 \end{bmatrix}}_{A} \mathbf{x} + \underbrace{\begin{bmatrix} N_1 \\ N_2 \\ \vdots \\ N_r \end{bmatrix}}_{B} u,$$

$$\mathbf{y} = \underbrace{\begin{bmatrix} I_m & 0 & \cdots & 0 \end{bmatrix}}_{C} \mathbf{x} + \underbrace{H_\infty}_{D} u.$$

Example 10.18 (*OCF*) Following Example 10.17, let

$$H(s) = \frac{Y(s)}{U(s)} = \frac{s^3 + 5s^2 + 10s + 15}{s^3 + 3s^2 + 6s + 9} = 1 + \frac{2s^2 + 4s + 6}{s^3 + 3s^2 + 6s + 9}$$

where $n = mr = 3$. Thus, $H(s) = H_\infty + H_{\text{sp}}(s)$, with $H_\infty = 1$ and $H_{\text{sp}}(s) = \frac{2s^2+4s+6}{s^3+3s^2+6s+9}$.

$$Y(s)(s^3 + 3s^2 + 6s + 9) = U(s)(s^3 + 5s^2 + 10s + 15)$$

This expands to

$$s^3(Y - U) + s^2(3Y - 5U) + s(6Y - 10U) + (9Y - 15U) = 0$$

which implies

$$Y(s) = U(s) + \frac{1}{s}(5U - 3Y) + \frac{1}{s^2}(10U - 6Y) + \frac{1}{s^3}(15U - 9Y).$$

Define

$$\begin{aligned} X_1(s) &= \frac{1}{s}(5U - 3Y) + \frac{1}{s^2}(10U - 6Y) + \frac{1}{s^3}(15U - 9Y), \\ X_2(s) &= \frac{1}{s}(10U - 6Y) + \frac{1}{s^2}(15U - 9Y), \\ X_3(s) &= \frac{1}{s}(15U - 9Y) \end{aligned}$$

so that $Y = U + X_1$ and

$$y(t) = \begin{bmatrix} 1 & 0 & 0 \end{bmatrix} \begin{bmatrix} x_1 \\ x_2 \\ x_3 \end{bmatrix} + u(t)$$

From this, we have the following relations:

$$\begin{aligned} (1)\quad sX_1(s) &= 5U(s) - 3Y(s) + \frac{1}{s}(10U - 6Y) + \frac{1}{s^2}(15U - 9Y) \\ &= 5U - 3(U + X_1) + X_2 \\ &= -3X_1 + X_2 + 2U \\ (2)\quad sX_2(s) &= 10U(s) - 6Y(s) + \frac{1}{s}(15U - 9Y) \\ &= 10U - 6(U + X_1) + X_3 \\ &= -6X_1 + X_3 + 4U \end{aligned}$$

$$\begin{aligned}(3)\quad sX_3(s) &= 15U(s) - 9Y(s)\\ &= 15U - 9(U + X_1)\\ &= -9X_1 + 6U\end{aligned}$$

Thus,

$$s\begin{bmatrix}x_1\\x_2\\x_3\end{bmatrix} = \begin{bmatrix}-3 & 1 & 0\\-6 & 0 & 1\\-9 & 0 & 0\end{bmatrix}\begin{bmatrix}x_1\\x_2\\x_3\end{bmatrix} + \begin{bmatrix}2\\4\\6\end{bmatrix}u \quad \xrightarrow{\mathcal{L}^{-1}} \quad \text{yields OCF.}$$

□

10.5 Modal Canonical Form

The modal canonical form (MCF) is concerned with representing the system matrix using a (block-)diagonal matrix A. We've seen this before in our discussion with characteristic modes. For simplicity, we will discuss SISO cases (i.e., $H(s)$ is not a matrix), since the MIMO cases are similar.

$$H_{\text{sp}}(s) = \frac{N(s)}{D(s)} \quad \text{with} \quad D(s) = s^r + \alpha_1 s^{r-1} + \cdots + \alpha_{r-1}s + \alpha_r$$

10.5.1 Case: D(s) has Distinct Real Roots

In this case, we can invoke partial fraction decomposition:

$$H_{\text{sp}}(s) = \frac{N(s)}{D(s)} = \frac{N(s)}{(s-\lambda_1)(s-\lambda_2)\cdots(s-\lambda_r)} = \frac{N_1}{s-\lambda_1} + \frac{N_2}{s-\lambda_2} + \cdots + \frac{N_r}{s-\lambda_r}$$

Here, the $\{\ \lambda_1, \lambda_2, \ldots, \lambda_r\ \}$ are called the *poles* of the transfer function $H_{\text{sp}}(s)$. Correspondingly, the roots of $N(s)$ are called the *zeros* of $H_{\text{sp}}(s)$.

Since $H(s) = \frac{Y(s)}{U(s)}$, we also have

$$Y(s) = N_1 \underbrace{\frac{U(s)}{s-\lambda_1}}_{X_1(s)} + N_2 \underbrace{\frac{U(s)}{s-\lambda_2}}_{X_2(s)} + \cdots + N_r \underbrace{\frac{U(s)}{s-\lambda_r}}_{X_r(s)} + H_\infty U(s) \quad (r = n)$$

$$\Longrightarrow X_i(s)(s-\lambda_i) = U(s) \quad \xrightarrow{\mathcal{L}^{-1}} \quad \dot{x}_i(t) = \lambda_i x_i(t) + u(t)$$

$$\begin{bmatrix} \dot{x}_1 \\ \dot{x}_2 \\ \vdots \\ \dot{x}_n \end{bmatrix} = \begin{bmatrix} \lambda_1 & 0 & \cdots & 0 \\ 0 & \lambda_2 & \ddots & 0 \\ \vdots & \vdots & \ddots & \vdots \\ 0 & 0 & \cdots & \lambda_n \end{bmatrix} \begin{bmatrix} x_1 \\ x_2 \\ \vdots \\ x_n \end{bmatrix} + \begin{bmatrix} 1 \\ 1 \\ \vdots \\ 1 \end{bmatrix} u(t)$$

$$y(t) = \begin{bmatrix} N_1 & N_2 & \cdots & N_n \end{bmatrix} \mathbf{x}(t) + H_\infty u(t)$$

Example 10.19 (*MCF with Distinct Real Roots*) Let

$$H(s) = \frac{Y(s)}{U(s)} = \frac{(s+5)(s+4)}{(s+1)(s+2)(s+3)} = \frac{N_1}{s+1} + \frac{N_2}{s+2} + \frac{N_3}{s+3}.$$

For partial fraction decomposition, we need

$$(s+5)(s+4) = N_1(s+2)(s+3) + N_2(s+1)(s+3) + N_3(s+1)(s+2). \tag{10.1}$$

Expanding this equation, we get

$$(1.1) = (N_1 + N_2 + N_3)s^2 + (5N_1 + 4N_2 + 3N_3)s + (6N_1 + 3N_2 + 2N_3).$$

This leads to the system of equations:

$$\begin{cases} N_1 + N_2 + N_3 = 1 \\ 5N_1 + 4N_2 + 3N_3 = 9 \\ 6N_1 + 3N_2 + 2N_3 = 20 \end{cases} \implies N_1 = 6, \quad N_2 = -6, \quad N_3 = 1.$$

Thus,

$$\begin{bmatrix} \dot{x}_1 \\ \dot{x}_2 \\ \dot{x}_3 \end{bmatrix} = \begin{bmatrix} -1 & 0 & 0 \\ 0 & -2 & 0 \\ 0 & 0 & -3 \end{bmatrix} \begin{bmatrix} x_1 \\ x_2 \\ x_3 \end{bmatrix} + \begin{bmatrix} 1 \\ 1 \\ 1 \end{bmatrix} u, \quad \mathbf{y} = \begin{bmatrix} 6 & -6 & 1 \end{bmatrix} \mathbf{x}.$$

A nice trick you can use to simplify solving N_1, N_2, N_3 here is to substitute in the poles for s.

$$N_1(s+2)(s+3) + N_2(s+1)(s+3) + N_3(s+1)(s+2) = s^2 + 9s + 20$$

- $s = -1$: $N_1(1)(2) + 0 + 0 = (-1)^2 + 9(-1) + 20 \implies 2N_1 = 12 \implies N_1 = 6$
- $s = -2$: $0 + N_2(-1)(1) + 0 = (-2)^2 + 9(-2) + 20 \implies N_2 = -6$
- $s = -3$: $0 + 0 + N_3(-2)(-1) = (-3)^2 + 9(-3) + 20 \implies N_3 = 1$

MCF can also be represented using a block diagram, as shown below:

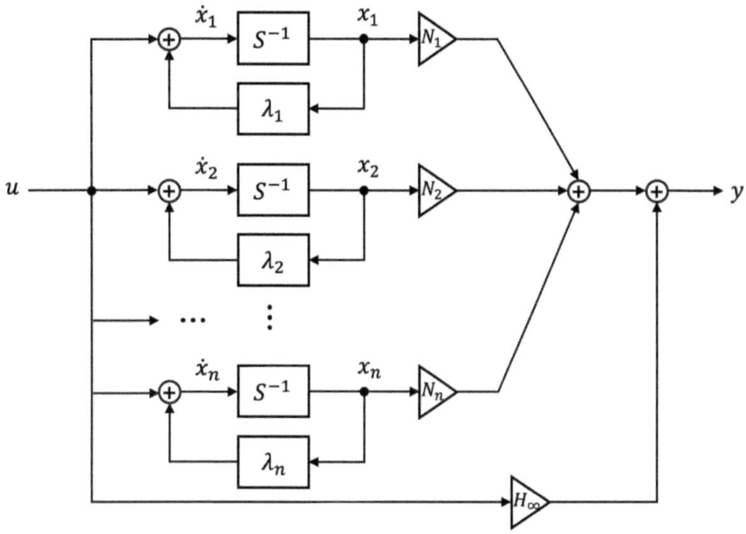

□

10.5.2 Case: $D(s)$ has Some Repeating Real Roots

In this case, we can invoke a special case of partial fractions, with repeating roots:

$$\begin{aligned} H_{\text{sp}}(s) &= \frac{N(s)}{D(s)} = \frac{N(s)}{(s-\lambda_1)^k(s-\lambda_2)\cdots(s-\lambda_r)} \\ &= \frac{N_{11}}{s-\lambda_1} + \frac{N_{12}}{(s-\lambda_1)^2} + \cdots + \frac{N_{1k}}{(s-\lambda_1)^k} + \frac{N_2}{s-\lambda_2} + \cdots + \frac{N_r}{s-\lambda_r} \end{aligned}$$

where $(r-1+k=n)$.

Then the MCF is obtained as

$$\dot{\mathbf{x}} = \underbrace{\left[\begin{array}{cccc|ccc} \lambda_1 & 1 & 0 & \cdots & 0 & \cdots & 0 \\ 0 & \lambda_1 & \ddots & 0 & 0 & \cdots & 0 \\ \vdots & \vdots & \ddots & 1 & \vdots & & \vdots \\ 0 & 0 & \cdots & \lambda_1 & 0 & \cdots & 0 \\ \hline 0 & 0 & \cdots & 0 & \lambda_2 & \ddots & 0 \\ \vdots & & & & & \ddots & \vdots \\ 0 & 0 & \cdots & 0 & 0 & \cdots & \lambda_r \end{array}\right]}_{A} \mathbf{x} + \underbrace{\left[\begin{array}{c} 0 \\ \vdots \\ 0 \\ 1 \\ \hline 1 \\ \vdots \\ 1 \end{array}\right]}_{B} u,$$

where the first $k-1$ elements of B are zeros and the last r elements are ones.

$$\mathbf{y} = \underbrace{\begin{bmatrix} N_{1k} \; N_{1k-1} \; \cdots \; N_{11} \; N_2 \; \cdots \; N_r \end{bmatrix}}_{C} \mathbf{x} + \underbrace{H_\infty}_{D} u.$$

Remark 10.17 This structure corresponds to the Jordan Canonical Form (JCF) for A.

Example 10.20 (*MCF with Some Repeated Real Roots*) Let

$$H(s) = \frac{Y(s)}{U(s)} = \frac{s^2 + 9s + 10}{(s+1)^2(s+3)}.$$

Using partial fractions:

$$H(s) = \frac{N_{11}}{s+1} + \frac{N_{12}}{(s+1)^2} + \frac{N_2}{s+3}.$$

Thus,

$$Y(s) = N_{11} \underbrace{\frac{U(s)}{s+1}}_{X_1(s)} + N_{12} \underbrace{\frac{U(s)}{(s+1)} \cdot \frac{1}{s+1}}_{X_2(s)} + N_2 \underbrace{\frac{U(s)}{s+3}}_{X_3(s)}.$$

Solve for coefficients:

$$\Longrightarrow N_{11}(s+1)(s+3) + N_{12}(s+3) + N_2(s+1)^2 \equiv s^2 + 9s + 10.$$

This leads to the system of equations:

$$\begin{cases} N_{11} + N_2 = 1 \\ 4N_{11} + N_{12} + 2N_2 = 9 \\ 3N_{11} + 3N_{12} + N_2 = 10 \end{cases} \Longrightarrow N_{11} = 3, \quad N_{12} = 1, \quad N_2 = -2.$$

Thus,

$$\begin{bmatrix} \dot{x}_2 \\ \dot{x}_1 \\ \dot{x}_3 \end{bmatrix} = \begin{bmatrix} -1 & 1 & 0 \\ 0 & -1 & 0 \\ 0 & 0 & -3 \end{bmatrix} \begin{bmatrix} x_2 \\ x_1 \\ x_3 \end{bmatrix} + \begin{bmatrix} 0 \\ 1 \\ 1s \end{bmatrix} u, \quad \mathbf{y} = \begin{bmatrix} 1 \; 3 \; -2 \end{bmatrix} \begin{bmatrix} x_2 \\ x_1 \\ x_3 \end{bmatrix}.$$

□

Chapter 11
Minimum-Energy Input

We now turn to the problem of computing the *minimum-energy control* (otherwise known as the *minimum-energy input*), which is the least amount of energy required to achieve a particular control task (e.g., reach a final state). Determining the minimum-energy input is important in many control problems because it helps optimize the efficiency of a system. For example, in many applications, reducing the energy input directly correlates to lower operating cost, such as power, wear and tear on system components, etc.

11.1 Preliminary Reviews and Background

11.1.1 Least-Squares Methods

We begin with a review of a key tool used to analyze minimum-energy control problems. *(Linear) least-squares* methods are typically used to solve a variety of *regression* problems: given a dataset $\{(\mathbf{x}_i, y_i)\}_{i=1}^n$, determine a model f such that $y_i = f(\mathbf{x}_i)$ for all $i = 1 \ldots, n$. The input to our function is a vector $\mathbf{x} \in \mathbb{R}^d$, where d is some number of features, and the output is (for simplicity) a scalar $y \in \mathbb{R}$.

Since we are given n of these input-output pairs, we accordingly stack them up into a matrix $X \in \mathbb{R}^{n \times d}$, where the i-th row is $\mathbf{x}_i^\top$, and vector $\mathbf{y} = (y_1, \ldots, y_n)^\top \in \mathbb{R}^n$. Further define the mean $\overline{\mathbf{x}} := \frac{1}{n}\sum_{i=1}^n \mathbf{x}_i \in \mathbb{R}^d$. We will assume that we've preprocessed the data such that any input $\mathbf{x}$ is centered, i.e., $\overline{\mathbf{x}} := \frac{1}{n}\sum_{i=1}^n \mathbf{x}_i = 0$. Let us similarly define $\overline{y} := \frac{1}{n}\sum_{i=1}^n y_i$.

We would like to fit a *linear* model of the form $f(X) = X\mathbf{w} + w_0$ over this data, and determine the values of weights $\mathbf{w}$ and w_0. Typically, we cannot find a model f which perfectly satisfies $\mathbf{y} = f(X)$ with equality. Thus, we take a *least-squares* approach, in which we aim to minimize the combined mean-squared error norm between the output of the model and the given labels:

SJ. Han, *Linear Systems*, KAIST Research Series,
https://doi.org/10.1007/978-981-96-9774-8_11

$$J(w_0, \mathbf{w}) = \|\mathbf{y} - (X\mathbf{w} + w_0\mathbb{1})\|_2^2$$

where $\mathbb{1}$ is the n-dimensional vector of ones.

Deriving the optimal values of w_0 and $\mathbf{w}$ is straightforward using methods from vector calculus. First, expand out the cost:

$$\begin{aligned} J(w_0, \mathbf{w}) &= (\mathbf{y}^\top - \mathbf{w}^\top X^\top - w_0\mathbb{1}^\top)(\mathbf{y} - X\mathbf{w} - w_0\mathbb{1}) \\ &= \mathbf{y}^\top - \mathbf{y}^\top X\mathbf{w} - w_0\mathbf{y}^\top\mathbb{1} - \mathbf{w}^\top X^\top\mathbf{y} + \mathbf{w}^\top X^\top X\mathbf{w} + w_0\mathbf{w}^\top X^\top\mathbb{1} \\ &\quad - w_0\mathbb{1}^\top\mathbf{y} + w_0\mathbb{1}^\top X\mathbf{w} + w_0^2\underbrace{\mathbb{1}^\top\mathbb{1}}_{n} \end{aligned}$$

Differentiate this cost with respect to w_0 and $\mathbf{w}$ separately, set each to 0, and solve a system of equations for the minimizing $\mathbf{w}$ and w_0.

$$\begin{aligned} \frac{\partial J}{\partial w_0} &= -\mathbf{y}^\top\mathbb{1} + \mathbf{w}^\top X^\top\mathbb{1} - \mathbb{1}^\top\mathbf{y} + \mathbb{1}^\top X\mathbf{w} + 2w_0 n \\ &= -2\sum_{i=1}^{n} y_i + 2\mathbf{w}^\top X^\top\mathbb{1} + 2w_0 n \triangleq 0 \\ \implies w_0 &= \frac{1}{n}\sum_{i=1}^{n} y_i + \frac{1}{n}\mathbf{w}\underbrace{X^\top\mathbb{1}}_{=n\overline{x}=0} = \overline{y} \end{aligned}$$

$$\begin{aligned} \nabla_{\mathbf{w}} J &= -X^\top\mathbf{y} - X^\top\mathbf{y} + 2X^\top X\mathbf{w} + w_0 X^\top\mathbb{1} + w_0 X^\top\mathbb{1} \\ &= -2X^\top\mathbf{y} + 2X^\top X\mathbf{w} + 2w_0 X^\top\mathbb{1} \triangleq 0 \\ \implies \mathbf{w} &= (X^\top X)^{-1}X^\top\mathbf{y} \end{aligned}$$

Including Constraints Now suppose that the model f must satisfy some other conditions

$$\min_{\tilde{\mathbf{w}}} J(\tilde{\mathbf{w}}) \text{ s.t. } M\tilde{\mathbf{w}} = \mathbf{c}$$

where $M \in \mathbb{R}^{p\times(n+1)}$ and $\mathbf{c} \in \mathbb{R}^p$ are known constants. For notation simplicity, let $\tilde{\mathbf{w}} \triangleq (w_0, \mathbf{w}^\top)^\top \in \mathbb{R}^{n+1}$ and $\tilde{X} \triangleq [\mathbb{1}, X] \in \mathbb{R}^{n\times(n+1)}$.

We can write a matrix equation which solves the optimal weights of the constrained linear least squares problem, then solve it using Lagrange multipliers. Construct

$$\mathcal{L}(\tilde{\mathbf{w}}; \lambda) \triangleq J(\tilde{\mathbf{w}}) + 2\lambda^\top (M\tilde{\mathbf{w}} - \mathbf{c})$$

for some $\lambda \in \mathbb{R}^p$. Differentiate it and set it equal to zero.

$$\nabla_{\tilde{\mathbf{w}}}\mathcal{L}(\tilde{\mathbf{w}};\lambda) = 2\left(\tilde{X}^\top\tilde{X}\tilde{\mathbf{w}} - \tilde{X}^\top\mathbf{y} + M^\top\lambda\right) = 0$$

We have two vectors of unknowns, $\tilde{\mathbf{w}}$ and λ. The matrix equation to solve these unknowns comes from putting the above equation together with the constraints.

$$\begin{bmatrix} \tilde{X}^\top\tilde{X} & M^\top \\ M & 0 \end{bmatrix}\begin{bmatrix} \tilde{\mathbf{w}} \\ \lambda \end{bmatrix} = \begin{bmatrix} \tilde{X}^\top\mathbf{y} \\ \mathbf{c} \end{bmatrix}$$

Note that when $\lambda = 0$, we obtain the original *normal equations* from the prior unconstrained problem:

$$\tilde{\mathbf{w}} = (\tilde{X}^\top\tilde{X})^{-1}\tilde{X}^\top\mathbf{y}$$

For general $\lambda \neq 0$, there are several algorithms to solve this matrix equation, including *LU factorization* and *QR factorization*. We will not go into the details of these algorithms here.

Constrained Least-Norm Problem Let's take a look at the special case of the constrained linear least-squares problem where $\mathbf{y} \equiv 0$ and $X \equiv I$. We will also ignore the bias term (set $w_0 = 0$) for simplicity.

$$\min_{\mathbf{w}} \|\mathbf{w}\|_2^2 \text{ s.t. } M\mathbf{w} = \mathbf{c}$$

In most applications, we have $M \in \mathbb{R}^{m\times n}$ with linearly independent rows and $m < n$ (i.e., an underdetermined system). Hence, there are infinitely-many solutions to $M\mathbf{w} = \mathbf{c}$. Since the goal is to find a solution $\mathbf{w}$ of this equation with the least norm, this is sometimes called the *least-norm problem*.

Since M has linearly-independent rows, the following *right pseudoinverse* exists:

$$M^\dagger \triangleq M^\top(MM^\top)^{-1}$$

We can show that $\mathbf{w}^* \triangleq M^\dagger\mathbf{c}$ is the optimal solution to the least-norm problem above by verifying two things.

1. $\mathbf{w}^*$ satisfies the equation: $M\mathbf{w}^* = MM^\top(MM^\top)^{-1}\mathbf{c} = \mathbf{c}$.
2. $\mathbf{w}^*$ achieves the least norm: for any other $\mathbf{w} \in \mathbb{R}^n$ which satisfies $M\mathbf{w} = \mathbf{c}$, note that

$$\|\mathbf{w}\|_2^2 = \left\|\mathbf{w}^* + (\mathbf{w} - \mathbf{w}^*)\right\|_2^2 = \left\|\mathbf{w}^*\right\|_2^2 + 2\mathbf{w}^{*\top}(\mathbf{w} - \mathbf{w}^*) + \left\|\mathbf{w} - \mathbf{w}^*\right\|_2^2 \quad (11.1)$$

Note that

$$\mathbf{w}^{*\top}(\mathbf{w} - \mathbf{w}^*) = \mathbf{c}^\top(M^\dagger)^\top(\mathbf{w} - \mathbf{w}^*) = \mathbf{c}^\top(MM^\top)^{-1}M(\mathbf{w} - \mathbf{w}^*) = 0$$

Therefore

$$(2.1) = \|\mathbf{w}^*\|_2^2 + \|\mathbf{w} - \mathbf{w}^*\|_2^2 \geq \|\mathbf{w}^*\|_2^2 \text{ by nonnegativity of norms.}$$

How would the solution change if we have $M \in \mathbb{R}^{m\times n}$ with linearly independent columns and $m > n$ (i.e., an overdetermined system)? It turns out that the optimal solution is still given by $\mathbf{w}^* \triangleq M^\dagger \mathbf{c}$. However, now we have to use the *left pseudoinverse* (also called the *Moore-Penrose inverse*)

$$M^\dagger \triangleq (MM^\top)^{-1}M^\top$$

which exists because M has full column rank.

Remark 11.18 In the literature, there are various versions of least squares problems depending on the types of constraints you have (e.g., quadratic, linear inequality, etc.) In machine learning literature, it is also common to add a *regularization* term. For our purposes, however, the linear equality constraint suffices.

11.1.2 $\mathcal{L}^2$ Space and Norm

We are most familiar with norms of vectors in $\mathbb{R}^n$. For a vector $\mathbf{x} = \begin{bmatrix} x_1 & x_2 & \cdots & x_n \end{bmatrix}^\top \in \mathbb{R}^n$, the discrete ℓ_p norms for finite-dimensional vectors are defined as follows:

- The ℓ_p norm:

$$\|\mathbf{x}\|_{\ell_p} = \left(\sum_{i=1}^{n} |x_i|^p\right)^{\frac{1}{p}}, \quad p \in [1, \infty)$$

- The ℓ_∞ norm:

$$\|\mathbf{x}\|_{\ell_\infty} = \max_{1\leq i\leq n} |x_i|$$

Similarly, for infinite sequences $\mathbf{x} = (x_1, x_2, x_3, \ldots)$, the ℓ_p norms are defined as:

$$\|\mathbf{x}\|_{\ell_p} = \left(\sum_{i=1}^{\infty} |x_i|^p\right)^{\frac{1}{p}}, \quad p \in [1, \infty)$$

We say that $\mathbf{x} \in \ell_p$ if $\|\mathbf{x}\|_{\ell_p} < \infty$.

On the other hand, when dealing with functions $f : I \to \mathbb{R}$ defined over a continuous domain I, we use the continuous $\mathcal{L}_p$ norms:

- The $\mathcal{L}_p$ norm:

$$\|f\|_{\mathcal{L}_p} = \left(\int_I |f(t)|^p \, dt \right)^{\frac{1}{p}}, \quad p \in [1, \infty)$$

- The $\mathcal{L}_\infty$ norm:

$$\|f\|_{\mathcal{L}_\infty} = \text{ess sup}_{t \in I} |f(t)|$$

Here, ess sup denotes the essential supremum, representing the maximum value of a function across its domain while ignoring any irregular values on sets of measure zero. And remark that $x(t)$, like $f(t)$, can also be expressed in terms of the $\mathcal{L}_p$ norm.

For vector-valued functions $\boldsymbol{f} : I \to \mathbb{R}^n$, where $\boldsymbol{f}(t) = [f_1(t), f_2(t), \ldots, f_n(t)]^\top$, we define the $\mathcal{L}_p$ norms by integrating the ℓ_p norms of $\boldsymbol{f}(t)$ over I:

$$\|\boldsymbol{f}\|_{\mathcal{L}_p} = \left(\int_D \|\boldsymbol{f}(t)\|_{\ell_p}^p \, dt \right)^{\frac{1}{p}}, \quad p \in [1, \infty).$$

Similarly, the $\mathcal{L}_\infty$ norm is defined as:

$$\|\boldsymbol{f}\|_{\mathcal{L}_\infty} = \text{ess sup}_{t \in I} \|\boldsymbol{f}(t)\|_{\ell_\infty} = \text{ess sup}_{t \in I} \max_{1 \le i \le n} |f_i(t)|.$$

We use $\mathcal{L}_p(I; \mathbb{R}^n)$ to denote the space of vector-valued functions with finite $\mathcal{L}_p$ norms:

$$\mathcal{L}_p(I; \mathbb{R}^n) = \left\{ \boldsymbol{f} : I \to \mathbb{R}^n \;\middle|\; \|\boldsymbol{f}\|_{\mathcal{L}_p} < \infty \right\}.$$

In particular, for $p = 2$:

$$\mathcal{L}_2(I; \mathbb{R}^n) = \left\{ \boldsymbol{f} : I \to \mathbb{R}^n \;\middle|\; \|\boldsymbol{f}\|_{\mathcal{L}_2} < \infty \right\}.$$

Note: The distinction between the discrete ℓ_p norms and the continuous $\mathcal{L}_p$ norms is important:

- ℓ_p norms are used for finite-dimensional vectors and also for infinite sequences (discrete case).
- $\mathcal{L}_p$ norms are used for functions defined over continuous domains (continuous case).

Note: $\mathcal{L}_p(I; \mathbb{R}^n)$ is a Banach space for any $1 \le p < \infty$, and $\mathcal{L}_2(I; \mathbb{R}^n)$ is a Hilbert space.

Definition 11.75 (*Banach Space*) A *Banach space* is a complete normed vector space; that is, a vector space equipped with a norm such that every Cauchy sequence converges within the space.

Definition 11.76 (*Hilbert Space*) A *Hilbert space* is a complete inner product space; that is, a vector space equipped with an inner product that induces a norm, and that is complete with respect to the metric induced by the norm.

All Hilbert spaces are Banach spaces (since the norm induced by the inner product makes it a normed space, and completeness ensures it is a Banach space), but not all Banach spaces are Hilbert spaces (since they may lack an inner product).

Example 11.21 (*A Banach Space that is not a Hilbert Space*) The space $\mathcal{L}_p(D)$ for $p \neq 2$, $1 \leq p < \infty$, is a Banach space but not a Hilbert space. This is because there is no inner product that induces the $\mathcal{L}_p$ norm when $p \neq 2$. For instance, consider $\mathcal{L}_1(D)$, the space of absolutely integrable functions on D. It is complete with respect to the $\mathcal{L}_1$ norm, making it a Banach space. However, it lacks an inner product structure compatible with the $\mathcal{L}_1$ norm, so it is not a Hilbert space. □

Definition 11.77 (*Admissible Control*) Let $U \subseteq \mathbb{R}^m$ be a set of control values for a system, $\dot{\mathbf{x}}(t) = f(t, \mathbf{x}(t), \mathbf{u}(t))$ where $\mathbf{u}(t) \in U$ is the control input. The set

$$\mathcal{U} \triangleq \left\{\mathbf{u} : \mathbb{R}^{\geq 0} \to U \;\middle|\; \mathbf{u} \text{ is measurable}\right\} \tag{11.2}$$

is then called the set of all *admissible control inputs*.

11.1.3 Linear Mapping

A linear mapping $\mathcal{M} : \mathcal{V} \to \mathcal{W}$ between two vector spaces $\mathcal{V}$, $\mathcal{W}$ over the same field $\mathbb{F}$ satisfies

$$\begin{aligned} \mathcal{M}(\alpha_1\mathbf{v}_1 + \alpha_2\mathbf{v}_2) &= \alpha_1\mathcal{M}(\mathbf{v}_1) + \alpha_2\mathcal{M}(\mathbf{v}_2) \\ &= \alpha_1\mathbf{w}_1 + \alpha_2\mathbf{w}_2, \quad \forall\mathbf{v}_1, \mathbf{v}_2 \in \mathcal{V},\ \forall\mathbf{w}_1, \mathbf{w}_2 \in \mathcal{W},\ \alpha_1, \alpha_2 \in \mathbb{F} \end{aligned}$$

In other words, a linear mapping preserves vector addition and scalar multiplication.

Since $\mathcal{V}$ and $\mathcal{W}$ are vector spaces, they have bases. Suppose $\{\mathbf{v}_i\}_{i=1}^m$ is a basis for $\mathcal{V}$ and $\{\mathbf{w}_j\}_{j=1}^n$ is a basis for $\mathcal{W}$. Any vector $\mathbf{x} \in \mathcal{V}$ can be expressed as

$$\mathbf{x} = \sum_{i=1}^{m} \alpha_i\mathbf{v}_i, \quad \alpha_i \in \mathbb{F}$$

and its image under $\mathcal{M}$ is

$$\mathcal{M}(\mathbf{x}) = \sum_{i=1}^{m} \alpha_i\mathcal{M}(\mathbf{v}_i) = \sum_{i=1}^{m} \alpha_i\mathbf{w}_i'$$

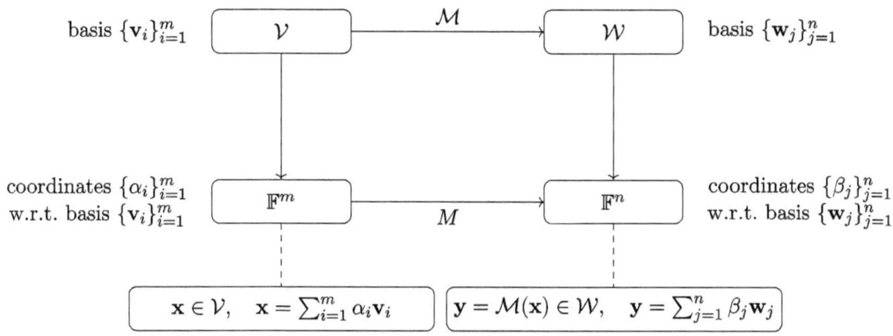

Fig. 11.1 Relationship between a linear mapping $\mathcal{M}$ between vector spaces $\mathcal{V}$ and $\mathcal{W}$, and its matrix representation M. This is a diagram that is commonly seen in linear algebra, but is useful for us to visualize too, while studying linear systems

where $\mathbf{w}_i' = \mathcal{M}(\mathbf{v}_i) \in \mathcal{W}$. Each $\mathbf{w}_i'$ can be expressed in terms of the basis of $\mathcal{W}$:

$$\mathbf{w}_i' = \sum_{j=1}^{n} M_{ji} \mathbf{w}_j$$

where $M_{ji} \in \mathbb{F}$ are the components of the matrix representation M of $\mathcal{M}$.

Note that as long as $\dim \mathcal{V} < \infty$ and $\dim \mathcal{W} < \infty$, there exists a matrix representation M for any linear mapping $\mathcal{M}$. One common diagram used to visualize these relationships, especially in linear algebra courses and references, is Fig. 11.1.

11.1.4 Adjoint Operators

Definition 11.78 (*Adjoint Operator*) The *adjoint operator* $\mathcal{M}^* : \mathcal{W} \to \mathcal{V}$ of a linear mapping $\mathcal{M} : \mathcal{V} \to \mathcal{W}$, for two Hilbert spaces $\mathcal{V}$ and $\mathcal{W}$ over the same field, satisfies

$$\langle \mathcal{M}\mathbf{v}, \mathbf{w} \rangle_{\mathcal{W}} = \langle \mathbf{v}, \mathcal{M}^*\mathbf{w} \rangle_{\mathcal{V}}, \quad \forall \mathbf{v} \in \mathcal{V}, \ \mathbf{w} \in \mathcal{W}$$

where $\langle \cdot, \cdot \rangle_{\mathcal{W}}$ and $\langle \cdot, \cdot \rangle_{\mathcal{V}}$ denote inner products on $\mathcal{W}$ and $\mathcal{V}$, respectively.

Example 11.22 Let $\mathcal{V} = \mathbb{R}^n$, $\mathcal{W} = \mathbb{R}^m$, and let $\mathcal{M} \in \mathbb{R}^{m \times n}$ be a matrix representing a linear mapping $\mathcal{M} : \mathcal{V} \to \mathcal{W}$. Then the adjoint operator is $\mathcal{M}^* = \mathcal{M}^\top \in \mathbb{R}^{n \times m}$, because

$$\langle \mathcal{M}\mathbf{v}, \mathbf{w} \rangle_{\mathcal{W}} = (\mathcal{M}\mathbf{v})^\top \mathbf{w} = \mathbf{v}^\top \mathcal{M}^\top \mathbf{w} = \langle \mathbf{v}, \mathcal{M}^\top \mathbf{w} \rangle_{\mathcal{V}}$$

□

Example 11.23 (*Integral Operator*) Let $K : [t_0, t_f] \times [t_0, t_f] \to \mathbb{R}$ be a measurable function that satisfies

$$\int_{t_0}^{t_f} \int_{t_0}^{t_f} |K(t,s)|^2 \, ds \, dt < \infty$$

Define a linear operator $\mathcal{M} : L_2(I; \mathbb{R}) \to L_2(I; \mathbb{R})$ by

$$(\mathcal{M}u)(t) \triangleq \int_{t_0}^{t_f} K(t,s)u(s) \, ds$$

for $u \in L_2(I; \mathbb{R})$, where $I = [t_0, t_f]$. Then, for $u, v \in L_2(I; \mathbb{R})$, the inner product $\langle \cdot, \cdot \rangle_{L_2(I;\mathbb{R})}$ satisfies:

$$\begin{aligned}
\langle \mathcal{M}u, v \rangle_{L_2} &= \int_{t_0}^{t_f} (\mathcal{M}u)(t) \, v(t) \, dt \\
&= \int_{t_0}^{t_f} \left(\int_{t_0}^{t_f} K(t,s)u(s) \, ds \right) v(t) \, dt \\
&= \int_{t_0}^{t_f} u(s) \left(\int_{t_0}^{t_f} K(t,s)v(t) \, dt \right) ds \\
&= \langle u, \mathcal{M}^* v \rangle_{L_2}
\end{aligned}$$

where the adjoint operator $\mathcal{M}^*$ is given by

$$(\mathcal{M}^* v)(s) = \int_{t_0}^{t_f} K(t,s)v(t) \, dt$$

Note that $\mathcal{M}$ is self-adjoint (i.e., $\mathcal{M} = \mathcal{M}^*$) if and only if $K(t,s) = K(s,t)$ for all $s, t \in I$. □

11.2 Minimum Energy Input: Continuous-Time Case

We are now ready to pose the minimum-energy input problem using mathematical terms. We begin with the CT case, and discuss the DT case in the following Section 11.3.

Given linear system $\dot{\mathbf{x}}(t) = A(t)\mathbf{x}(t) + B(t)\mathbf{u}(t)$ over $I \triangleq [t_0, t_f]$, the control input $\mathbf{u}(t)$ is used to steer or transfer the state from an initial state $\mathbf{x}_0 \triangleq \mathbf{x}(t_0)$ to a desired final state $\mathbf{x}_f = \mathbf{x}(t_f)$. Towards designing the control input $\mathbf{u}(t)$ for the purpose of using minimum energy, there are a few important questions we ask:

1. How can we compute a control $\mathbf{u}$ that successfully transfers the system from $\mathbf{x}_0$ to $\mathbf{x}_f$?

2. How quickly can we perform this transfer from $\mathbf{x}_0$ to $\mathbf{x}_f$?
3. How can we determine the most efficient control $\mathbf{u}$ to minimize energy usage?

The minimum energy control problem is easier to pose mathematically and solve for LTI systems, so we will restrict our focus to the LTI case. For a given controllable CT LTI (A, B) with the initial state $\mathbf{x}(t_0) = \mathbf{x}_0$, we aim to find an input $\mathbf{u}(t)$ that steers the system to the final state while minimizing energy usage. Here, energy usage is quantified by the squared $\mathcal{L}_2$-norm of the control, defined as:

$$\|\tilde{\mathbf{u}}\|_{\mathcal{L}_2}^2 \triangleq \int_{t_0}^{t_f} \tilde{\mathbf{u}}^\top(t)\tilde{\mathbf{u}}(t)\,dt \leq \|\mathbf{u}\|_{\mathcal{L}_2}^2 \triangleq \int_{t_0}^{t_f} \mathbf{u}^\top(t)\mathbf{u}(t)\,dt \quad \forall \mathbf{u} : \mathbb{R}^{\geq 0} \to \mathbb{R}^m \tag{11.3}$$

In our context, we define the space $\mathcal{U}$ of admissible control signals (see Definition 11.77) as the set of all input signals that are "allowed" for the system.

$$\mathcal{U} = \left\{\mathbf{u} : \mathbb{R}^{\geq 0} \to \mathbb{R}^m \;\middle|\; \mathcal{L}\{\mathbf{u}(t)\} \text{ exists for all } t\right\}$$

where $\mathcal{L}\{\cdot\}$ is the Laplace transform.

For a controllable LTI system (A, B), we can pose the problem of finding a minimum-energy input $\tilde{\mathbf{u}}(t)$ as a *constrained least-squares (CLS)* optimization problem:

$$\tilde{\mathbf{u}} = \arg\min_{\mathbf{u}\in\mathcal{U}} \|\mathbf{u}\|_{\mathcal{L}_2}^2 \quad \text{s.t.} \quad 0 = e^{A(t_f - t_0)}\mathbf{x}_0 + \int_{t_0}^{t_f} e^{A(t_f-\tau)}B\mathbf{u}(\tau)\,d\tau$$

The constraint is a linear equality constraint:

$$-e^{A(t_f - t_0)}\mathbf{x}_0 = \int_{t_0}^{t_f} e^{A(t_f-\tau)}B\mathbf{u}(\tau)\,d\tau$$

Multiplying $e^{-A(t_f - t_0)}$ to both sides:

$$\underbrace{-\mathbf{x}_0}_{\triangleq \mathbf{c}} = \int_{t_0}^{t_f} e^{A(t_0-\tau)}B\mathbf{u}(\tau)\,d\tau \triangleq \mathcal{M}\mathbf{u}$$

Here, $\mathcal{M} : \mathcal{L}_2(I; \mathbb{R}^m) \to \mathbb{R}^n$, and $\mathcal{M}(\cdot) \triangleq \int_{t_0}^{t_f} e^{A(t_0-\tau)}B(\cdot)\,d\tau$ is a linear mapping.

The minimum-energy input problem solves the following CLS problem:

$$\min_{\mathbf{u}\in\mathcal{U}} \|\mathbf{u}\|_{\mathcal{L}_2}^2 \quad \text{s.t.} \quad \mathcal{M}\mathbf{u} = \mathbf{c} \tag{11.4}$$

The optimal solution to generic CLS can be obtained using the Lagrangian method. Introduce the Lagrangian multiplier vector $\boldsymbol{\lambda} \in \mathbb{R}^n$. Then, the Lagrangian can be formulated as follows:

$$\mathcal{L}(\mathbf{u}, \boldsymbol{\lambda}) = \int_{t_0}^{t_f} \mathbf{u}^\top(t)\mathbf{u}(t)\,dt + \boldsymbol{\lambda}^\top(\mathcal{M}\mathbf{u} - \mathbf{c})$$

To find the optimal $\mathbf{u}$, we take the derivative of $\mathcal{L}$ with respect to $\mathbf{u}$ and set it to zero:

$$\nabla_{\mathbf{u}}\mathcal{L} = \int_{t_0}^{t_f} \frac{\partial}{\partial \mathbf{u}} \left(\mathbf{u}^\top(t)\mathbf{u}(t)\right) dt + \frac{\partial}{\partial \mathbf{u}} \left(\boldsymbol{\lambda}^\top(\mathcal{M}\mathbf{u} - \mathbf{c})\right) = 0$$

Since $\mathcal{M}$ is an operator mapping from one Hilbert space to another, we can define its adjoint operator $\mathcal{M}^*$.

$$\nabla_{\mathbf{u}}\mathcal{L} = \int_{t_0}^{t_f} 2\mathbf{u}(t)\,dt + \mathcal{M}^*\boldsymbol{\lambda} = 2\mathbf{u}(t) + \mathcal{M}^*\boldsymbol{\lambda} = 0$$

Solving for $\mathbf{u}$, we get:

$$\mathbf{u}(t) = -\frac{1}{2}\mathcal{M}^*\boldsymbol{\lambda}$$

Using the constraint, $\mathcal{M}\mathbf{u} = \mathbf{c}$, we can get $\boldsymbol{\lambda}$:

$$\mathcal{M}\left(-\frac{1}{2}\mathcal{M}^*\boldsymbol{\lambda}\right) = \mathbf{c} \quad \Longrightarrow \quad \boldsymbol{\lambda} = -2(\mathcal{M}\mathcal{M}^*)^{-1}\mathbf{c}$$

Thus, the optimal solution to generic CLS can be described as follows:

$$\mathbf{u} = \mathcal{M}^*(\mathcal{M}\mathcal{M}^*)^{-1}\mathbf{c} \tag{11.5}$$

Now we need to derive the form of the adjoint operator $\mathcal{M}^*$. Before we do so, let's determine what the interpretation of the adjoint operator is in the context of control theory.

Example 11.24 (*Adjoint System*) The concept of the adjoint operator is related to the adjoint system in control theory. Consider the system

$$\dot{\mathbf{x}}(t) = A(t)\mathbf{x}(t)$$

where $A(t)$ is a time-varying matrix. The adjoint system is given by

$$\dot{\mathbf{z}}(t) = -A^\top(t)\mathbf{z}(t)$$

Here, the negative transpose of $A(t)$ appears because, in the context of inner products, the adjoint of the differential operator involves the negative transpose. The state transition matrices (STMs) for $\mathbf{x}(t)$ and $\mathbf{z}(t)$ are related through transposition and inversion. □

Next, let's derive the adjoint operator, $\mathcal{M}^*$:

$$\begin{aligned}
\mathbf{c} = \mathcal{M}\mathbf{u} \Longrightarrow \mathbf{c}^\top \mathbf{c} = \mathbf{c}^\top (\mathcal{M}\mathbf{u}) &= \int_{t_0}^{t_f} \left(\mathcal{M}^*\mathbf{c}\right)^\top (\tau)\mathbf{u}(\tau)\, d\tau \\
\Longrightarrow \mathbf{x}_0^\top \mathbf{x}_0 = -\mathbf{x}_0^\top &\left(\int_{t_0}^{t_f} e^{A(t_0-\tau)} B\mathbf{u}(\tau)\, d\tau\right) \\
&= \int_{t_0}^{t_f} (\underbrace{B^\top e^{A^\top (t_0-\tau)}}_{\mathcal{M}^*} \underbrace{(-\mathbf{x}_0)}_{\mathbf{c}})^\top \mathbf{u}(\tau)\, d\tau
\end{aligned}$$

This tells us the adjoint operator should be $(\mathcal{M}^*\mathbf{c})(\tau) = B^\top e^{A^\top (t_0-\tau)}\mathbf{c}$. Thus, the CLS solution is:

$$\tilde{\mathbf{u}}(t) = \mathcal{M}^*(\mathcal{M}\mathcal{M}^*)^{-1}\mathbf{c}(t) = -B^\top \underbrace{e^{A^\top (t_0-t)}}_{\substack{\triangleq \Phi^\top (t_0,t) \\ \text{in LTI case}}} \underbrace{\left(\int_{t_0}^{t_f} e^{A(t_0-\tau)} B B^\top e^{A^\top (t_0-\tau)}\, d\tau\right)^{-1}}_{\triangleq W_C^{-1}(t_0,t_f) \text{ in LTI case}} \mathbf{x}_0 \tag{11.6}$$

Here, $W_C(t_0, t_f)$ is called the *controllability Gramian*, which we will see in Chapter 12.

Remark 11.19 The *CT controllability Gramian* and *CT reachability Gramian* are defined as follows:

$$\begin{aligned}
W_C(t_0, t_f) &\triangleq \int_{t_0}^{t_f} \Phi(t_0, s)B(s)B^\top (s)\Phi^\top (t_0, s)ds \\
W_R(t_0, t_f) &\triangleq \int_{t_0}^{t_f} \Phi(t_f, s)B(s)B^\top (s)\Phi^\top (t_f, s)ds
\end{aligned}$$

As their names suggest, these Gramians are often used the characterize the extent to which we can control a system and reach certain states. We will discuss more properties about these Gramians in Chapter 12.

In general linear (LTV) systems, the minimum energy control input can be described as follows:

$$\tilde{\mathbf{u}}(t) = -B^\top (t)\Phi^\top (t_0, t)W_C^{-1}(t_0, t_f)\mathbf{x}_0 \tag{11.7}$$

There is some relationship between the controllability Gramian and reachability Gramian (you can derive them easily). Since $W_R = \Phi(t_f, t_0)W_C\Phi^\top (t_f, t_0)$,

$$W_R^{-1} = \Phi^\top (t_0, t_f)W_C^{-1}\Phi(t_0, t_f), \quad W_C^{-1} = \Phi^\top (t_f, t_0)W_R^{-1}\Phi(t_f, t_0)$$

From (11.7) and the Gramian relation, we can derive the minimum energy input in different forms:

1. When $t_0 = 0$, and the system is LTI,

$$\tilde{\mathbf{u}}(t) = -B^\top e^{A^\top(t_f-t)} W_R^{-1}(0, t_f) e^{At_f} \mathbf{x}_0$$

2. When $t_0 = 0$, $\mathbf{x}_f \in \mathbb{R}^n$, $\mathbf{x}_f \neq 0$, and the system is LTI,

$$\tilde{\mathbf{u}}(t) = -B^\top e^{A^\top(t_f-t)} W_R^{-1}(0, t_f)(e^{At_f} \mathbf{x}_0 - \mathbf{x}_f)$$

3. When $t_0 = 0$, $\mathbf{x}_f \in \mathbb{R}^n$, $\mathbf{x}_f \neq 0$, and the system is LTV,

$$\tilde{\mathbf{u}}(t) = -B^\top(t)\Phi^\top(t_f, t) W_R^{-1}(0, t_f)(\Phi(t_f, 0)\mathbf{x}_0 - \mathbf{x}_f)$$

11.3 Minimum Energy Input: Discrete-Time Case

Now we will go over the derivation of the minimum-energy input in the DT case, which largely involves one of the simplest classes of linear mappings: *matrices*, which map from some $\mathbb{R}^n$ to some other $\mathbb{R}^m$.

Definition 11.79 Given controllable DT LTI (A, B), the *minimum energy input* is $\mathbf{u} \triangleq \{\mathbf{u}[0], \mathbf{u}[1], \ldots, \mathbf{u}[t_f - 1]\}$ (taking $t_0 = 0$), a sequence that steers from $\mathbf{x}_0$ to $\mathbf{x}_f$ at time $t_f \in \mathbb{N}$ using the smallest amount of "energy", given by the ℓ_2 norm $\|\mathbf{u}\|_2^2 \triangleq \sum_{s=0}^{t_f-1} \|\mathbf{u}[s]\|_2^2$.

As a constrained linear least-squares problem, we can write

$$\min_{\mathbf{u}} \|\mathbf{u}\|_2^2 \text{ s.t. } \mathbf{x}_f = A^{t_f}\mathbf{x}_0 + \sum_{s=1}^{t_f} A^{t_f-s} B\mathbf{u}[s-1]$$

Lemma 11.7 *The minimum energy input in the DT case is given by*

$$\mathbf{u}[t] = -B^\top(A^\top)^{t_f-1-t} \left(\sum_{s=0}^{t_f-1} A^s B B^\top (A^\top)^s \right)^{-1} (A^{t_f}\mathbf{x}_0 - \mathbf{x}_f) \tag{11.8}$$

Proof We follow basically the same steps as in the CT case.

1. Manipulate the constraint to get a linear equality of the form $M\mathbf{u} = \mathbf{c}$:

$$-(A^{t_f}\mathbf{x}_0 - \mathbf{x}_f) = \sum_{s=1}^{t_f} A^{t_f-s}B\mathbf{u}[s-1] = \underbrace{\begin{bmatrix}B & AB & \cdots & A^{t_f-1}B\end{bmatrix}}_{\triangleq \mathcal{C}_d(t_f)} \begin{bmatrix}\mathbf{u}[t_f-1] \\ \mathbf{u}[t_f-2] \\ \vdots \\ \mathbf{u}[0]\end{bmatrix}$$

where we will denote $\mathcal{C}_d(t_f)$ to be the discrete-time controllability matrix until time $t_f \in \mathbb{N}$. In this case, $M \equiv \mathcal{C}_d(t_f)$ and $\mathbf{c} \equiv \mathbf{x}_f - A^{t_f}\mathbf{x}_0$.

2. Note that M is an underdetermined system because there are more columns than rows. Furthermore, M is full row rank because the system is controllable.
3. We must use the right-pseudoinverse solution to the least-norm problem above:

$$M^{\dagger} \equiv \mathcal{C}_d(t_f)^{\dagger} = \mathcal{C}_d^{\top}(t_f)(\mathcal{C}_d(t_f)\mathcal{C}_d^{\top}(t_f))^{-1}$$

and so

$$\begin{aligned}
\begin{bmatrix}\mathbf{u}[t_f-1] \\ \mathbf{u}[t_f-2] \\ \vdots \\ \mathbf{u}[0]\end{bmatrix} &= -\begin{bmatrix}B^{\top} \\ B^{\top}A^{\top} \\ \vdots \\ B^{\top}(A^{t_f-1})^{\top}\end{bmatrix} \left(\begin{bmatrix}B & AB & \cdots & A^{t_f-1}B\end{bmatrix}\begin{bmatrix}B^{\top} \\ B^{\top}A^{\top} \\ \vdots \\ B^{\top}(A^{t_f-1})^{\top}\end{bmatrix}\right)^{-1} (A^{t_f}\mathbf{x}_0 - \mathbf{x}_f) \\
&= \begin{bmatrix}B^{\top} \\ B^{\top}A^{\top} \\ \vdots \\ B^{\top}(A^{t_f-1})^{\top}\end{bmatrix} \left(\sum_{s=0}^{t_f-1} A^s BB^{\top}(A^{\top})^s\right)^{-1} (A^{t_f}\mathbf{x}_0 - \mathbf{x}_f)
\end{aligned}$$

The desired formula (11.8) comes from simply matching the matrix entries on the LHS and RHS. ■

Chapter 12
Controllability

Now that we have discussed the canonical forms and the concepts of minimum-energy inputs, it is necessary to study whether or not a system can be controlled in the first place. As we will see in this chapter *controllability* refers to the ability to steer or control a system from any initial state to any desired final state using the system's inputs. If a system is controllable, other related questions can also be asked, such as "to what extent can the system be controlled"? Moreover, for systems that can be controlled, several different problems can be solved, such as reference-tracking, disturbance-rejection, regulation, and stabilization.

12.1 Controllability and Reachability

Definition 12.80 (*Controllable*) System $\dot{\mathbf{x}} = A\mathbf{x} + B\mathbf{u}$, $\mathbf{x}(0) = \mathbf{x}_0$ is *controllable* until time $t_f > 0$ for any $\mathbf{x}_0 \in \mathbb{R}^n$ if $\exists \mathbf{u}$ which can steer the system from $\mathbf{x}_0$ to $\mathbf{x}_f \triangleq \mathbf{x}(t_f) = 0$ over time interval $[0, t_f]$.

We note that there is no requirement on the system's behavior after time t_f.

Definition 12.81 (*Controllable Subspace*) The *controllable subspace* $\mathcal{C}[t_0, t_f]$ consist of all $\mathbf{x}_0$ for which $\exists \mathbf{u} : [t_0, t_f] \to \mathbb{R}^m$ which transfers the state to $\mathbf{x}_f = 0$.

$$\mathcal{C}[t_0, t_f] \triangleq \{\mathbf{x}_0 \in \mathbb{R}^n \mid \exists\, \mathbf{u},\ 0 = \Phi(t_f, t_0)x_0 + \int_{t_0}^{t_f} \Phi(t_f, s)B(s)\mathbf{u}(s)ds\}$$

Definition 12.82 (*Reachable Subspace*) Analogous to Definition 12.81, the *reachable subspace* $R[t_0, t_f]$ consists of all $\mathbf{x}_f$ which can be steered to from $\mathbf{x}_0 \equiv 0$ (shifting coordinates for simplicity)

$$\mathcal{R}[t_0, t_f] \triangleq \{\mathbf{x}_f \in \mathbb{R}^n \mid \exists\, \mathbf{u},\ \mathbf{x}(t_f) = \mathbf{x}_f = \int_{t_0}^{t_f} \Phi(t_f, s)B(s)\mathbf{u}(s)ds\}$$

SJ. Han, *Linear Systems*, KAIST Research Series,
https://doi.org/10.1007/978-981-96-9774-8_12

Definition 12.83 (*Reachable*) A system is *reachable* at time $t_f > 0$ if $\mathcal{R}[t_0, t_f] = \mathbb{R}^n$.

In summary, the *controllability problem* is posed as: given final state $\mathbf{x}_f$ at time T, find the set of all initial states $\mathbf{x}_0$, such that $\exists$ trajectory $\mathbf{x}(t)$ which steers $\mathbf{x}_0$ at $t = t_0$ to $\mathbf{x}_f$ at $t = T$. On the other hand, the *reachability problem* is posed as: given initial state $\mathbf{x}_0$ at time t_0, find the set of all final states $\mathbf{x}_f$, such that $\exists$ trajectory $\mathbf{x}(t)$ which steers $\mathbf{x}_0$ at $t = t_0$ to $\mathbf{x}_f$ at $t = T$. From these problem statements alone, you may guess that controllability and reachability are like related properties. In fact, for LTI systems they are equivalent, as shown in the following proposition.

Proposition 12.3 (Controllability-Reachability Equivalence) *The LTI system is controllable if and only if it is reachable.*

We will come back to prove this proposition later. First, we will discuss simpler ways of verifying controllability/reachability beyond using the direct definitions above.

The following matrices are useful in characterizing the reachable and controllable subspaces.

Definition 12.84 (*Controllability and Reachability Gramians*) Given times t_0 and $t_f > t_0$, the *reachability Gramian* is:

$$W_R(t_0, t_f) \triangleq \int_{t_0}^{t_f} \Phi(t_f, s)B(s)B^\top(s)\Phi^\top(t_f, s)ds$$

and the *controllability Gramian* is:

$$W_C(t_0, t_f) \triangleq \int_{t_0}^{t_f} \Phi(t_0, s)B(s)B^\top(s)\Phi^\top(t_0, s)ds$$

We've previously defined the Gramians in Remark 11.19, during the discussion of minimum-energy inputs.

Theorem 12.27 *A linear system (either LTI or LTV) is controllable iff* $\forall t > t_0$, $W_c(t_0, t_f)$ *is nonsingular.*

Proof (Sufficiency.) Suppose $W_c(t_0, t_f)$ is nonsingular. Choose specific control input

$$\mathbf{u}(t) = -B^\top(t)\Phi^\top(t_0, t)W_C^{-1}(t_0, t_f)\mathbf{x}_0,$$

then

$$\begin{aligned}
\mathbf{x}_f &= \Phi(t_f, t_0)\mathbf{x}_0 + \int_{t_0}^{t_f} \Phi(t_f, \tau)B(\tau)\mathbf{u}(\tau)d\tau \\
&= \Phi(t_f, t_0)\mathbf{x}_0 - \int_{t_0}^{t_f} \Phi(t_f, \tau)B(\tau)B^\top(\tau)\Phi^\top(t_0, \tau)W_C^{-1}(t_0, t_f)\mathbf{x}_0 d\tau \\
&= \Phi(t_f, t_0)\mathbf{x}_0 - \Phi(t_f, t_0)W_C(t_0, t_f)W_C^{-1}(t_0, t_f)\mathbf{x}_0
\end{aligned}$$

(Necessity:) Now, suppose the system is controllable. Suppose for the sake of contradiction that W_C is singular. Therefore, $\exists \tilde{\mathbf{x}} \in \mathbb{R}^n$, $\tilde{\mathbf{x}} \neq 0$ such that

$$0 = \tilde{\mathbf{x}}^\top W_C(t_0, t_f)\tilde{\mathbf{x}} = \int_{t_0}^{t_f} \tilde{\mathbf{x}}^\top \Phi(t_0, \tau)B(\tau)B^\top(\tau)\Phi^\top(t_0, \tau)\tilde{\mathbf{x}}d\tau.$$

The integrated is $\|\tilde{x}^\top \Phi(t_o, \tau)B(\tau)\|^2 \geq 0$ for all τ. So the integral being 0 means the integrated is 0.

$$\|\tilde{\mathbf{x}}^\top \Phi(t_o, \tau)B(\tau)\| = 0 \quad \forall \tau \in [t_0, t_f]$$

Since the system is controllable, there exists continuous input u(t) s.t.

$$0 = \Phi(t_f, t_0)\tilde{\mathbf{x}} + \int_{t_0}^{t_f} \Phi(t_f, \tau)B(\tau)\mathbf{u}(\tau)d\tau.$$

Thus we can obtain an expression for $\tilde{\mathbf{x}}$,

$$\tilde{\mathbf{x}} = -\Phi^{-1}(t_f, t_0)\int_{t_0}^{t_f} \Phi(t_f, \tau)B(\tau)\mathbf{u}(\tau)d\tau = -\int_{t_0}^{t_f} \Phi(t_0, \tau)B(\tau)\mathbf{u}(\tau)d\tau.$$

And,

$$\|\tilde{\mathbf{x}}\|_2^2 = \tilde{\mathbf{x}}^\top\tilde{\mathbf{x}} = -\int_{t_0}^{t_f} \tilde{\mathbf{x}}^\top \Phi(t_0, \tau)B(\tau)\mathbf{u}(\tau)d\tau = 0$$

This contradicts $\tilde{\mathbf{x}} \neq 0$. Thus no such $\tilde{\mathbf{x}}$ can exist. ∎

For LTI case, additional tests for controllability can be performed.

Theorem 12.28 (Equivalent Conditions for Controllability) *Given LTI (A, B), the following statements are equivalent:*

CTRB1. *(A, B) is controllable.*
CTRB2. *$W_C(t_0, t_f) \triangleq \int_{t_0}^{t_f} \Phi(t_0, s)BB^\top\Phi^\top(t_0, s)ds$ is nonsingular.*
CTRB3. *The controllability matrix*

$$\mathcal{C} \triangleq [B \; AB \; A^2B \; \cdots \; A^{n-1}B] \in \mathbb{R}^{n\times nm} \tag{12.1}$$

has rank n (full row rank).

CTRB4. *The matrix $[A - \lambda I \;\; B] \in \mathbb{R}^{n\times(n+m)}$ has full row rank $\forall \lambda \in \mathbb{C}$ (note: we only need to check this for $\{\lambda_i\}$ eigenvalues of A).*
CTRB5. *$\mathbf{w}^\top B \neq 0 \; \forall$ left eigenvectors $\mathbf{w} \in \mathbb{C}^n$ of A.*

Additional properties:

1. Controllability is preserved under equivalence transformations: (A, B) is controllable if and only if (PAP^{-1}, PB) is controllable for some equivalent transformation $\tilde{\mathbf{x}}(t) = P\mathbf{x}(t)$.

A proof sketch of this fact can be seen by simply taking a transformation of the controllability matrix

$$[PB\ PAP^{-1}PB\ \cdots\ (PAP^{-1})^{n-1}PB] = P[B\ AB\ \cdots\ A^{n-1}B] = P\mathcal{C}$$

which has the same rank as the original $\mathcal{C}$ since P is nonsingular.

2. Controllability can be related back to canonical forms. In particular, the *Kalman controllable form (KCF)* is given as

$$\dot{\mathbf{x}}(t) = \begin{bmatrix} \dot{\mathbf{x}}^{(1)}(t) \\ \mathbf{x}^{(2)}(t) \end{bmatrix} = \begin{bmatrix} A_{11} & A_{12} \\ & A_{22} \end{bmatrix} \begin{bmatrix} \mathbf{x}^{(1)}(t) \\ {}^{(2)}(t) \end{bmatrix} + \begin{bmatrix} B_1 \end{bmatrix} \mathbf{u}(t)$$

We can get $\dot{\mathbf{x}}_2(t) = A_{22}\mathbf{x}_2(t)$ that is the uncontrollable part of the system.

Definition 12.85 (*Stabilizability*) LTI system (A, B) is *stabilizable* if all uncontrollable modes are asymptotically stable, meaning that $\lim_{t\to\infty} \|\mathbf{x}^{(2)}(t)\| = 0$, for $\mathbf{x}^{(2)}$ in the KCF. Alternatively, another mathematical condition is to say (A, B) is stabilizable if $\exists K \in \mathbb{R}^{m\times n}$ s.t. $A - BK$ is stable.

Stabilizability is weaker than controllability, because controllability requires the final state $\mathbf{x}_f$ to be exactly 0, but stabilizability only requires $\mathbf{x}_f$ to approach 0. We will discuss more about the KCF in future chapters. A similar PBH test can be used to check stabilizability.

Theorem 12.29 (Equivalent Statements for Stabilizability) *The following statements are equivalent:*

1. (A, B) *is stabilizable.*
2. A_{22} *(the uncontrollable part of the CCF) is Hurwitz.*
3. $\operatorname{rank}[A - \lambda I\ B] = n$ *holds* $\forall\lambda \in \bar{\mathbb{C}}^+$.

Conditions (CTRB4.) and (CTRB5.) in Theorem 12.28 are called the *Popov-Belevitch-Hautus (PBH) test.* We will prove it in Sect. 12.2.2.

Example 12.25 Consider CT LTI system with system matrices

$$A = \begin{bmatrix} 0 & 1 & 0 \\ 0 & 0 & 1 \\ -6 & -11 & -6 \end{bmatrix}, B = \begin{bmatrix} 0 \\ 1 \\ -3 \end{bmatrix}$$

Let's use one of the conditions in Theorem 12.28 to check its controllability. First, the eigenvalues of A are $\lambda_1 = -1, \lambda_2 = -2, \lambda_3 = -3$, left eigenvectors are $\mathbf{w}_1 = \begin{bmatrix} 6 \\ 5 \\ 1 \end{bmatrix}$, $\mathbf{w}_2 = \begin{bmatrix} 3 \\ 4 \\ 1 \end{bmatrix}$, $\mathbf{w}_3 = \begin{bmatrix} 2 \\ 3 \\ 1 \end{bmatrix}$. Check $\mathbf{w}_3^\top B = \begin{bmatrix} 2 & 3 & 1 \end{bmatrix} \begin{bmatrix} 0 \\ 1 \\ -3 \end{bmatrix}$ =0. By the PBH eigenvector test, the system is not controllable. □

12.2 Proofs of Equivalence of the Controllability Tests

In the following subsections, we will prove the equivalence of some of the statements in Theorem 12.28.

12.2.1 Proof: (CTRB1.) and (CTRB2.) are Equivalent to (CTRB3.)

CT LTI Case Before we prove this equivalence, let's see the following lemma about the matrix exponential property.

Lemma 12.8 *For $A \in \mathbb{R}^{n\times n}$, there exists scalar, analytic functions $\{\alpha_k(t)\}_{k=0}^{n-1}$ s.t.*

$$e^{At} = \sum_{k=0}^{n-1} \alpha_k(t) A^k.$$

Proof Recall $\Phi(t,0) = e^{At}$ is the unique solution to matrix ODE $\dot{X}(t) = AX(t)$, $X(0) = I$.

Verify $\sum_{k=0}^{n-1} \alpha_k(t) A^k$ satisfies ODE by using the following two steps.

1. Show $X(0) = I$.

$$\sum_{k=0}^{n-1} \alpha_k(0) A^k = I$$

   Then

$$\alpha_0(0) = 1, \alpha_1(0) = \alpha_2(0) = \ldots = \alpha_{n-1}(0) = 0$$

   by matching the coefficients of A^k.
2. Show $\dot{X}(t) = AX(t)$.
   We have

$$\dot{X}(t) = \sum_{k=0}^{n-1} \dot{\alpha}_k(t) A^k$$

   and

$$\begin{aligned} AX(t) &= \sum_{k=0}^{n-1} \alpha_k(t) A^{k+1} \\ &= \sum_{k=1}^{n-1} \alpha_{k-1}(t) A^k + \alpha_{n-1}(t) A^n \end{aligned}$$

$$= \sum_{k=1}^{n-1} \alpha_{k-1}(t) A^k - \sum_{k=0}^{n-1} c_k \alpha_{n-1}(t) A^k$$

$$= -c_0 \alpha_{n-1}(t) + \sum_{k=1}^{n-1} (\alpha_{k-1}(t) - c_k \alpha_{n-1}(t)) A^k,$$

where the third equality is due to Cayley-Hamilton.
In order for $\dot{X}(t) = AX(t)$, we need $\alpha_k(t)$ to satisfy

$$\begin{cases} \dot{\alpha}_0(t) = -c_o \alpha_{n-1}(t) \\ \alpha_k(t) = \alpha_{k-1}(t) - c_k \alpha_{n-1}(t) \quad \forall k = 1, \ldots, n-1 \end{cases}$$

by matching the coefficients. In the state-space form, we get

$$\begin{bmatrix} \dot{\alpha}_0 \\ \alpha_1 \\ \alpha_2 \\ \\ \alpha_{n-1} \end{bmatrix} = \begin{bmatrix} 0\ 0 \cdots 0 & -c_0 \\ 0 \cdots 0 & -c_1 \\ 1 \cdots 0 & -c_2 \\ \vdots \ \ddots \ \vdots & \vdots \\ 0 \cdots 1 & -c_{n-1} \end{bmatrix} \begin{bmatrix} \alpha_0 \\ n-1 \end{bmatrix}, \begin{bmatrix} \alpha_0(0) \\ (0) \\ (0) \\ \\ n-1(0) \end{bmatrix} = \begin{bmatrix} 1 \end{bmatrix},$$

which looks exactly like the observable canonical form (OCF).

As we proved the existence and uniqueness of the solutions to autonomous LTI systems, there clearly exists a solution $\{\alpha_k(t)\}_{k=0}^{n-1}$ to above state-space form. ■

In addition, recall the solution form for LTI (A, B):

$$\mathbf{x}(t_f) = e^{A(t_f - t_0)} \mathbf{x}(t_0) + \int_{t_0}^{t_f} e^{A(t_f - \tau)} B\mathbf{u}(\tau) d\tau$$

$$e^{-At_f} \mathbf{x}_f - e^{-At_0} \mathbf{x}_0 = \int_{t_0}^{t_f} e^{-A\tau} B\mathbf{u}(\tau) d\tau$$

Using Lemma 12.8, we get

$$e^{-A\tau} = \sum_{k=0}^{n-1} \alpha_k(\tau) A^k.$$

Then

$$e^{-At_f}\mathbf{x}_f - e^{-At_0}\mathbf{x}_0 = \int_{t_0}^{t_f}\sum_{k=0}^{n-1}\alpha_k(\tau)A^k B\mathbf{u}(\tau)d\tau = \begin{bmatrix} B & AB & \cdots & A^{n-1}B \end{bmatrix}\begin{bmatrix} \int_{t_0}^{t_f}\alpha_0(\tau)\mathbf{u}(\tau)d\tau \\ \int_{t_0}^{t_f}\alpha_1(\tau)\mathbf{u}(\tau)d\tau \\ \int_{t_0}^{t_f}\alpha_2(\tau)\mathbf{u}(\tau)d\tau \\ \vdots \\ \int_{t_0}^{t_f}\alpha_{n-1}(\tau)\mathbf{u}(\tau)d\tau \end{bmatrix}.$$

$\therefore \forall \mathbf{x}_0 \in \mathbb{R}^n$, the solution exists iff $\text{rank}(\mathcal{C}) = \text{rank}(\begin{bmatrix} B & AB & \cdots & A^{n-1}B \end{bmatrix}) = n$.

Now, let's prove that (CTRB1.) and (CTRB2.) are indeed equivalent to (CTRB3.).

Theorem 12.30 *LTI system* (A, B) *is controllable on* $[t_0, t_f]$ *iff* $\text{rank}(\mathcal{C}) = n$, *where* $\mathcal{C}$ *is the controllability matrix* (12.1).

Proof First, note that for all $p \geq n$, $\text{rank}(\begin{bmatrix} B & AB & \cdots & A^{p-1}B \end{bmatrix}) = \text{rank}(\begin{bmatrix} B & AB & \cdots & A^{n-1}B \end{bmatrix})$ because of Cayley-Hamilton.

($\Longrightarrow$) Suppose $\text{rank}(\mathcal{C}) < n$. That means that there exists $\tilde{\mathbf{x}} \neq 0$, $\tilde{\mathbf{x}} \in \mathbb{R}^n$ s.t. $\tilde{\mathbf{x}}^\top A^k B = 0$ for all $k = 0, \ldots, n-1$. Then

$$\begin{aligned} \tilde{\mathbf{x}}^\top W_c(t_0, t_f) &= \int_{t_0}^{t_f} \tilde{\mathbf{x}}^\top e^{A(t_0-t)} BB^\top e^{A^\top(t_0-t)} dt \\ &= \int_{t_0}^{t_f} \tilde{\mathbf{x}}^\top \left(\sum_{k=0}^{n-1}\alpha_k(t_0-t)A^k\right) BB^\top e^{A^\top(t_0-t)} dt \\ &= \int_{t_0}^{t_f} \left(\sum_{k=0}^{n-1}\alpha_k(t_0-t)\tilde{\mathbf{x}}^\top A^k B\right) B^\top e^{A^\top(t_0-t)} dt = 0, \end{aligned}$$

where the second equality results from Lemma 12.8, and the last equality is due to the assumption $\tilde{\mathbf{x}}^\top A^k B = 0$ for all $k = 0, \ldots, n-1$. Therefore, there exists $\tilde{\mathbf{x}} \neq 0$ s.t. $\tilde{\mathbf{x}}^\top W_c(t_0, t_f) = 0 \implies W_c(t_0, t_f)$ is noninvertible. By contrapositive, we can get the following result:

$$\therefore W_c(t_0, t_f) \text{ is nonsingular } \implies \text{rank}(\mathcal{C}) = n.$$

($\Longleftarrow$) Suppose $W_c(t_0, t_f)$ is noninvertible. That means that there exists $\tilde{\mathbf{x}} \neq 0$ s.t.

$$0 = \tilde{\mathbf{x}}^\top W_c(t_0, t_f)\tilde{\mathbf{x}} = \int_{t_0}^{t_f} \tilde{\mathbf{x}}^\top e^{A(t_0-t)} BB^\top e^{A^\top(t_0-t)}\tilde{\mathbf{x}} dt.$$

Using the fact that $\tilde{\mathbf{x}}^\top e^{A(t_0-t)} BB^\top e^{A^\top(t_0-t)}\tilde{\mathbf{x}} = \|\tilde{\mathbf{x}}^\top e^{A(t_0-t)}B\|^2 \geq 0$ for all t, we can get

$$\|\tilde{\mathbf{x}}^\top e^{A(t_0-t)}B\|^2 = 0 \implies \tilde{\mathbf{x}}^\top e^{A(t_0-t)}B = 0 \quad \forall t \in [t_0, t_f]. \tag{12.2}$$

When $t = t_0$ in (12.2), we have $\tilde{\mathbf{x}}^\top B = 0$. Differentiating (12.2) with $t = t_0$,

$$-\tilde{\mathbf{x}}^\top A e^{A(t_0-t)}B = 0 \implies \tilde{\mathbf{x}}^\top AB = 0. \tag{12.3}$$

Similarly, differentiating (12.2) twice with $t = t_0$, gives us $\tilde{\mathbf{x}}^\top A^2 B = 0$. In general, differentiating (12.2) k times with $t = t_0$ yields

$$\tilde{\mathbf{x}}^\top A^k B = 0 \;\; \forall\, k \in \mathbb{N}.$$

Thus, we have

$$\tilde{\mathbf{x}}^\top \begin{bmatrix} B \; AB \; \cdots \; A^{n-1}B \end{bmatrix} = 0,$$

which means $\text{rank}(\mathcal{C}) < n$.

By contrapositive, we can get the following result:

$$\therefore \text{rank}(\mathcal{C}) = n \implies W_c(t_0, t_f) \text{ is nonsingular.} \qquad \blacksquare$$

CT LTV Case. For LTV systems, we can create a similar controllability matrix. First, a brief detour into *adjoint systems*. Uncontrolled linear system $\dot{\mathbf{x}}(t) = A(t)\mathbf{x}(t)$ with state transition matrix (STM) $\Phi(t, t_0)$ has an adjoint system given by

$$\dot{\mathbf{z}}(t) = -A^\top(t)\mathbf{z}(t)$$

with STM $\Phi^\top(t, t_0)$.

As we saw from Chap. 3, $\Phi(t, t_0)$ satisfies

$$\frac{\partial \Phi}{\partial t}(t, t_0) = A(t)\Phi(t, t_0) \text{ and } \Phi(t_0, t_0) = I.$$

But $\Phi(t_0, t)$ satisfies

$$\frac{\partial \Phi}{\partial t}(t_0, t) = -\Phi(t_0, t)A(t) \text{ and } \Phi(t_0, t_0) = I. \tag{12.4}$$

This is easily verifiable for LTI case, $\Phi(t, t_0) = e^{A(t-t_0)}$.

Definition 12.86 (*Controllability Matrix Sequence for LTV systems*) For LTV $(A(t), B(t))$ with $A \in C^{N-1}$, $B \in C^N$, and $N \in \mathbb{N}$, define the *controllability matrix sequence* $\{K_i(t)\}_{i=0}^N$ s.t.

$$\begin{cases} K_0(t) = B(t) \\ K_i(t) = -A(t)K_{i-1}(t) + \dot{K}_{i-1}(t) \quad i = 1, 2, \ldots, N \end{cases}$$

Lemma 12.9 *The sequence* $\{K_i(t)\}_{i=0}^N$ *has the following property:*

$$\forall t, \tau,\; \frac{\partial^i}{\partial \tau^i}(\Phi(t, \tau)B(\tau)) = \Phi(t, \tau)K_i(\tau) \quad i = 1, 2, \ldots \tag{12.5}$$

Proof Lemma 12.9 has a simple proof by induction. First, when $i = 0$, the formula clearly holds since $K_0(\tau) = B(\tau)$. Now suppose the formula holds when $i = j$. If $i = j + 1$,

$$\begin{aligned}\frac{\partial^i}{\partial \tau^i}(\Phi(t,\tau)B(\tau)) &= \frac{\partial}{\partial \tau}\left(\frac{\partial^i}{\partial \tau^i}(\Phi(t,\tau)B(\tau))\right) \\ &= \frac{\partial}{\partial \tau}(\Phi(t,\tau)K_j(\tau)) \\ &= \dot{\Phi}(t,\tau)K_j(\tau) + \Phi(t,\tau)\dot{K}_j(\tau) \\ &= -\Phi(t,\tau)A(\tau)K_j(\tau) + \Phi(t,\tau)\dot{K}_j(\tau) \\ &= \Phi(t,\tau)K_{j+1}(\tau),\end{aligned}$$

where the second equality comes from the induction case, and the fourth equality is due to (12.4). By induction, this concludes our proof. ■

Lemma 12.10 *Suppose that there exists $N \in \mathbb{N}$ s.t. $A \in C^{N-1}$, $B \in C^N$. Then LTV system $(A(t), B(t))$ is controllable on $[t_0, t_f]$ if for some $\tau^* \in [t_0, t_f]$,*

$$\operatorname{rank}(\begin{bmatrix} K_0(\tau^*) & K_1(\tau^*) & \cdots & K_N(\tau^*) \end{bmatrix}) = n.$$

Proof Suppose $(A(t), B(t))$ is not controllable on $[t_0, t_f]$. That means that $W_c(t_0, t_f)$ is noninvertable, and there exists $\tilde{\mathbf{x}} \neq 0$ s.t. $0 = \tilde{\mathbf{x}}^\top W_c(t_0, t_f)\tilde{\mathbf{x}} = \int_{t_0}^{t_f} \tilde{\mathbf{x}}^\top \Phi(t_0, t)BB^\top \Phi^\top(t_0, t)\tilde{\mathbf{x}}dt$.

Using the fact that $\tilde{\mathbf{x}}^\top \Phi(t_0, t)BB^\top \Phi^\top(t_0, t)\tilde{\mathbf{x}} = \|\tilde{\mathbf{x}}^\top \Phi(t_0, t)B\|^2 \geq 0$ for all t, we can get

$$\|\tilde{\mathbf{x}}^\top \Phi(t_0, t)B\|^2 = 0 \Longrightarrow \tilde{\mathbf{x}}^\top \Phi(t_0, t)B = 0 \quad \forall t \in [t_0, t_f].$$

Let $\hat{\mathbf{x}} = \Phi^\top(t_0, \tau^*)\tilde{\mathbf{x}}$, $\hat{\mathbf{x}} \neq 0$ so that

$$\hat{\mathbf{x}}^\top \Phi(\tau^*, t)B(t) = \tilde{\mathbf{x}}^\top \Phi(t_0, \tau^*)\Phi(\tau^*, t)B(t) = \tilde{\mathbf{x}}^\top \Phi(t_0, t)B(t) = 0 \quad \forall t \in [t_0, t_f]. \tag{12.6}$$

When $t = \tau^*$ in (12.6),

$$\hat{\mathbf{x}}^\top \Phi(\tau^*, \tau^*)B(\tau^*) = \hat{\mathbf{x}}^\top B(\tau^*) = \hat{\mathbf{x}}^\top K_0(\tau^*) = 0.$$

Differentiating (12.6), we get

$$\begin{aligned}0 &= \frac{d}{dt}(\hat{\mathbf{x}}^\top \Phi(\tau^*, t)B(t)) \\ &= \hat{\mathbf{x}}^\top(\dot{\Phi}(\tau^*, t)B(t) + \Phi(\tau^*, t)\dot{B}(t)) \\ &= \hat{\mathbf{x}}^\top(-\Phi(\tau^*, t)A(t)B(t) + \Phi(\tau^*, t)\dot{B}(t)) \\ &= \hat{\mathbf{x}}^\top \Phi(\tau^*, t)(-A(t)B(t) + \dot{B}(t)) \\ &= \hat{\mathbf{x}}^\top \Phi(\tau^*, t)K_1(t),\end{aligned}$$

where the third equality results from (12.4), and the last equality is due to Definition 12.86. Then with $t = \tau^*$,

$$\hat{\mathbf{x}}^\top K_1(\tau^*) = 0.$$

Differentiate (12.6) k times with $t = \tau^*$,

$$\hat{\mathbf{x}}^\top K_j(\tau^*) = 0.$$

Thus, we have

$$\tilde{\mathbf{x}}^\top \left[K_0(\tau^*)\ K_1(\tau^*)\ \cdots\ K_N(\tau^*)\right] = 0,$$

then $\text{rank}(\mathcal{C}) < n$.

By contrapositive, we can get the following result:

$$\therefore \text{rank}(\mathcal{C}) = n \implies \text{LTV } (A(t), B(t)) \text{ is controllable.}$$ ■

DT LTI Case. Consider the DT LTI system (A, B)

$$\mathbf{x}[t+1] = A\mathbf{x}[t] + B\mathbf{u}[t].$$

Applying these system dynamics recursively through values of $t \in \mathbb{N}$, we get

$$\begin{aligned}
\mathbf{x}[1] &= A\mathbf{x}_0 + B\mathbf{u}[0] \\
\mathbf{x}[2] &= A\mathbf{x}[1] + B\mathbf{u}[1] = A^2\mathbf{x}_0 + AB\mathbf{u}[0] + B\mathbf{u}[1] \\
&\vdots \\
\mathbf{x}[n] &= A\mathbf{x}[n-1] + B\mathbf{u}[n-1] \\
&= A^n\mathbf{x}_0 + A^{n-1}B\mathbf{u}[0] + A^{n-2}B\mathbf{u}[1] + \cdots + B\mathbf{u}[n-1].
\end{aligned}$$

In matrix equation form, we have

$$\mathbf{x}[n] - A^n\mathbf{x}_0 = \mathcal{C}_d \begin{bmatrix} \mathbf{u}[n-1] \\ \mathbf{u}[n-2] \\ \vdots \\ \mathbf{u}[0] \end{bmatrix}$$

with $\mathcal{C}_d \triangleq \left[B\ AB\ \cdots\ A^{n-1}B\right]$. Here, the DT controllability matrix is basically the same as its CT counterpart.

Corollary 12.4 *DT LTI* (A, B) *is controllable iff* $\text{rank}(\mathcal{C}_d) = n$.

Sometimes, $\mathcal{C}_d$ is called the *reachability matrix* because the range $Im(\mathcal{C}_d)$ is just the reachable subspace (setting $x_0 = 0$).

1. $\begin{bmatrix} B \; AB \; \cdots \; A^{n-1}B \end{bmatrix}$ with $k < n$ is the reachability matrix in k steps.
2. The largest possible reachable set (maximum reachable set) is attained in at most $k = n$ steps.

12.2.2 Proof: (CTRB1.) and (CTRB2.) are Equivalent to (CTRB4.)

First, we prove the PBH rank test described by (CTRB4.).

Proof of PBH rank test. Note this condition is automatically satisfied when λ is not an eigenvalue of A. Thus, we only need to check it when λ is an eigenvalue.

(Sufficiency.) Suppose (A, B) is not controllable. Transforming to CCF yields a form where $\tilde{A}_{22}$ has a nonzero dimension.

Let λ be an eigenvalue of $\tilde{A}_{22}$. Then the matrix $[\tilde{A}_{22} - \lambda I \;\; \tilde{B}]$ has rank less than n. Consequently, the matrix $[A - \lambda I \;\; B] = P^{-1}[\tilde{A}_{22} - \lambda I \;\; \tilde{B}] \begin{bmatrix} P & 0 \\ 0 & I \end{bmatrix}$ also has rank less than n. By contrapositive argument, the rank condition holds, which means the system is controllable.

(Necessity.) Suppose $\text{rank}[\lambda I - A \;\; B] < n$, Then $\exists \; \mathbf{x} \neq 0 \; \mathbf{x} \in \mathbb{R}^n$, such that $\mathbf{x}^\top [A - \lambda I \;\; B] = 0$, then we can get

$$\mathbf{x}^\top (A - \lambda I) = 0, \quad \mathbf{x}^\top B = 0$$

Note $\mathbf{x}^\top (A - \lambda I) = 0$ implies $\mathbf{x}^\top A^k = \lambda^k \mathbf{x}^\top \forall k \in \mathbb{N}$, so that $\mathbf{x}^\top [B \; AB \; \cdots \; A^{n-1}B]$ =0. This doesn't satisfy the controllability matrix test.

Again, by contrapositive argument, the system is controllable when the rank condition holds. ■

Theorem 12.31 *LTI* (A, B) *is controllable on* $[t_0, t_f]$ *iff* $\text{rank}(\begin{bmatrix} A - \lambda I & B \end{bmatrix}) = n \;\; \forall \lambda \in \mathbb{C}$.

Proof ($\Longleftarrow$) Condition is satisfied when λ is not an eigenvalue of A, so we will check the condition only when λ is eigenvalue of A.

Suppose that $\text{rank}(\begin{bmatrix} A - \lambda I & B \end{bmatrix}) < n$. That means that there exists $\tilde{\mathbf{x}} \neq 0, \tilde{\mathbf{x}} \in \mathbb{R}^n$ s.t. $\tilde{\mathbf{x}}^\top \begin{bmatrix} A - \lambda I & B \end{bmatrix} = 0$. Then

$$\tilde{\mathbf{x}}^\top (A - \lambda I) = 0, \tag{12.7a}$$

$$\tilde{\mathbf{x}}^\top B = 0 \tag{12.7b}$$

(12.7a) implies that

$$\tilde{\mathbf{x}}^\top A = \lambda \tilde{\mathbf{x}}^\top$$

$$\implies \tilde{\mathbf{x}}^\top A^2 = \lambda \tilde{\mathbf{x}}^\top A = \lambda^2 \tilde{\mathbf{x}}^\top$$

$$\implies \tilde{\mathbf{x}}^\top A^3 = \lambda^2 \tilde{\mathbf{x}}^\top A = \lambda^3 \tilde{\mathbf{x}}^\top$$

$$\vdots$$

$$\implies \tilde{\mathbf{x}}^\top A^k = \lambda^k \tilde{\mathbf{x}}^\top \quad \forall k \in \mathbb{N}.$$

Using the above results and (12.7b), we get

$$\tilde{\mathbf{x}}^\top \begin{bmatrix} B \ AB \ \cdots \ A^{n-1}B \end{bmatrix} = \begin{bmatrix} \tilde{\mathbf{x}}^\top B \ \lambda \tilde{\mathbf{x}}^\top B \ \cdots \ \lambda^{n-1} \tilde{\mathbf{x}}^\top B \end{bmatrix} = 0.$$

Therefore, it doesn't satisfies $\text{rank}(\mathcal{C}) = n$ test in (12.28).
By contrapositive, we can get the following result:

$$\therefore \text{ LTI } (A, B) \text{ is controllable } \implies \text{rank}(\begin{bmatrix} A - \lambda I \ B \end{bmatrix}) = n.$$

($\Longrightarrow$) The proof of this direction follows similarly to the proof of the PBH rank test shown at the beginning of this section.

12.2.3 *Proof: (CTRB1.) and (CTRB2.) are Equivalent to (CTRB5.)*

Theorem 12.32 *LTI (A, B) is controllable on $[t_0, t_f]$ iff $\mathbf{w}^\top B \neq 0$ for all left eigenvectors $\mathbf{w} \in \mathbb{C}^n$ of A.*

Proof ($\Longleftarrow$) Suppose that there exists $\mathbf{w} \in \mathbb{C}^n$, $\mathbf{w} \neq 0$ s.t.

$$\mathbf{w}^\top A = \lambda \mathbf{w}^\top, \mathbf{w}^\top B = 0.$$

Then

$$\mathbf{w}^\top \begin{bmatrix} B \ AB \ \cdots \ A^{n-1}B \end{bmatrix} = \begin{bmatrix} \mathbf{w}^\top B \ \lambda \mathbf{w}^\top B \ \cdots \ \lambda^{n-1} w^\top B \end{bmatrix} = 0,$$

which means that $\text{rank}(\begin{bmatrix} B \ AB \ \cdots \ A^{n-1}B \end{bmatrix}) < n$.
Therefore, it doesn't satisfies $\text{rank}(\begin{bmatrix} B \ AB \ \cdots \ A^{n-1}B \end{bmatrix}) = n$ test in (CTRB3.).
By contrapositive, we can get the following result:

$$\therefore \text{ LTI } (A, B) \text{ is controllable } \implies \mathbf{w}^\top B \neq 0 \text{ for all left eigenvectors } \mathbf{w} \in \mathbb{C}^n \text{ of } A.$$

($\Longrightarrow$) Suppose that $\text{rank}(\begin{bmatrix} B \ AB \ \cdots \ A^{n-1}B \end{bmatrix}) < n$.
Then by changing coordinates, we get

$$\tilde{A} = T^{-1}AT = \begin{bmatrix} \tilde{A}_{11} & \tilde{A}_{12} \\ 0 & \tilde{A}_{22} \end{bmatrix}, \tilde{B} = T^{-1}B = \begin{bmatrix} \tilde{B}_{11} \\ 0 \end{bmatrix}.$$

Let λ be an eigenvalue of $\tilde{A}_{22}$, and w_{22} is an left eigenvector of $\tilde{A}_{22}$. By defining

$$\mathbf{w} = (T^{-1})^\top \begin{bmatrix} 0 \\ \mathbf{w}_{22} \end{bmatrix},$$

we have

$$\begin{aligned} \mathbf{w}^\top A &= \begin{bmatrix} 0 \\ \mathbf{w}_{22} \end{bmatrix}^\top T^{-1} \left(T \begin{bmatrix} \tilde{A}_{11} & \tilde{A}_{12} \\ 0 & \tilde{A}_{22} \end{bmatrix} T^{-1} \right) \\ &= \left[0 \ \mathbf{w}_{22}^\top \tilde{A}_{22} \right] T^{-1} \\ &= \left[0 \ \lambda \mathbf{w}_{22}^\top \right] T^{-1} \\ &= \lambda \left[0 \ \mathbf{w}_{22}^\top \right] T^{-1} \\ &= \lambda \mathbf{w}^\top \end{aligned}$$

and

$$\begin{aligned} \mathbf{w}^\top B &= \begin{bmatrix} 0 \\ \mathbf{w}_{22} \end{bmatrix}^\top T^{-1} \left(T \begin{bmatrix} \tilde{B}_{11} \\ 0 \end{bmatrix} \right) \\ &= \left[0 \ \mathbf{w}_{22}^\top \right] \begin{bmatrix} \tilde{B}_{11} \\ 0 \end{bmatrix} \\ &= 0. \end{aligned}$$

This shows that there exists $w \in \mathbb{C}^n$, $w \neq 0$ s.t.

$$\mathbf{w}^\top A = \lambda \mathbf{w}^\top, \mathbf{w}^\top B = 0.$$

By contrapositive, we can get the following result:

$$\begin{aligned} &\therefore \mathbf{w}^\top B \neq 0 \text{ for all left eigenvectors } \mathbf{w} \in \mathbb{C}^n \text{ of } A \\ &\implies \text{LTI } (A, B) \text{ is controllable.} \end{aligned}$$ ■

Chapter 13
Observability

A related system characteristic is *observability*, which refers to the ability to determine the internal state of a system just by looking at its outputs. If a system is observable, you can figure out what's going on inside the system based only on what you can observe from the outside, like inferring the condition of an engine by listening to its sound or reading its temperature. In essence, controllability is about controlling the system, and observability is about being able to "see" the state of the system from the outside.

Before delving into the intricacies of observability, let's set the stage by defining the concept and its significance in control theory. Observability plays a crucial role in determining the extent to which we can extract information about the internal state of a system solely from its output. In this section, we'll explore the definition of observability, key metrics such as the observability Gramian. First consider the linear system without any control input.

Definition 13.87 (*Observable*) LTI or LTV system $\dot{\mathbf{x}} = A\mathbf{x}, \mathbf{x}(t_0), \mathbf{y} = C\mathbf{x}$ is *observable* in time interval $[t_0, t_f]$, $t_f > t_0$ if any initial state $\mathbf{x}_0 \in \mathbb{R}$ can be uniquely determined from $\mathbf{y}(t)$, $t \in [t_0, t_f]$.

If the initial-state $\mathbf{x}_0$, can be found from $\mathbf{u}$ and $\mathbf{y}$ measured over a finite interval of time t_0, the system is said to be observable; otherwise the system is said to be unobservable. More intuitively, it asks how well can the internal state $\mathbf{x}(t)$ be estimated given only output information $\{\mathbf{y}(s) : s \in [0, t)\}$.

Remark 13.20 Unlike controllability, observability is not affected by the control input $\mathbf{u}$. We will see that conditions for observability only depend on system matrix A and output matrix C.

Many of the definitions and concepts introduced for controllability and reachability are also applicable here for observability.

Definition 13.88 (*Observability Gramian*) The *observability Gramian* for linear system $(A(t), C(t))$ is

SJ. Han, *Linear Systems*, KAIST Research Series,
https://doi.org/10.1007/978-981-96-9774-8_13

$$W_O(t_0, t_f) \triangleq \int_{t_0}^{t_f} \Phi^\top(t, t_0) C^\top(t) C(t) \Phi(t, t_0)\, dt \in \mathbb{R}^{n\times n} \tag{13.1}$$

Theorem 13.33 *$(A(t), C(t))$ is observable iff $W_O(t_0, t_f)$ is nonsingular.*

Proof (*Sufficiency.*) Suppose $W_O(t_0, t_f)$ is nonsingular. Then $\mathbf{x}_0 = W_O^{-1}(t_0, t_f) \int_{t_0}^{t} \Phi^\top(s, t_0) C^\top(t) \mathbf{y}(s)\, ds$ is constructed from $\{\mathbf{y}(s) : t_0 \leq s \leq t\}$ for any $t \in [t_0, t_f]$.
Check that it satisfies the solution equation $\mathbf{y}(t) = C(t)\Phi(t, t_0)\mathbf{x}_0 \ \forall t \in [t_0, t_f]$.

$$\implies \underbrace{\int_{t_0}^{t} \Phi^\top(s, t_0) C^\top(t) C(t) \Phi(t, t_0)\, dt}_{\triangleq W_O(t_0, t_f)}\, x_0 = \int_{t_0}^{t} \Phi^\top(s, t_0) C^\top(t) \mathbf{y}(t)\, dt \tag{13.2}$$

Multiply both sides by $W_O^{-1}(t_0, t)$ to get the desired result.
((*Necessity.*) Suppose $W_O(t_0, t_f)$ singular. Then

$$\begin{aligned} \implies \exists \tilde{\mathbf{x}} \in \mathbb{R}^n, \tilde{\mathbf{x}} \neq \mathbf{0} \text{ such that } \tilde{\mathbf{x}}^\top W_O(t_0, t_f)\tilde{\mathbf{x}} &= 0 \\ &= \int_{t_0}^{t_f} \tilde{\mathbf{x}} \Phi^\top(s, t_0) C^\top(s) C(s) \Phi(s, t_0)\tilde{\mathbf{x}}\, ds \\ &= \int_{t_0}^{t_f} \|C(s)\Phi(s, t_0)\tilde{\mathbf{x}}\|_2^2\, ds \\ &\implies C(s)\Phi(s, t_0)\tilde{\mathbf{x}} = 0 \ \forall s \in [t_0, t_f] \end{aligned} \tag{13.3}$$

The solution equation is $\mathbf{y}(t) = C(t)\Phi(t, t_0)\mathbf{x}_0$. Hence, both $\mathbf{x}_0 = 0$ and $\mathbf{x}_0 = \tilde{\mathbf{x}}$ yield the same output $\mathbf{y}(t) = 0$. Therefore, we cannot uniquely determine the input $\mathbf{x}$ when we observe a zero output, and so the system is not observable. By contrapositive, observable $\implies W_O(t_0, t_f)$ is nonsingular. ∎

Just as how stabilizability was considered to be a weaker version of controllability (see Definition 12.85), there is also a weaker version of observability called detectability.

Definition 13.89 (*Detectable*) LTI system (A, C) is *detectable* if all unobservable modes of the system are asymptotically stable. Alternatively, a more mathematical condition is to say (A, C) is detectable if $\exists L \in \mathbb{R}^{n\times k}$ s.t. $A - LC$ is stable.

CT LTI Case. For the CT LTI case (i.e. $\dot{\mathbf{x}}(t) = A\mathbf{x}(t), \mathbf{y}(t) = C\mathbf{x}(t)$), we can check for observability using tests similar to controllability.

Theorem 13.34 (Equivalent Conditions for Observability) *Given LTI (A, C), the following statements are equivalent:*

OBSV1. (A, C) *is observable.*
OBSV2. $W_O(t_0, t_f) \triangleq \int_{t_0}^{t_f} \Phi^\top(t, t_0) C^\top(t) C(t) \Phi(t, t_0)\, dt$ *is nonsingular.*
OBSV3. *The observability matrix* $\mathcal{O}(A, C) \triangleq \begin{bmatrix} C \\ CA \\ \vdots \\ CA^{n-1} \end{bmatrix}$ *has full column rank* $(= n)$.
OBSV4. *PBH rank test:* $\begin{bmatrix} A-\lambda I \\ C \end{bmatrix}$ *has full column rank* $\forall \lambda \in \mathbb{C}$.
OBSV5. *PBH eigenvector test:* $C\mathbf{v} \neq 0\ \forall$ *right eigenvectors* $\mathbf{v} \in \mathbb{C}^n$ *of* A.

The proofs of equivalence for these conditions follow very similarly to the controllability versions, and so we will not discuss them here.

Example 13.26 Consider the system

$$\begin{aligned} \dot{\mathbf{x}} &= \begin{bmatrix} -\frac{8}{45} & \frac{1}{30} \\ -\frac{1}{45} & -\frac{1}{10} \end{bmatrix} \mathbf{x} + \begin{bmatrix} 2 \\ 1 \end{bmatrix} \mathbf{u}, \quad \mathbf{x}_0 = \begin{bmatrix} 5 \\ 2 \end{bmatrix} \\ \mathbf{y} &= \begin{bmatrix} 3 & 4 \end{bmatrix} \mathbf{x} \end{aligned} \tag{13.4}$$

Let's use one of the conditions in Theorem 13.34 to determine whether this system is observable or not. Since

$$\text{rank} \begin{bmatrix} C \\ CA \end{bmatrix} = \text{rank} \begin{bmatrix} 3 & 4 \\ -\frac{28}{45} & -\frac{3}{10} \end{bmatrix} = 2 \tag{13.5}$$

has full column rank, the system is observable. □

Definition 13.90 (*Unobservable Subspace*) For LTI (A, C), the *unobservable subspace* is $\ker(\mathcal{O}(A, C))$, i.e. the nullspace of observability matrix.

Theorem 13.35 *LTI* (A, C) *is observable iff* $\ker(\mathcal{O}(A, C)) = \{\mathbf{0}\}$.

DT LTI Case. Observability for DT LTI systems follows similarly to the DT LTI conditions for controllability. Given a DT LTI system $\mathbf{x}[t+1] = A\mathbf{x}[t]$, $\mathbf{y}[t] = C\mathbf{x}[t]$, recursively iterate through time:

$$\left\{ \begin{array}{l} \mathbf{y}[0] = C\mathbf{x}_0 \\ \mathbf{y}[1] = C\mathbf{x}[1] = CA\mathbf{x}_0 \\ \mathbf{y}[2] = C\mathbf{x}[2] = CA^2\mathbf{x}_0 \\ \vdots \end{array} \right\} \Longrightarrow \begin{bmatrix} \mathbf{y}[0] \\ \mathbf{y}[1] \\ \vdots \\ \mathbf{y}[n-1] \end{bmatrix} = \underbrace{\begin{bmatrix} C \\ CA \\ \vdots \\ CA^{n-1} \end{bmatrix}}_{\triangleq \mathcal{O}} \mathbf{x}_0$$

Again, by Cayley-Hamilton, we do not need to consider powers A^k for $k \geq n$. The above system of equations has a unique solution $\mathbf{x}_0$ to any output measurement sequence $\{\mathbf{y}[0] \cdots \mathbf{y}[n-1]\}$ iff $\mathcal{O}$ is full column rank.

CT LTV Case. For $A \in \mathcal{C}^{N-1}$, $C \in \mathcal{C}^{N}$ for some $N \in \mathbb{N}$, define the sequence $\{L_i\}_{i=0}^{N}$ such that

$$\begin{cases} L_0(t) &= C(t) \\ L_i(t) &= L_{i-1}A(t) + \dot{L}_{i-1}(t) \end{cases} \tag{13.6}$$

Theorem 13.36 *LTV $(A(t), C(t))$ with $A \in \mathcal{C}^{N-1}$, $C \in \mathcal{C}^{N}$ for some $N \in \mathbb{N}$ is observable on $[t_0, t_f]$ if for some $\tau^* \in [t_0, t_f]$,*

$$rank \begin{bmatrix} L_0(\tau^*) \\ L_1(\tau^*) \\ \vdots \\ L_N(\tau^*) \end{bmatrix} = n \tag{13.7}$$

Chapter 14
Minimal Realizations

14.1 The Kalman Decomposition

In Sect. 12.1, the Kalman decomposition was posed as follows

$$C(\tilde{A}, \tilde{B}) \triangleq PC(A, B) = \begin{bmatrix} C_1(A, B) \\ 0_{(n-r)\times nm} \end{bmatrix},$$

$$\tilde{A} \triangleq PAP^{-1} = \begin{bmatrix} \tilde{A}_{11} & \tilde{A}_{12} \\ 0 & \tilde{A}_{22} \end{bmatrix}, \quad \tilde{B} \triangleq PB = \begin{bmatrix} \tilde{B}_1 \\ 0 \end{bmatrix}$$

Here, we'll look at how equivalence transformation P should be constructed such that the original system (A, B, C) can be transformed to the Kalman decomposition form. First, we will recall a few mathematical definitions from linear algebra.

Definition 14.91 (*Invariant Subspace*) Let $T : \mathcal{V} \to \mathcal{V}$ be a linear operator. Then subspace $\mathcal{W} \subseteq \mathcal{V}$ is a *T-invariant subspace* if $\mathbf{v} \in \mathcal{W} \Longrightarrow T\mathbf{v} \in \mathcal{W}$ (i.e. $T\mathcal{W} \subseteq \mathcal{W}$).

Remark 14.21 We've already introduced the notion of invariant sets while discussing LaSalle's principle in Chap. 7. These invariant sets are directly related to Definition 14.91.

Definition 14.92 (*Orthogonal Complement*) Let $\mathcal{W}$ be subspace of $\mathbb{R}^n$. Then $\mathcal{W}^\perp$ is the *orthogonal complement* of $\mathcal{W}$, defined by

$$\mathcal{W}^\perp \triangleq \left\{\mathbf{v} \in \mathbb{R}^n \mid \langle \mathbf{w}, \mathbf{v} \rangle = 0 \quad \forall \mathbf{w} \in \mathcal{W}\right\}, \quad \mathcal{W} \cap \mathcal{W}^\perp = \{0\}, \quad \mathcal{W} \oplus \mathcal{W}^\perp = \mathbb{R}^n.$$

Note that our linear operator is $A : \mathbb{R}^n \to \mathbb{R}^n$ (matrix case: $\mathcal{V} = \mathbb{R}^n$ in Definition 14.91).

Suppose $\mathcal{W} \subseteq \mathbb{R}^n$ with $\dim \mathcal{W} = r(< n)$ is A-invariant and $\mathcal{W}$ has basis $\{\mathbf{w}_1, \mathbf{w}_2, \dots \mathbf{w}_r\}, \mathbf{w}_i \in \mathcal{W}$. Choose subspace $\mathcal{U} = \mathcal{W}^\perp$ and denote its basis as $\{\mathbf{u}_{r+1}, \dots \mathbf{u}_n\}, \mathbf{u}_i \in \mathcal{U}$. Essentially, $\{\mathbf{u}_{r+1}, \dots \mathbf{u}_n\}$ completes $\{\mathbf{w}_1, \mathbf{w}_2, \dots \mathbf{w}_r\}$ to form basis of $\mathbb{R}^n$. We write the coordinate transformation P^{-1} as below.

SJ. Han, *Linear Systems*, KAIST Research Series,
https://doi.org/10.1007/978-981-96-9774-8_14

$$P^{-1} \triangleq \begin{bmatrix} | & & | & | & & | \\ \mathbf{w}_1 & \dots & \mathbf{w}_r & \mathbf{u}_{r+1} & \dots & \mathbf{u}_n \\ | & & | & | & & | \end{bmatrix} = \left[\mathcal{W} | \mathcal{U} \right].$$

Now, we can write equations for A as block matrices:

$$\begin{aligned} &\tilde{A} = PAP^{-1} \\ &\Longrightarrow P^{-1}\tilde{A} = AP^{-1} \\ &\Longrightarrow \left[\mathcal{W} | \mathcal{U} \right] \left[\begin{array}{c|c} \tilde{A}_{11} & \tilde{A}_{12} \\ \hline \tilde{A}_{21} & \tilde{A}_{22} \end{array} \right] = A \left[\mathcal{W} | \mathcal{U} \right] \\ &\Longrightarrow \left[\mathcal{W}\tilde{A}_{11} + \mathcal{U}\tilde{A}_{21} | \mathcal{W}\tilde{A}_{12} + \mathcal{U}\tilde{A}_{22} \right] = \left[A\mathcal{W} | A\mathcal{U} \right] \end{aligned}$$

where $\mathcal{W} \in \mathbb{R}^{n\times r}$, $\mathcal{U} \in \mathbb{R}^{n\times(n-r)}$, $\tilde{A}_{11} \in \mathbb{R}^{r\times r}$, $\tilde{A}_{12} \in \mathbb{R}^{r\times(n-r)}$, $\tilde{A}_{21} \in \mathbb{R}^{(n-r)\times r}$, $\tilde{A}_{22} \in \mathbb{R}^{(n-r)\times(n-r)}$.

Since every column $\mathbf{w}_i$ of $\mathcal{W}$ is A-invariant, we know $A\mathbf{w}_i$ can still be written as a linear combination of $\{\mathbf{w}_1, \dots \mathbf{w}_r\}$. This, combined with $\mathcal{W} \cap \mathcal{U} = \{0\}$ means $\tilde{A}_{21} = 0$. Thus, the "completely controllable" part of (A, B) is A-invariant subspace (Similarly, in equation $\tilde{B} = PB$, $\tilde{B}_2 = 0$.).

Note that if $\mathcal{U}$ also happens to be A-invariant, then we have $\tilde{A}_{12} = 0$ too.

Then, we can check the results of Kalman Decomposition with specific equivalence transformation P:

$$\begin{aligned} C(\tilde{A}, \tilde{B}) \triangleq PC(A, B) &= \left[PB \; PAB \; \dots \; PA^{n-1}B \right] \\ &= \left[\tilde{B} \; \tilde{A}\tilde{B} \; \dots \; \tilde{A}^{n-1}\tilde{B} \right] \\ &= \begin{bmatrix} \tilde{B}_1 & \tilde{A}_{11}\tilde{B}_1 & \dots & \tilde{A}_{11}^{n-1}\tilde{B}_1 \\ 0_{(n-r)\times m} & 0_{(n-r)\times m} & \dots & 0_{(n-r)\times m} \end{bmatrix} \\ &= \begin{bmatrix} C_1(A, B) \\ 0_{(n-r)\times nm} \end{bmatrix} \end{aligned}$$

where rank $C(A, B) = r$, $\tilde{B}_1 \in \mathbb{R}^{r\times m}$, $C_1(A, B) \in \mathbb{R}^{r\times nm}$.

14.2 Duality Principle

It is important to recognize the close connection between controllability and observability via one fundamental principle of control theory: the *duality principle*. This principle highlights the complementarity of controllability and observability and illustrates how the ability to control a system (controllability) corresponds to the ability to fully distinguish its internal dynamics from its output (observability). Here, we will look at the duality principle and its implications for systems analysis and design.

Definition 14.93 (*Dual System*) The *dual system* of linear $(A(t), B(t), C(t), D(t))$ is the linear system $(A^\top(t), C^\top(t), B^\top(t), D^\top(t))$. The dimension of the internal state is the same, but the control inputs/measurement outputs are switched.

Theorem 14.37 *LTI (A, B) is controllable (stabilizable) iff $(A^\top, B^\top)$ is observable (detectable).*

This can be proven by seeing $\mathcal{C}(A, B) = \mathcal{O}(A^\top, B^\top)$. Moreover: Im $\mathcal{C}(A, B) = \ker \mathcal{O}(A^\top, B^\top)^\perp$. Let us now finish our discussion on Kalman decomposition from the previous lecture.

1. So far: by choosing $\tilde{\mathbf{x}} \triangleq P_c\mathbf{x}(t)$, where P_c yields an equivalent system

$$\tilde{A} \triangleq P_c A P_c^{-1} = \begin{bmatrix} \tilde{A}_c & \tilde{A}_{12} \\ \mathbf{0} & \tilde{A}_{\bar{C}} \end{bmatrix}, \tilde{B} \triangleq P_c B = \begin{bmatrix} \tilde{B}_C \\ \mathbf{0} \end{bmatrix}, \tilde{C} \triangleq C P_c^{-1} = \begin{bmatrix} \tilde{C}_C & \tilde{C}_{\bar{C}} \end{bmatrix}, \tag{14.1}$$

   where $(\tilde{A}_C, \tilde{B}_C)$ form the controllable subspace (of dimension r) and subspace $(\tilde{A}_{\bar{C}}, \mathbf{0})$ is not affected at all by input u.
2. We can alternatively choose $\tilde{\mathbf{x}}(t) = P_0\mathbf{x}(t)$, where P_0 such that

$$\tilde{A} \triangleq P_0 A P_0^{-1} = \begin{bmatrix} \tilde{A}_0 & \mathbf{0} \\ \tilde{A}_{21} & \tilde{A}_{\bar{O}} \end{bmatrix}, \tilde{B} \triangleq P_0 B = \begin{bmatrix} \tilde{B}_O \\ \tilde{B}_{\bar{O}} \end{bmatrix}, \tilde{C} \triangleq C P_O^{-1} = \begin{bmatrix} \tilde{C}_O & \mathbf{0} \end{bmatrix}, \tag{14.2}$$

   where $(\tilde{A}_O, \tilde{C}_O)$ form the observable subspace.

To construct the controllable and unobservable subspaces, we will use the concept of invariant subspaces (see Definition 14.91). Consider Im $B \triangleq \{Bu | u \in \mathbb{R}^m\}$ and $\ker C \triangleq \{\mathbf{x} \in \mathbb{R}^n | C\mathbf{x} = 0\}$ and define

$$\underbrace{\mathcal{V}_C \triangleq \text{Im } B + A \text{ Im } B + \cdots + A^{n-1} \text{Im } B}_{\text{reachable subspace}}, \quad \underbrace{\mathcal{V}_{\bar{O}} \triangleq \bigcap_{i=0}^{n-1} \ker C * A^i}_{\text{unobservable subspace}} \tag{14.3}$$

Lemma 14.11 *$A\mathcal{V}_C \subseteq A\mathcal{V}_C$ and $A\mathcal{V}_{\bar{O}} \subseteq \mathcal{V}_{\bar{O}}$ ($\mathcal{V}_C$ and $\mathcal{V}_{\bar{O}}$ are both A-invariant)*

Proof We will progress through the following steps.

1. Any vector $\mathbf{v} \in \mathcal{V}_C$ can be expressed as $\mathbf{v} = B\mathbf{u}_0 + AB\mathbf{u}_1 + \cdots + A^{n-1}B\mathbf{u}_{n-1}$ for some $\mathbf{u}_i \in \mathbb{R}^m$.

$$\begin{aligned} \implies A\mathbf{v} &= AB\mathbf{u}_0 + A^2 B\mathbf{u}_1 + \cdots + A^n B\mathbf{u}_{n-1} \\ &= B\tilde{\mathbf{u}}_0 + AB\tilde{\mathbf{u}}_1 + \cdots + A^{n-1}B\tilde{\mathbf{u}}_{n-1} \text{ by Cayley-Hamilton theorem} \\ &\in \mathcal{V}_C \end{aligned} \tag{14.4}$$

   Note by using the Cayley-Hamilton theorem, $\tilde{\mathbf{u}}_i \in \mathbb{R}^m$, which is expressed in terms of $\mathbf{u}_i \in \mathbb{R}^m$ and the coefficients of $\mathcal{X}_A(\lambda)$.

2. Any vector $\mathbf{v} \in \mathcal{V}_{\bar{O}}$ satisfies $CA^i\mathbf{v} = \mathbf{0}\,\forall i = 0, 1, \ldots, n-1. \Longrightarrow A\mathbf{v}$ also satisfies $CA^i(A\mathbf{v}) = \mathbf{0}\,\forall i = 0, 1, \ldots, n-2$.
   When $i = n-1$, use Cayley-Hamilton theorem to show it is still equal to $\mathbf{0}$. Thus, $A\mathbf{v} \in \mathcal{V}_{\bar{O}}$. ■

This tells us both the reachable subspace and unobservable subspace are A-invariant.

Lemma 14.12 *The following relationships hold:*

$$A(\mathcal{V}_C \cap \mathcal{V}_{\bar{O}}),\quad A(\mathcal{V}_C + \mathcal{V}_{\bar{O}}) \subseteq \mathcal{V}_C + \mathcal{V}_{\bar{O}}$$

Decompose the state as:

$$\mathbb{R}^n = \mathcal{V}_1 \oplus \mathcal{V}_2 \oplus \mathcal{V}_3 \oplus \mathcal{V}_4,$$

where $\mathcal{V}_1 \triangleq \mathcal{V}_C \cap \mathcal{V}_{\bar{O}}$ and $\mathcal{V}_2, \mathcal{V}_3$ satisfy

1. $\mathcal{V}_2 \oplus \mathcal{V}_1 = \mathcal{V}_C$,
2. $\mathcal{V}_3 \oplus \mathcal{V}_1 = \mathcal{V}_{\bar{O}}$,
3. $\mathcal{V}_4$ completes the basis.

Note that in order for the dimensions to match, we require

$$n_i \triangleq \dim \mathcal{V}_i \text{ (i.e., } \sum_{i=1}^{4} n_i = n)$$

Create transformation $P^{-1} \triangleq [P_1 \,|\, P_2 \,|\, P_3 \,|\, P_4] \in \mathbb{R}^{n\times n}$, $P_i \in \mathbb{R}^{n\times n_i}$ and let $\tilde{\mathbf{x}} = P\mathbf{x}$. Then

$$\tilde{A} = \left[\begin{array}{cc|cc} A_{C\bar{O}} & A_{12} & A_{13} & A_{14} \\ \mathbf{0} & A_{CO} & \mathbf{0} & A_{24} \\ \hline \mathbf{0} & \mathbf{0} & A_{\bar{C}\bar{O}} & A_{34} \\ \mathbf{0} & \mathbf{0} & \mathbf{0} & A_{\bar{C}O} \end{array}\right],\ \tilde{B} = \left[\begin{array}{c} B_{C\bar{O}} \\ \hline B_{CO} \\ \hline \mathbf{0} \\ \hline \mathbf{0} \end{array}\right],\ \tilde{C} = \left[\mathbf{0}\,\middle|\,C_{CO}\,\middle|\,\mathbf{0}\,\middle|\,C_{\bar{C}O}\right] \tag{14.5}$$

We have essentially decomposed a general LTI system (A, B, C) into a form which readily gives us the controllable and observable parts:

- Subsystem (A_{CO}, B_{CO}, C_{CO}) is both controllable and observable
- Subsystem $\left(\begin{bmatrix} A_{C\bar{O}} & A_{12} \\ \mathbf{0} & A_{CO} \end{bmatrix}, \begin{bmatrix} B_{C\bar{O}} \\ B_{CO} \end{bmatrix}, \begin{bmatrix} \mathbf{0} & C_{CO} \end{bmatrix}\right)$ is controllable
- Subsystem $\left(\begin{bmatrix} A_{CO} & A_{24} \\ \mathbf{0} & A_{\bar{C}O} \end{bmatrix}, \begin{bmatrix} B_{CO} \\ \mathbf{0} \end{bmatrix}, \begin{bmatrix} C_{CO} & C_{\bar{C}O} \end{bmatrix}\right)$ is observable

You can also interpret the transformation P:

$$P^{-1} \triangleq [P_1 \,|\, P_2 \,|\, P_3 \,|\, P_4] \equiv \left[P_{C\bar{O}}\,\middle|\,P_{CO}\,\middle|\,P_{\bar{C}\bar{O}}\,\middle|\,P_{\bar{C}O}\right] \tag{14.6}$$

- $P_{C\bar{O}}$ columns from basis for states that are both reachable and observable
- P_{CO} chosen so that columns $\left[P_{C\bar{O}}\ P_{CO}\right]$ form basis for reachable subspace
- $P_{\bar{C}\bar{O}}$ chosen so that columns of $\left[P_{C\bar{O}}\ P_{\bar{C}\bar{O}}\right]$ form basis for unobservable subspace
- $P_{\bar{C}O}$ chosen to make $\left[P_{C\bar{O}}\,|\,P_{CO}\,|\,P_{\bar{C}\bar{O}}\,|\,P_{\bar{C}O}\right]$ invertible

Remark 14.22 Some submatrix dimensions may be zero! In fact, for systems that are both controllable and observable, $P^{-1} = P_{CO}$.

14.3 Controllability and Observability Properties in Realizations

How do controllability and observability properties appear in input-output descriptions of LTI systems? Recall a realization of proper, rational $H(s)$ is a collection of matrices $\{A, B, C, D\}$ such that $H(s) = C(sI - A)^{-1}B + D$.

Definition 14.94 (*Order of Transfer Function*) The *order* of a transfer function $H(s)$ is the highest exponent of $H(s)$. For proper and rational $H(s) = \frac{N(s)}{D(s)}$, the order is the degree of the denominator polynomial.

Example 14.27 Consider the following SISO example:

$$\dddot{y}(t) + 6\ddot{y}(t) + 11\dot{y}(t) + 6y(t) = \dot{u}(t) + 3\mathbf{u}(t)$$

$$\implies y(t) = \frac{s+3}{(s+1)(s+2)(s+3)} = \frac{1}{(s+1)(s+2)}$$

There is a pole-zero cancellation at $s = -3$. □

Theorem 14.38 *Let $H(s)$ be an input-output transfer function description of a system.*

- *If there are no pole-zero cancellations in $H(s)$, then the system is both controllable and observable.*
- *If there is a pole-zero cancellation in $H(s)$, then the system is either uncontrollable, unobservable, or both.*

Example 14.28 Suppose we are given that the CCF of a LTI system is

$$A = \begin{bmatrix} -6 & -11 & -6 \\ 1 & 0 & 0 \\ 0 & 1 & 0 \end{bmatrix}, \quad B = \begin{bmatrix} 1 \\ 0 \\ 0 \end{bmatrix}, \quad C = \begin{bmatrix} 0\ 1\ 3 \end{bmatrix}, \quad D = \begin{bmatrix} 0 \end{bmatrix}.$$

The controllability matrix $\mathcal{C}(A, B)$ is given by:

$$\mathcal{C}(A, B) = \begin{bmatrix} B & AB & A^2B \end{bmatrix} = \begin{bmatrix} 1 & -6 & 25 \\ 0 & 1 & -6 \\ 0 & 0 & 1 \end{bmatrix}$$

Hence $rank\ \mathcal{C} = 3$, thus the system is controllable.

The observability matrix $\mathcal{O}(A, C)$ is given by:

$$\mathcal{O}(A, C) = \begin{bmatrix} C \\ CA \\ CA^2 \end{bmatrix} = \begin{bmatrix} 0 & 1 & 3 \\ 1 & 3 & 0 \\ -3 & -11 & -6 \end{bmatrix} \sim \begin{bmatrix} 1 & 0 & -9 \\ 0 & 1 & 3 \\ 0 & 0 & 0 \end{bmatrix}$$

Hence $rank\ \mathcal{O} < 3$, thus the system is unobservable.

Now let's try the CCF of the reduced order version $\hat{H}(s) = \frac{1}{(s+1)(s+2)}$, and define the matrices for the reduced order state-space representation:

$$\hat{A} = \begin{bmatrix} -2 & -3 \\ 1 & 0 \end{bmatrix}, \ \hat{B} = \begin{bmatrix} 1 \\ 0 \end{bmatrix}, \ \hat{C} = \begin{bmatrix} 0 & 1 \end{bmatrix}, \ \hat{D} = D$$

Then the controllability and observability matrices for $(\hat{A}, \hat{B}, \hat{C}, \hat{D})$ are:

$$\mathcal{C}(\hat{A}, \hat{B}) = \begin{bmatrix} 1 & -2 \\ 0 & 1 \end{bmatrix}, \ \mathcal{O}(\hat{A}, \hat{C}) = \begin{bmatrix} 0 & 1 \\ 1 & 0 \end{bmatrix}$$

Both are full rank, which implies the reduced order system is controllable and observable. □

14.4 Minimal Realizations

We are now ready to define minimal realizations, and investigate its relationship with the Kalman decomposition.

Definition 14.95 (*Minimal Realization*) A realization of $H(s)$ is *minimal* if there is no other realization of $H(s)$ of a smaller order. Thus, minimal realizations don't have pole-zero cancellations.

Remark 14.23 For proper rational $H(s) = \frac{N(s)}{D(s)}$, this means the polynomials $N(s)$ and $D(s)$ are *coprime*. Recall that integers $a, b \in \mathbb{Z}$ are coprime if their greatest common divisor (GCD) is 1. In proper rational transfer functions $H(s) = \frac{N(s)}{D(s)}$, if there exist polynomials $\tilde{N}(s)$, $\tilde{D}(s)$ such that $\frac{N(s)}{D(s)} = \frac{\tilde{N}(s)}{\tilde{D}(s)}$ with deg $\tilde{D}(s) <$ deg $D(s)$, then $N(s)$ and $D(s)$ are not coprime. Later, in Chap. 17, we'll come back to the concept of coprime factorization.

The Kalman decomposition especially gives us the realization:

$$H(z) = \tilde{C}(sI - \tilde{A})^{-1}\tilde{B} + \tilde{D} = C_{co}(sI - A_{co})^{-1}B_{co} + D_{co}$$

where $(\tilde{A}, \tilde{B}, \tilde{C}, \tilde{D})$ is the Kalman decomposed version of the system (A, B, C, D) and $(A_{co}, B_{co}, C_{co}, D_{co})$ is the controllable and observable part. This can be shown by using the transformation P which created (14.5) and the zero-state equivalence property

$$D = \tilde{D} \quad \text{and} \quad CA^kB = \tilde{C}\tilde{A}^k\tilde{B} \quad \forall k \geq 0$$

Theorem 14.39 (Kalman 1965) *Realization (A, B, C, D) of proper, rational transfer matrix $H(s)$ is minimal if and only if it is both controllable and observable.*

The intuition here is that if (A, B) not controllable, the order can be reduced. Similarly, if (A, C) not observable, the order can be reduced.

Instead of proving this theorem directly, let's treat the two cases—controllability and observability—separately. We invoke the same notation used in Sect. 14.2.

Controllability from a Minimal Realization. If we choose $\tilde{\mathbf{x}}(t) = P_c\mathbf{x}(t)$ where P_c yields equivalent system, we have:

$$\tilde{A} = P_cAP_c^{-1}, \quad \tilde{B} = P_cB, \quad \tilde{C} = CP_c^{-1}$$

Then note that:

$$\tilde{C}(sI - \tilde{A})^{-1}\tilde{B} = \begin{bmatrix}\tilde{C}_c & \tilde{C}_{\bar{c}}\end{bmatrix}\begin{bmatrix} sI_r - \tilde{A}_c & \tilde{A}_{12} \\ 0 & sI_{n-r} - \tilde{A}_{\bar{c}}\end{bmatrix}^{-1}\begin{bmatrix}\tilde{B}_c \\ 0\end{bmatrix}$$

Inverse of 2×2 block triangular matrix is

$$\begin{bmatrix} A & B \\ 0 & D\end{bmatrix}^{-1} = \begin{bmatrix} A^{-1} & B^* \\ 0 & D^{-1}\end{bmatrix} \quad \text{where} \quad B^* = -A^{-1}BD^{-1}$$

and we have:

$$\tilde{C}(sI - \tilde{A})^{-1}\tilde{B} = \begin{bmatrix}\tilde{C}_c & \tilde{C}_{\bar{c}}\end{bmatrix}\begin{bmatrix} (sI_r - \tilde{A}_c)^{-1} & * \\ 0 & (sI_{n-r} - \tilde{A}_{\bar{c}})^{-1}\end{bmatrix}\begin{bmatrix}\tilde{B}_c \\ 0\end{bmatrix} = \tilde{C}_c(sI - \tilde{A}_c)^{-1}\tilde{B}_c$$

System order becomes reduced from n to rank $\mathbf{C}(A, B)$.

Observability from a Minimal Realization. Similarly, if we choose $\tilde{\mathbf{x}}(t) = P_o\mathbf{x}(t)$ where P_o yields equivalent system, then

$$\tilde{A} = P_o A P_o^{-1} = \begin{bmatrix} \tilde{A}_o & 0 \\ \tilde{A}_{21} & \tilde{A}_{\bar{o}} \end{bmatrix}, \quad \tilde{B} = P_o B = \begin{bmatrix} B_o \\ \tilde{B}_{\bar{o}} \end{bmatrix}, \quad \tilde{C} = C P_o^{-1} = \begin{bmatrix} \tilde{C}_o & 0 \end{bmatrix}$$

and

$$\tilde{C}(sI - \tilde{A})^{-1}\tilde{B} = \tilde{C}_o(sI - \tilde{A}_o)^{-1}\tilde{B}_o$$

System order becomes reduced from n to rank $\mathcal{O}(A, C)$.

Essentially, by combining the two pieces above, Theorem 14.39 suggests the combined joint reduction (a realization that is both controllable and observable) is possible.

Lemma 14.13 *The reachable subspace* $Im(\mathcal{C}(A, B))$ *is the smallest* A*-invariant subspace containing* B*, the column span of* B*.*

Proof Suppose $\exists\ S$ which is another A-invariant subspace such that:

$$B \subseteq S \subseteq \text{Im}(\mathcal{C}(A, B))$$

Apply A repeatedly across the entire relation:

$$\begin{aligned} AB &\subseteq AS \subseteq S \subseteq \text{Im}(\mathcal{C}(A, B)) \\ A^2B &\subseteq AS \subseteq S \subseteq \text{Im}(\mathcal{C}(A, B)) \\ &\vdots \\ A^{n-1}B &\subseteq AS \subseteq S \subseteq \text{Im}(\mathcal{C}(A, B)) \end{aligned}$$

But $\text{Im}(\mathcal{C}(A, B)) = \{B, AB, \ldots, A^{n-1}B\} \subseteq S$, hence $S = \text{Im}(\mathcal{C}(A, B))$. ■

Lemma 14.14 *Unobservable subspace* $Ker(\mathcal{O}(A, C))$ *is the largest* A*-invariant subspace contained in the kernel of* C*.*

An A-invariant subspace contained in the $\text{Ker}(C)$ follows similarly to above. The overall relationships between these null and range spaces is shown in Fig. 14.1.

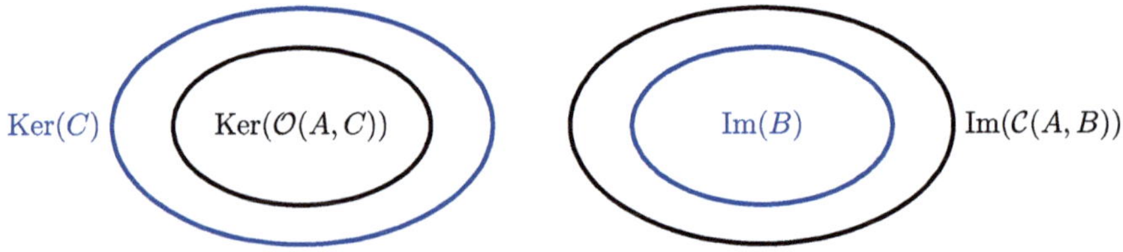

Fig. 14.1 The relationships among the reachable and unobservable subspaces, and the null and range spaces of C and B respectively. This visualizes the results we obtained in Lemmas 14.13 and 14.14

14.5 Gilbert Realization

A common type of minimal realization is the *Gilbert realization*. It's used only when $H(s)$ has distinct poles. (basically looks like MCF).

The transfer function $H(s)$ is decomposed via partial fraction decomposition:

$$H(s) = \frac{N_1}{s - \lambda_1} + \frac{N_2}{s - \lambda_2} + \cdots + \frac{N_r}{s - \lambda_r}$$

where $N_i \in \mathbb{R}^{k \times m}$.

Total size of realization is $\sum_{i=1}^{r} \operatorname{rank} N_i$. Find $B_i \in \mathbb{R}^{l_i \times m}$, $C_i \in \mathbb{R}^{k \times l_i}$ s.t. $C_i B_i = N_i$ for $i = 1, \ldots, r$ and $l_i = \operatorname{rank} N_i$. The form of A, B, C will be:

$$A = \begin{bmatrix} \lambda_1 I_{l_1} & & 0 \\ & \ddots & \\ 0 & & \lambda_r I_{l_r} \end{bmatrix}, \quad B = \begin{bmatrix} B_1 \\ \vdots \\ B_r \end{bmatrix}, \quad C = [C_1 \ldots C_r]$$

Chapter 15
Controller and Observer Design

15.1 Pole-Placement

In this section, let's consider the CT LTI system with system matrices (A, B, C) and initial condition $\mathbf{x}_0 \in \mathbb{R}^n$. That is:

$$\mathcal{H} \triangleq \begin{cases} \dot{\mathbf{x}}(t) = A\mathbf{x}(t) + B\mathbf{u}(t) \\ \mathbf{y}(t) = C\mathbf{x}(t) \\ \mathbf{x}(0) = \mathbf{x}_0. \end{cases}$$

We saw, in Chap. 3, that the explicit solution to this system can be expressed using the STM:

$$\mathbf{x}(t) = e^{At}\mathbf{x}_0 + \int_0^t e^{A(t-s)} B\mathbf{u}(s)ds$$

In particular, for open-loop system (with $\mathbf{u} \equiv 0$), the state $\mathbf{x}(t) = e^{At}\mathbf{x}_0$ will decay to zero asymptotically (i.e., exponentially, since they are equivalent for linear systems, as we saw in II) for any $\mathbf{x}_0$ if the eigenvalues of A all have negative real part. In the case where not all eigenvalues of A have negative real part (i.e., the system is not open-loop stable), the control input $\mathbf{u}$ is valuable to turn the system into a stable one. Starting from this chapter, we will see several approaches to control design.

15.1.1 State-Feedback Control

One of the simplest forms of control law that can be used to close the loop (i.e., obtain a *closed-loop response*) is the following constant-gain state-feedback law

SJ. Han, *Linear Systems*, KAIST Research Series,
https://doi.org/10.1007/978-981-96-9774-8_15

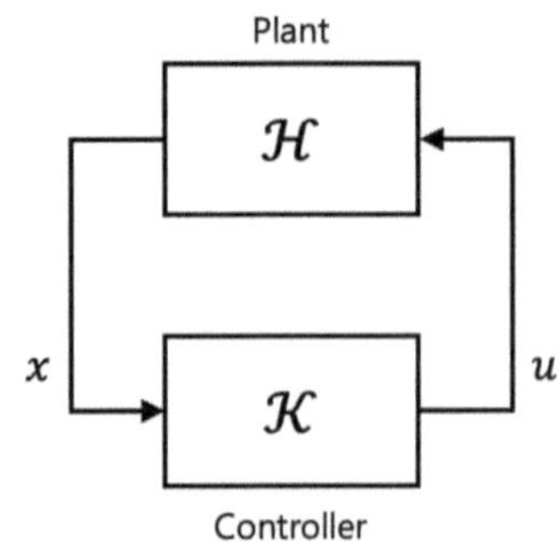

Fig. 15.1 A feedback configuration, where $\mathcal{H}$ describes the open-loop system and $\mathcal{K}$ describes the controller (i.e., $\mathbf{u} = \mathcal{K}\{\mathbf{y}\}$). In the case of constant-gain state-feedback, the $\mathcal{K}$ is a constant matrix multiplication operation (i.e., $\mathcal{K}\{\mathbf{x}\} = K\mathbf{x}$). Due to the loop-like structure of the system, it is also easy to see why the entire configuration is called "closed-loop"

$$\mathbf{u}(t) = -K\mathbf{x}(t), \tag{15.1}$$

where $K \in \mathbb{R}^{m\times n}$ is a *constant gain* matrix. We call this *state feedback* because the control input $\mathbf{u}$ is directly proportional to the state $\mathbf{x}$, and is being *fed back* into the LTI system. State-feedback is usually implemented when the full state is available for use (i.e., $C = I$ and $\mathbf{y} = \mathbf{x}$).

Figure 15.1 modifies the general feedback interconnection from Chap. 1 into a version with only the components and signals that are most relevant to our discussion so far. The closed-loop system dynamics are obtained by substituting (15.1) into the open-loop dynamics.

$$\dot{\mathbf{x}}(t) = (A - BK)\mathbf{x}(t), \quad \mathbf{y}(t) = \mathbf{x}(t), \quad \mathbf{x}(0) = \mathbf{x}_0$$

which is a fully autonomous system. Its system matrices are then given by $(A - BK, 0, I)$. Furthermore, the solution trajectory of the system is described by

$$\mathbf{x}(t) = e^{(A-BK)t}\mathbf{x}_0 \tag{15.2}$$

Similar to what we saw before, $\mathbf{x}(t)$ will decay asymptotically/exponentially to 0 if the eigenvalues of $A - BK$ have all negative real parts. This, of course, is dependent on the choice of K. Thus, by appropriately designing the gain K in the state-feedback law, we can convert a potentially unstable open-loop system A into a stable closed-loop system $A - BK$. This procedure is called the *eigenvalue placement problem.* As we've seen in Chap. 10, the eigenvalues of a system also correspond to its poles, and so the eigenvalue placement problem is often interchangeably called the *pole placement problem.*

Sometimes, even if A is stable, it may be desired to move its poles to different locations on the complex plane. This may be to achieve other properties (e.g., faster decay rate to the steady-state value). We see one such example as follows.

Example 15.29 (*Pole/Eigenvalue Placement*) Suppose we are given 2D LTI system with scalar control input,

$$\dot{\mathbf{x}}(t) = \begin{bmatrix} -15 & 8 \\ -15 & 7 \end{bmatrix} \mathbf{x}(t) + \begin{bmatrix} 1 \\ 1 \end{bmatrix} u(t).$$

Note that the eigenvalues of the A matrix, -5 and -3, are already stable in open-loop. Let's find a suitable state-feedback gain matrix K which replaces the closed-loop eigenvalues at $-10 \pm j$.

Since u is one dimensional and $\mathbf{x}$ is 2D, we have $K \in \mathbb{R}^{1\times 2}$. Let's specifically denote $K = [k_1, k_2]$. Then, the characteristic polynomial of the closed-loop system matrix is computed as $\det(\lambda I - (A - BK)) = \lambda^2 + (k_1 + k_2 + 8)\lambda + (k_1 + 15)$.

Now, we want this to equal $(\lambda + 10 + j)(\lambda + 10 - j)$. Comparing the coefficients, we have $k_1 + k_2 + 8 = 20$ and $k_1 + 15 = 101$. Therefore, we have the unique solution $k_1 = 86$, $k_2 = -74$, i.e., $K = [86, \ -74]$.

Substituting it back to obtain the closed-loop system $\dot{\mathbf{x}}(t) = (A - BK)\mathbf{x}(t)$, you can indeed verify that the eigenvalues are at $-10 \pm j$. □

Remark 15.24 Here, we used the feedback control law $\mathbf{u} = -K\mathbf{x}$. Actually, the sign convention of the state feedback law (i.e., $\mathbf{u}(t) = -K\mathbf{x}(t)$ versus $\mathbf{u}(t) = K\mathbf{x}(t)$) varies by reference. In the case the feedback law is given by $\mathbf{u}(t) = K\mathbf{x}(t)$, the closed-loop system matrix is given by $A + BK$ instead of $A - BK$. Regardless of the convention, the theory behind pole or eigenvalue placement remains the same, and in future subsections, we will often use $\mathbf{u} = \pm K\mathbf{x}$ interchangeably. But it is still important to be careful of which convention you are using in your calculations, and to make sure to stick to it.

15.1.2 Ackermann's Formula

Ackermann's formula gives us a more principled approach for performing pole-placement, especially when the control input is scalar-valued ($u(t) \in \mathbb{R}$ for all t). Assume $\mathcal{H}$ is controllable, with controllability matrix $C(A, B)$. Choose $u(t) = k^\top x(t)$ where $k \triangleq \begin{bmatrix} k_1 & k_2 & \dots & k_n \end{bmatrix}$ so that $A_{\text{cl}} = A + Bk^\top$. Write its characteristic polynomial as follows:

$$\begin{aligned} &\chi_{A_{\text{cl}}}(\lambda) \triangleq \lambda^n + c_{n-1}\lambda^{n-1} + \cdots + c_l\lambda + c_0 \\ \Longrightarrow\ &\chi_{A_{\text{cl}}}(A_{\text{cl}}) = A_{\text{cl}}^n + c_{n-1}A_{\text{cl}}^{n-1} + \cdots + c_1A_{\text{cl}} + c_0 I = 0 \text{ by Cayley-Hamilton} \end{aligned} \tag{15.3}$$

Compute powers of A_{cl} in terms of A, B, k:

$$A_{\mathrm{cl}}^2 = (A + Bk^\top)^2 = A^2 + ABk^\top + \underbrace{Bk^\top A + (Bk^\top)^2}_{=Bk^\top A_{\mathrm{cl}}}$$

$$\begin{aligned} A_{\mathrm{cl}}^3 &= (A^2 + ABk^\top + Bk^\top A_{\mathrm{cl}})(A + Bk^\top) \\ &= A^3 + A^2Bk^\top + \underbrace{ABk^\top A + A(Bk^\top)^2}_{=ABk^\top A_{\mathrm{cl}}} + Bk^\top A_{\mathrm{cl}}^2 \\ &\vdots \\ A_{\mathrm{cl}}^n &= A^n + A^{n-1}Bk^\top + A^{n-2}Bk^\top A_{\mathrm{cl}} + \cdots + ABk^\top A_{\mathrm{cl}}^{n-2} + Bk^\top A_{\mathrm{cl}}^{n-1} \end{aligned}$$

Substituting into (15.3) and simplifying yields

$$\begin{aligned} \chi_{A_{\mathrm{cl}}}(A_{c1}) = \chi_{A_{\mathrm{cl}}}(A) \\ &+ (A^{n-1}Bk^\top + A^{n-2}Bk^\top A_{\mathrm{cl}} + \cdots + ABk^\top A_{\mathrm{cl}}^{n-2} + Bk^\top A_{\mathrm{cl}}^{n-1}) \\ &+ \cdots \\ &+ C_2(ABk^\top + Bk^\top A_{\mathrm{cl}}) + C_1 Bk^\top \triangleq 0 \end{aligned}$$

$$\implies \chi_{A_{\mathrm{cl}}}(A) + \underbrace{\begin{bmatrix} B & AB & \ldots & A^{n-1}B \end{bmatrix}}_{C(A,B)} \begin{bmatrix} * \\ * \\ \vdots \\ k^\top \end{bmatrix} = 0$$

$$\therefore \begin{bmatrix} * \\ * \\ \vdots \\ k^\top \end{bmatrix} = -C(A, B)^{-1}\chi_{A_{\mathrm{cl}}}(A).$$

where the terms that are irrelevant to our calculations are marked by $*$.

To solve for $k^\top$, multiply $\begin{bmatrix} 0 & 0 & \ldots & 0 & 1 \end{bmatrix}$ to both sides:

$$k^\top = -\begin{bmatrix} 0 & 0 & \ldots & 0 & 1 \end{bmatrix} C(A, B)^{-1}\chi_{A_{\mathrm{cl}}}(A). \tag{15.4}$$

While we presented Ackermann's formula for scalar u only, a similar argument can be made for vector-valued $\mathbf{u}(t) \in \mathbb{R}^m$, $m \geq 2$.

15.2 General Feedback Interconnection System

More generic feedback control systems consist of system $\mathcal{H}$ being *feedback-interconnected* with another system $\mathcal{K}$; we've actually seen this before in I (and also Fig. 15.1). In the previous section, we assumed $\mathcal{K}$ to be a constant-gain system, in which $\mathcal{K}\{\mathbf{x}\}$ simply multiplies a constant gain K to the input signal $\mathbf{x}$. More gen-

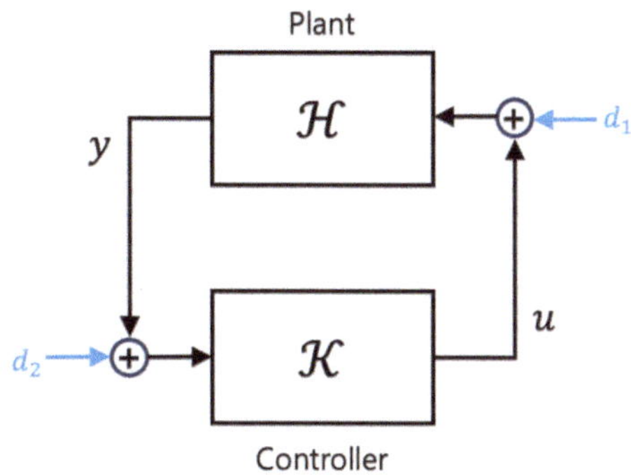

Fig. 15.2 A more general configuration of Fig. 15.1 with output-feedback (note the $\mathbf{y}$ instead of $\mathbf{x}$), and external disturbances $\mathbf{d}_1$ and $\mathbf{d}_2$

erally, the controller $\mathcal{K}$ could have its own system dynamics with its own *internal state* $\boldsymbol{\xi}(t)$):

$$\mathcal{K} \triangleq \begin{cases} \dot{\boldsymbol{\xi}}(t) = A_k \boldsymbol{\xi}(t) + B_k y(t) \\ u(t) = C_k \boldsymbol{\xi}(t) + D_k y(t) \end{cases} \quad (A_k, B_k, C_k) \text{ stabilizable, detectable} \tag{15.5}$$

In fact, note that the constant-gain state-feedback controller $u = -K\mathbf{x}(t)$ is equivalent to (15.5) when $A_k = 0$, $B_k = 0$, $C_k = 0$, $D_k = -K$ with $\mathbf{y}(t) = \mathbf{x}(t)$ (i.e., $C = I$). A more general feedback interconnection with $\mathcal{K}$ of the form (15.5) is shown in Fig. 15.2, including some disturbances.

In II, we saw that internal stability of a system is concerned with the stability of its *internal state* $\mathbf{x}(t)$. And for uncontrolled systems $\dot{\mathbf{x}}(t) = A\mathbf{x}(t)$ or $\dot{\mathbf{x}}(t) = A(t)\mathbf{x}(t)$, we characterized internal stability using Lyapunov, asymptotic, exponential, uniform, etc. stability notions. One question is: how should we characterize the stability of more general feedback interconnected systems like Fig. 15.2?

One way (loosely-speaking) is to say that $(\mathcal{H}, \mathcal{K})$ is *internally stable* if for all initial states $\mathbf{x}_0$, $\boldsymbol{\xi}_0$, and all bounded disturbance signals $\mathbf{d}_1$ and $\mathbf{d}_2$, all states and other signals (i.e. $\mathbf{x}$, $\boldsymbol{\xi}$, $\mathbf{y}$, $\mathbf{u}$) remain bounded $\forall t$. We will characterize the stability of general feedback interconnected systems in more detail in the following IV. For the rest of this chapter, however, we will focus on how the observability plays a role in the design of feedback control laws, since observer-based control design can be viewed as one type of system which abides by the configuration of Fig. 15.2.

15.3 State Observers (State Estimators)

In many practical control problems, we cannot directly observe the true internal state $\mathbf{x}(t)$ of a system, so a state-feedback law $\mathbf{u}(t) = \mathcal{K}\{\mathbf{x}\}(t)$ cannot be implemented. We thus require an additional component to develop an *estimate* $\hat{\mathbf{x}}(t)$ of the internal state $\mathbf{x}(t)$ given inputs $\{\mathbf{u}(s) : s \in [0, t]\}$ and measurements $\{\mathbf{y}(s) : s \in [0, t]\}$ (which we can observe). This additional component is called a *state observer*. We especially wish to design an observer such that asymptotic state observation is satisfied i.e.:

$$\lim_{t\to\infty} \|\mathbf{x}(t) - \hat{\mathbf{x}}(t)\| = 0. \tag{15.6}$$

There are two main types of state observers in the literature which satisfy (15.6):

1. *Full-order Observer*: observes all state variables in the system, both variables available for direct measurement (via $\mathbf{y} = C\mathbf{x}$) and variables which are *hidden*.
2. *Reduced-order Observer*: observes only state variables which are *hidden*.

Here, hidden state variables are components x_i of the state $\mathbf{x}$ that cannot be directly measured using the measurement equation $\mathbf{y} = C\mathbf{x}$.

15.3.1 Full-Order Observers

It is simpler to design a full-order observer rather than a reduced-order one, and so we will begin our discussion from here. The typical approach to satisfy (15.6) is to design another linear dynamics with internal state $\hat{\mathbf{x}}$:

$$\dot{\hat{\mathbf{x}}}(t) = \underbrace{A\hat{\mathbf{x}}(t) + B\mathbf{u}(t)}_{\text{linear dynamics term}} + \underbrace{L(\mathbf{y}(t) - \hat{\mathbf{y}}(t))}_{\text{correction term}}, \qquad \hat{\mathbf{y}}(t) = C\hat{\mathbf{x}}(t) \tag{15.7}$$

The observer (15.7) is composed of two main parts:

- the *linear dynamics term* propagates the observer's internal state $\hat{\mathbf{x}}$ forward in time.
- the *correction term* accounts for possible measurement errors by comparing the true measurement $\mathbf{y}$ against the *estimated measurement* $\hat{\mathbf{y}}$, which is created by substituting the state estimate $\hat{\mathbf{x}}$ into the measurement equation $\mathbf{y} = C\mathbf{x}$.

The *observer gain* $L \in \mathbb{R}^{n\times p}$ is another feedback gain matrix to be designed by the user. Its role is similar to the gain K we've seen in pole placement problems from Sect. 15.1.

Define $\mathbf{e}(t) \triangleq \mathbf{x}(t) - \hat{\mathbf{x}}(t)$ to be the *state error*. Then the observer's *error dynamics* is given by:

$$\begin{aligned}
\dot{\mathbf{e}}(t) &= \dot{\mathbf{x}}(t) - \dot{\hat{\mathbf{x}}}(t) \\
&= A\mathbf{x}(t) + B\mathbf{u}(t) - A\hat{\mathbf{x}}(t) - B\mathbf{u}(t) + L(C\mathbf{x}(t) - C\hat{\mathbf{x}}(t)) \\
&= (A - LC)\mathbf{e}(t)
\end{aligned} \tag{15.8}$$

This is precisely the design for a *full-order observer*. A block diagram representation of the observer, together with the plant, is shown in Fig. 15.3.

Remark 15.25 Note that this structure resembles the general feedback interconnection shown in Fig. 15.2. Here, $\mathcal{H}$ is the original LTI system with matrices (A, B, C),

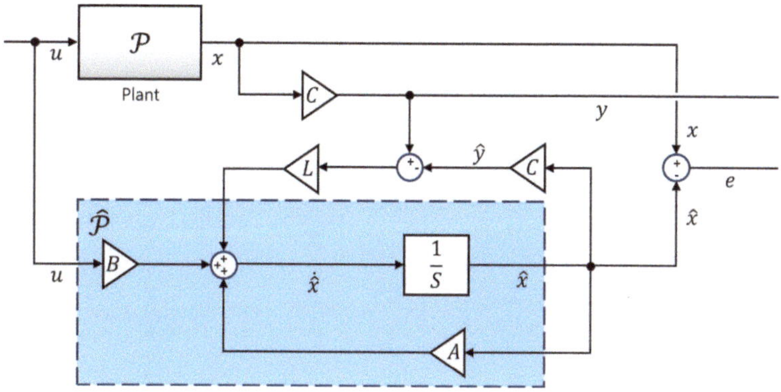

Fig. 15.3 The full-order observer, attached to the plant in block-diagram form

while $\mathcal{K}$ is precisely the observer we designed, (15.7), with internal state $\boldsymbol{\xi} \equiv \hat{\mathbf{x}}$ and system matrices $(A - LC, B, C)$.

Lemma 15.15 *An LTI observer exists if and only if (A, C) is detectable.*

We won't prove this lemma, but the intuition is simple. Note that the LTI version of the observer error dynamics (15.8) is:

$$\dot{\mathbf{e}}(t) = (A - LC)\mathbf{e}(t),$$

which is recognized from the definition of detectability (see Definition 13.89).

Choosing the constant gain $L \in \mathbb{R}^{n\times p}$ to place the eigenvalues of $A - LC$ on the open left-half plane allows for asymptotic stability of the error dynamics. This is analogous to the pole placement problem we saw in Sect. 15.1, where we chose controller gain $K \in \mathbb{R}^{n\times m}$ to ensure that $A - BK$ is Hurwitz. In fact, to design the gain for the observer, we can use the duality principle (see Sect. 14.2) and apply the same pole placement technique on the following *dual system*:

$$\begin{cases} \dot{\tilde{\mathbf{x}}}(t) = A^\top \tilde{\mathbf{x}}(t) + C^\top \tilde{\mathbf{u}}(t) \\ \tilde{\mathbf{y}}(t) = B^\top \tilde{\mathbf{x}}(t) \\ \tilde{\mathbf{u}}(t) \triangleq -L\tilde{\mathbf{x}}(t) \end{cases}$$

where all the system matrices are transposed due to the duality principle.

Remark 15.26 In the case where the measurement is scalar ($y(t) \in \mathbb{R}$), Ackermann's formula (Sect. 15.1) can also be performed to design the poles for an observer, where the gain L is obtained from

$$L^\top = -\begin{bmatrix} 0 \; 0 \ldots 0 \; 1 \end{bmatrix} \mathcal{O}(A, C)^{-\top} \chi_{A_{\text{cl}}}(A^\top). \tag{15.9}$$

Note the resemblance between (15.4) and (15.9).

15.3.2 Reduced-Order Observers

Now, suppose that only the first $k < n$ components of the state $\mathbf{x}(t)$ can be measured. How do we design an observer in this case?

First, partition $\mathbf{x}(t) = [\mathbf{x}_m^\top, \mathbf{x}_{\bar{m}}^\top]^\top$, where subscript m denotes the components which can be measured, and the complement $\bar{m}$ denotes the components which are hidden.

$$\begin{bmatrix} \mathbf{x}_m(t) \\ \mathbf{x}_{\bar{m}}(t) \end{bmatrix} = \begin{bmatrix} A_{11} & A_{12} \\ A_{21} & A_{22} \end{bmatrix} \begin{bmatrix} \mathbf{x}_m(t) \\ \mathbf{x}_{\bar{m}}(t) \end{bmatrix} + \begin{bmatrix} B_1 \\ B_2 \end{bmatrix} \mathbf{u}(t), \qquad \mathbf{y}(t) = \begin{bmatrix} C_m & C_{\bar{m}} \end{bmatrix} \begin{bmatrix} \mathbf{x}_m(t) \\ \mathbf{x}_{\bar{m}}(t) \end{bmatrix}$$

Since k components can be measured, $C \in \mathbb{R}^{k \times n}$ is full row rank. We arrange the C matrix to be in the form $[C_m \; C_{\bar{m}}]$ so that $C_m^{-1} \in \mathbb{R}^{k \times k}$ exists. Then we can immediately compute

$$\mathbf{x}_m(t) = C_m^{-1}\mathbf{y}(t) \Longrightarrow \hat{\mathbf{x}}_m(t) = \mathbf{x}_m(t)$$

Remark 15.27 If the measured states are not ordered, we can apply equivalence transformation to the state equation. i.e. we can choose equivalence transformation P such that $\tilde{C} = CP = \begin{bmatrix} I & 0 \end{bmatrix}$.

In order to observe the hidden (unmeasured) states $\mathbf{x}_{\bar{m}}$, consider designing a $\mathbf{z}(t)$ such that $\hat{\mathbf{x}}_{\bar{m}}(t) = L\mathbf{y}(t) + \mathbf{z}(t)$. Then we can rewrite $\mathbf{z}(t)$ as follows:

$$\mathbf{z} = \hat{\mathbf{x}}_{\bar{m}} - L\mathbf{y} = \hat{\mathbf{x}}_{\bar{m}} - LC_m\mathbf{x}_m$$

Then, $\dot{\mathbf{z}}(t)$ is given as follows:

$$\begin{aligned}
\dot{\mathbf{z}} &= \dot{\hat{\mathbf{x}}}_{\bar{m}} - LC_m\dot{\mathbf{x}}_m \\
&= A_{21}\mathbf{x}_m + A_{22}\hat{\mathbf{x}}_{\bar{m}} + B_2\mathbf{u} - LC_mA_{11}\mathbf{x}_m - LC_mA_{12}\hat{\mathbf{x}}_{\bar{m}} - LC_mB_1\mathbf{u} \\
&= (A_{22} - LC_mA_{12})\hat{\mathbf{x}}_{\bar{m}} + (A_{21} - LC_mA_{11})\mathbf{x}_m + (B_2 - LC_mB_1)\mathbf{u} \\
&= (A_{22} - LC_mA_{12})(L\mathbf{y} + \mathbf{z}) + (A_{21} - LC_mA_{11})C_m^{-1}\mathbf{y} + (B_2 - LC_mB_1)\mathbf{u} \\
&= (A_{22} - LC_mA_{12})\mathbf{z} + \left[(A_{21} - LC_mA_{11})C_m^{-1} + (A_{22} - LC_mA_{12})L\right]\mathbf{y} + (B_2 - LC_mB_1)\mathbf{u} \\
&= \hat{A}\mathbf{z} + G\mathbf{y} + H\mathbf{u}
\end{aligned}$$

where

$$\hat{A} \triangleq A_{22} - LC_mA_{12}, \quad G \triangleq (A_{21} - LC_mA_{11})C_m^{-1} + \hat{A}L, \quad H \triangleq B_2 - LC_mB_1$$

We need to choose $\hat{A}$, G, H, L to ensure (15.6) is satisfied.

1. We already have $\mathbf{e}_m(t) = \mathbf{x}_m(t) - \hat{\mathbf{x}}_m(t) = 0$.
2. If we choose $\mathbf{e}_{\bar{m}}(t) = \mathbf{x}_{\bar{m}}(t) - \hat{\mathbf{x}}_{\bar{m}}(t)$, then we can get $\dot{\mathbf{e}}_{\bar{m}}(t) = \hat{A}\mathbf{e}_{\bar{m}}$ because

$$\begin{aligned}
\dot{\mathbf{e}}_{\bar{m}} &= \dot{\mathbf{x}}_{\bar{m}} - \dot{\hat{\mathbf{x}}}_{\bar{m}} \\
&= (A_{21}\mathbf{x}_m + A_{22}\mathbf{x}_{\bar{m}} + B_2\mathbf{u}) - (\dot{\mathbf{z}} + LC_m\dot{\mathbf{x}}_m) \\
&= (A_{21}\mathbf{x}_m + A_{22}\bar{\mathbf{x}}_m + B_2\mathbf{u}) - \hat{A}z - G\mathbf{y} - H\mathbf{u} - LC_m(A_{11}\mathbf{x}_m + A_{12}\mathbf{x}_{\bar{m}} + B_1\mathbf{u}) \\
&= (A_{21} - LC_mA_{11})\mathbf{x}_m + \underbrace{(A_{22} - LC_mA_{12})}_{\hat{A}}\mathbf{x}_{\bar{m}} + \underbrace{(B_2 - LC_mB_1)}_{H}\mathbf{u} - H\mathbf{u} - \hat{A}z \\
&\quad - (A_{21} - LC_mA_{11})\underbrace{C_m^{-1}\mathbf{y}}_{\mathbf{x}_m} - \underbrace{(A_{22} - LC_mA_{12})}_{\hat{A}}L\mathbf{y} \\
&= \hat{A}\mathbf{x}_{\bar{m}} - \hat{A}z - \hat{A}L\mathbf{y} = \hat{A}x_{\bar{m}} - \hat{A}(\hat{\mathbf{x}}_{\bar{m}} - L\mathbf{y}) - \hat{A}L\mathbf{y} \\
&= \hat{A}(\mathbf{x}_{\bar{m}} - \hat{\mathbf{x}}_{\bar{m}}) = \hat{A}\mathbf{e}_{\bar{m}}.
\end{aligned}$$

Thus, to create this *reduced-order observer*, we use pole placement to find L that makes $\hat{A}$ Hurwitz.

Note that both full-order observers and reduced-order observers can be constructed for systems even if not all state components x_i are measured. Which one you choose to implement in order to estimate your state is dependent upon the application you are using it for. Although the reduced-order observer might require more complex design techniques, it is more efficient than full-order observers as it estimates only hidden states, reducing computation load.

15.4 Observer-Based Controllers

As before, a state-feedback control law $\mathbf{u}(t) = \pm K\mathbf{x}(t)$ relies on knowing $\mathbf{x}(t)$ precisely, but in many practical scenarios, this might not be the case. This motivates the construction of *observer-based controllers*, where observers (discussed in the previous section) are first used to construct $\hat{\mathbf{x}}$, then used in the state-feedback control law. We are effectively combining everything together to create a observer and controller architecture for general linear systems.

Theorem 15.40 *Given LTI system*

$$\mathcal{H} \triangleq \begin{cases} \dot{\mathbf{x}}(t) = A\mathbf{x}(t) + B\mathbf{u}(t) \\ \mathbf{y}(t) = C\mathbf{x}(t) + D\mathbf{u}(t) \\ \mathbf{x}(0) = \mathbf{x}_0 \end{cases},$$

the controller

$$\mathcal{K} \triangleq \begin{cases} \dot{\hat{\mathbf{x}}}(t) = (A + LC + BK)\hat{\mathbf{x}}(t) - L\mathbf{y}(t) \\ \mathbf{u}(t) = K\hat{\mathbf{x}}(t) \end{cases}$$

(with $\hat{\mathbf{x}} = \boldsymbol{\xi}$ *as its internal state) yields a closed-loop system*

$$\begin{bmatrix} \dot{\mathbf{x}} \\ \dot{\hat{\mathbf{x}}} \end{bmatrix} = A_{cl} \begin{bmatrix} \mathbf{x} \\ \hat{\mathbf{x}} \end{bmatrix}$$

where A_{cl}*, the closed-loop system matrix, has eigenvalues the same as* $A + BK$ *and* $A + LC$*.*

Proof By substituting in the given equations, our closed-loop system is

$$\begin{bmatrix} \dot{\mathbf{x}} \\ \dot{\hat{\mathbf{x}}} \end{bmatrix} = \underbrace{\begin{bmatrix} A & BK \\ -LC & A + BK + LC \end{bmatrix}}_{\triangleq A_{\text{cl}}} \begin{bmatrix} \mathbf{x} \\ \hat{\mathbf{x}} \end{bmatrix}$$

Note that (as per Remark 15.24), we are now using the control law $\mathbf{u} = K\mathbf{x}$ instead of with a $-$ sign, which is why our closed-loop matrix is $A + BK$ rather than $A - BK$.

Use transformation $P = \begin{bmatrix} I & -I \\ 0 & I \end{bmatrix}$ to get equivalent system:

$$PA_{\text{cl}}P^{-1} = \begin{bmatrix} A + LC & 0 \\ -LC & A + BK \end{bmatrix}$$

which is clearly a block-diagonal matrix. Because the eigenvalues of a block-diagonal matrix are the combined eigenvalues of each of the matrices along the block-diagonal, we have that the eigenvalues of A_{cl} are eigenvalues of $A + LC$ and $A + BK$. ■

Remark 15.28 There are further important implications of Theorem 15.40. First, it tells us that controllability and observability of (A, B, C) suffice to be able to place closed-loop eigenvalues anywhere. Second, stabilizability and detectability of (A, B, C) suffice to be able to obtain an *internally stabilizing controller*, which we will discuss more about in Chap. 17.

Theorem 15.40 also gives us an important result in regards to choosing the poles for the observer and the controller.

Proposition 15.4 (Separation Principle) *The state-feedback controller and state observer can be designed separately. This is because the eigenvalues of* $A + BK$ *and* $A + LC$ *are separate (i.e., they lie on independent block matrices along the diagonal of the closed-loop system matrix* A_{cl}*)*

Remark 15.29 (*Rules of Thumb for Designing Estimator Poles*) In control systems design, selecting appropriate poles for observers is crucial for achieving desirable system dynamics. Estimator poles are typically selected to be 2 to 6 times faster than controller poles to ensure quicker decay of estimation errors and maintain control dynamics dominance in the system response. However, making the observer respond faster, which means more noise from sensors can interfere with the actuators. So, if there is a lot of sensor noise, it may be better to set the observer poles to be slower than twice the speed of the controller poles. This slows down the system's response but improves its ability to smooth out noise. In either case, the overall system behavior is more influenced by the observer than the controller.

Additional references on controllability, observability, and reachability are [1, 2].

References

1. John M Davis et al. "Controllability, observability, realizability, and stability of dynamic linear systems". In: *arXiv preprint* arXiv:0901.3764 (2009).
2. Eduardo D. Sontag. "Reachability and Controllability". In: *Mathematical Control Theory: Deterministic Finite Dimensional Systems*. New York, NY: Springer US, 1990, pp. 79 129.

Chapter 16
Problems and Exercises

System Realizations

Problem 1. Represent the following input-output transfer functions in (1) controllable canonical form, (2) observable canonical form, and (3) modal canonical form. You are welcome to use MATLAB (or some other program) to solve any linear matrix equations.

$$\text{(a) } H(s) = \frac{s^4 + 4s^3 + 6s^2 + 8s + 10}{(s+1)^2(s+2)^2(s+3)(s+4)}$$

$$\text{(b) } H(s) = \begin{bmatrix} \dfrac{s^3 + 3s^2 + 3s + 1}{s^3 + 6s^2 + 11s + 6} & \dfrac{s-1}{s^3 + 6s^2 + 11s + 6} \\ \dfrac{s+4}{s^3 + 6s^2 + 11s + 6} & \dfrac{s^2 + 3s + 2}{s^3 + 6s^2 + 11s + 6} \end{bmatrix}$$

Problem 2: Modal Canonical Form. Recall we discussed modal canonical form (MCF) for distinct poles and for nonsimple, repeating poles. In this problem, we will consider what happens in other cases.

(a) Consider a transfer function which, after partial fraction decomposition, yields a pair of complex conjugate poles:

$$H(s) = \frac{2}{s + (1 - j)} + \frac{2}{s + (1 + j)} + \frac{3}{s + 2} + \frac{5}{s + 3}$$

How would you represent this system in MCF? Sketch the corresponding block-diagram of the system.

(b) Suppose a strictly proper transfer function with denominator polynomial of degree n has one *simple* repeating pole λ_k with multiplicity n_k, $1 < n_k < n$. How would you represent this system in MCF?

Problem 3: LTV Equivalence Transformations. In this chapter, we discussed equivalence transformations for linear *time-invariant* (LTI) systems. In the analogous

SJ. Han, *Linear Systems*, KAIST Research Series,
https://doi.org/10.1007/978-981-96-9774-8_16

case for linear *time-varying* (LTV) systems, our constant coordinate transformation $P \in \mathbb{R}^n$, now becomes a time-varying matrix $P(t) \in \mathbb{R}^{n \times n}$ which is continuously-differentiable ($\mathcal{C}^1$) and nonsingular for all t.

We can define equivalence in a similar way for LTV systems too. Given *equivalence transformation* $P(t) \in \mathbb{R}^{n \times n}$ which is $\mathcal{C}^1$ and nonsingular for all t, the LTV system $\{A(t), B(t), C(t), D(t)\}$ is *(algebraically) equivalent* to $\{\tilde{A}(t), \tilde{B}(t), \tilde{C}(t), \tilde{D}(t)\}$, where

$$\tilde{A}(t) = (\dot{P}(t) + P(t)A(t))P^{-1}(t), \quad \tilde{B}(t) = P(t)B(t),$$
$$\tilde{C}(t) = C(t)P^{-1}(t), \quad \tilde{D}(t) = D(t)$$

Prove that there exists an equivalence transformation $P(t)$ which transforms $A(t)$ to a constant matrix $A_0 \in \mathbb{R}^{n \times n}$.
(*Hint:* Consider a transformation of the form $P(t) \triangleq e^{A_0 t}\Psi^{-1}(t)$, where $\Psi(t)$ is a fundamental matrix to the LTV system $\dot{\mathbf{x}} = A(t)\mathbf{x}$.)

Minimum-Energy Input

Problem 4. Consider the system

$$\dot{\mathbf{x}}(t) = \begin{bmatrix} 0 & 1 \\ -1 & 0 \end{bmatrix} \mathbf{x}(t) + \begin{bmatrix} 1 \\ 0 \end{bmatrix} u(t)$$

with initial condition $\mathbf{x}_0 \triangleq (1, 0)^\top$.

(a) Show that this system is controllable.
(b) Compute the minimum energy input which transfers the system from $\mathbf{x}_0$ to 0 in t_f timesteps.
(c) Compute the minimum energy input $u^*(t)$ which transfers the system from $\mathbf{x}_0$ to $\mathbf{x}_f = (2, 3)^\top$ in t_f timesteps.
(d) For both parts (b) and (c), plot the resulting optimal trajectories for each of the values $t_f \in \{\pi, 10, 10^{-3}\}$.
(e) For both parts (b) and (c), plot $u^*(t)$ as a function of t. Interpret your plot.

Controllability, Reachability, and Observability

Problem 5. This problem aims to derive equivalent conditions for *output controllability* of linear systems. First, the linear (LTI or LTV) system $(A(t), B(t), C(t))$, given by

$$\mathbb{R}^n \ni \dot{\mathbf{x}}(t) = A(t)\mathbf{x}(t) + B(t)\mathbf{u}(t), \mathbf{x}(t_0) = \mathbf{x}_0, \qquad \mathbb{R}^k \ni \mathbf{y}(t) = C(t)\mathbf{x}(t)$$

is called *output controllable* on $[t_0, t_f]$ if for any $\mathbf{x}_0$ (and corresponding $\mathbf{y}_0 \triangleq C\mathbf{x}_0$), there exists a $\mathbf{u} : \mathbb{R}^{\geq 0} \to \mathbb{R}^m$ which can steer $\mathbf{y}_0$ towards $\mathbf{y}_f \triangleq \mathbf{y}(t_f) = 0$.

(a) Define the "output version" of the controllability Gramian as

$$W_{c,y}(t_0, t_f) \triangleq \int_{t_0}^{t_f} C(t_f)\Phi(t_f, \tau)B(\tau)B^\top(\tau)\Phi^\top(t_f, \tau)C^\top(t_f)d\tau$$

Prove that $(A(t), B(t), C(t))$ if output controllable iff $W_{c,y}(t_0, t_f)$ is nonsingular. Does your proof work if rank $C(t_f) < k$?
(Hint: For the necessity proof, consider initial state $\mathbf{x}_0 = \Phi(t_0, t_f)C^\top(t_f)(C(t_f)C^\top(t_f))^{-1}\tilde{\mathbf{y}}$ *after defining* $\tilde{\mathbf{y}} \in \mathbb{R}^k$ *to satisfy some property.)*

(b) Now let us consider specifically the LTI case (A, B, C) with rank $C = k$. Define a suitable controllability matrix $\mathcal{C}$ to test output controllability. Prove that (A, B, C) if output controllable iff rank $\mathcal{C} = k$.

Problem 6: Cartpole Revisited. Recall, from the previous Parts II and I, the inverted pendulum on a cart system. You have previously derived linearized state-space models around two equilibria points $\theta = 0$ (pendulum down) and $\theta = \pi$ (pendulum up). We will simplify the dynamics further, and instead use the following two linearizations for this problem.

- Linearizing downwards (around $(x, \dot{x}, \theta, \dot{\theta}) = (0, 0, 0, 0)$) gives us

$$\dot{\mathbf{x}}(t) = \begin{bmatrix} 0 & 1 & 0 & 0 \\ 0 & -\frac{b}{M} & -\frac{mg}{M} & 0 \\ 0 & 0 & 0 & 1 \\ 0 & \frac{b}{ML} & \frac{(M+m)g}{ML} & 0 \end{bmatrix} \mathbf{x}(t) + \begin{bmatrix} 0 \\ \frac{1}{M} \\ 0 \\ -\frac{1}{ML} \end{bmatrix} F(t) \tag{16.1}$$

- Linearizing upwards (around $(x, \dot{x}, \theta, \dot{\theta}) = (0, 0, \pi, 0)$) gives us

$$\dot{\mathbf{x}}(t) = \begin{bmatrix} 0 & 1 & 0 & 0 \\ 0 & -\frac{b}{M} & -\frac{mg}{M} & 0 \\ 0 & 0 & 0 & 1 \\ 0 & -\frac{b}{ML} & -\frac{(M+m)g}{ML} & 0 \end{bmatrix} \mathbf{x}(t) + \begin{bmatrix} 0 \\ \frac{1}{M} \\ 0 \\ \frac{1}{ML} \end{bmatrix} F(t) \tag{16.2}$$

(a) Use MATLAB commands to convert each linearized state-space model into the three canonical forms (1) CCF, (2) OCF, and (3) MCF. In your implementation of `sys`, use measurement equation $\mathbf{y}(t) = \mathbf{x}(t)$.
(b) Apply each of the 3 controllability tests (rank of the controllability matrix, PBH rank test, PBH eigenvector test) to each of the two linearized systems.

Problem 7 [1]. Consider the LTI system with the state equation:

$$\begin{bmatrix} \dot{x}_1 \\ \dot{x}_2 \\ \dot{x}_3 \end{bmatrix} = \begin{bmatrix} 0 & 1 & 0 \\ 0 & 0 & 1 \\ -\alpha_3 & -\alpha_2 & -\alpha_1 \end{bmatrix} \begin{bmatrix} x_1 \\ x_2 \\ x_3 \end{bmatrix} + \begin{bmatrix} 0 \\ 0 \\ 1 \end{bmatrix} u$$

Insert a state-feedback input $u = v - kx$, where v is the overall closed loop system input, where k is a constant row vector. Show that given any polynomial $p(s) = \sum_{k=0}^{3} a_k s^{3-k}$ with $a_0 = 1$, then there exists a row vector k such that the closed-loop system has $p(s)$ as its characteristic equation.
(This representation naturally extends to n dimensions.)

Problem 8. Consider a single-input system $\dot{\mathbf{x}} = A\mathbf{x} + bu$, where $u \in \mathbb{R}$, with $A = \text{diag}(\lambda_1, \cdots, \lambda_n)$ and the λ_i not necessarily distinct. State the necessary and sufficient conditions for full controllability. Now, generalize this problem for the case that A cannot be diagonalized, but can be converted into a Jordan form.

Problem 9: Hankel Singular Values for Linear Systems [2]. Consider the controllability and observability Gramians W_c, W_o of a LTI system (A, B, C) over some time period $[0, T]$.

(a) Determine what happens to these Gramians under similarity transformations of the state-space (PAP^{-1}, PB, CP^{-1}).
(b) Prove that the eigenvalues of the produce $W_c W_o$ are constant under this similarity transform P.

(These eigenvalues are called the *Hankel singular values* of the linear operator representing the linear system with zero initial condition.)

Problem 10: Complete Controllability [1]. Given a LTI system R, show that if R is completely controllable on $[t_0, t_1]$, then R is completely controllable on any $[t_0', t_1']$, where $t_0' \leq t_0 < t_1 \leq t_1'$. Show that this is no longer true when the interval $[t_0, t_1]$ is not a subset of $[t_0', t_1']$.

Problem 11: Cartpole Revisited. Consider the following problems about eigenvalue placement.

(a) Given 2D LTI system

$$\dot{\mathbf{x}}(t) = \begin{bmatrix} -15 & 8 \\ -15 & 7 \end{bmatrix} \mathbf{x}(t) + \begin{bmatrix} 1 \\ 1 \end{bmatrix} u(t),$$

find a suitable state-feedback gain matrix K which places the closed-loop eigenvalues at $-10 \pm j$.
your calculations by hand.

(b) Recall the inverted pendulum on a cart problem from a previous problem, and from II. We will apply eigenvalue placement and design K to stabilize the pole around the upright pole position ($\theta = \pi$). Use the MATLAB command `place` to place the eigenvalues of the system at $(-1.1, -1.2, -1.3, -1.4)$. Use the linearized dynamics from (16.2) and choose parameter values $M = 3$, $m = 1$, $b = 0.1$, $L = 0.5$, $I = 0.6$, $g = 9.81$.

(c) Use `ode45` to plot the state trajectories (x vs t, $\dot{x}$ vs t, θ vs t, and $\dot{\theta}$ vs t) of the closed-loop system you designed in part b. Use the initial condition $\mathbf{x}_0 = (1, 0, \pi - 0.1, 0)^\top$. Are you able to successfully stabilize the system?

(d) Repeat part c with initial condition $\mathbf{x}_0 = (1, 0, \pi/2, 0)^\top$. What happens to the system? Why?

(e) Repeat parts b and c (with original initial condition $\mathbf{x}_0 = (1, 0, \pi - 0.1, 0)^\top$) to place the poles at $(-3.1, -3.2, -3.3, -3.4)$. How would you compare the behavior of the system against part c?

You may notice there are infinitely many ways to place the poles of your system. But how do you make the best choice? The *linear quadratic regulator (LQR)* is a common method for choosing a set of eigenvalues which optimizes a specific *linear quadratic cost*, which is a function of the state and control input. We will talk more about LQR in the following Part IV.

Problem 12: Pole-Placement [2]. Consider the following completely controllable and observable MIMO system with two inputs and two outputs:

$$\dot{x} = ax + \begin{bmatrix} b_1 & b_2 \end{bmatrix} \begin{bmatrix} u_1 \\ u_2 \end{bmatrix}$$

$$\begin{bmatrix} y_1 \\ y_2 \end{bmatrix} = \begin{bmatrix} c_1 \\ c_2 \end{bmatrix} x$$

Define α_i to be the number of times you have to differentiate the ith output before one of the two inputs appear on the right side. Show that we have

$$\begin{bmatrix} y_1^{\alpha_1} \\ y_2^{\alpha_2} \end{bmatrix} = \begin{bmatrix} a^{\alpha_1} c_1 \\ a^{\alpha_2} c_2 \end{bmatrix} x + \underbrace{\begin{bmatrix} c_1 a^{\alpha_1 - 1} b_1 & c_1 a^{\alpha_1 - 1} b_2 \\ c_2 a^{\alpha_2 - 1} b_1 & c_2 a^{\alpha_2 - 1} b_2 \end{bmatrix}}_{\triangleq M} \underbrace{\begin{bmatrix} u_1 \\ u_2 \end{bmatrix}}_{\triangleq \mathbf{u}}$$

Assume that M is nonsingular and show that the control law

$$\mathbf{u} = M^{-1} \left(\begin{bmatrix} c_1 a^{\alpha_1} x \\ c_2 a^{\alpha_2} x \end{bmatrix} + \mathbf{v} \right)$$

places $\alpha_1 + \alpha_2$ poles of the closed-loop system at the origin, and the remaining at the zeros of the system $C(sI - A)^{-1}B$.

Problem 13: Internal Model Principle [2]. Consider the completely-controllable and observable SISO system

$$\dot{\mathbf{x}} = A\mathbf{x} + bu + w$$
$$y = cx$$

Here, w is an unknown, constant disturbance.

(a) A choice of state-feedback controller $u = -\mathbf{k}^\top \mathbf{x}$ will stabilize the system, but it will not yield $y \to 0$ as $t \to \infty$. For this reason, we can add a new state variable q such that

$$\dot{q} = y$$

Give conditions on the transfer function of the open-loop system $c(sI - A)^{-1}$ so as to be able to stabilize the augmented system using the new state feedback law

$$u = -\mathbf{k}^\top \mathbf{x} - fq$$

Does this guarantee that $y(t) \to 0$ as $t \to \infty$? Why or why not?

(b) Generalize this example to a disturbance of the form $we^{\lambda t}$, where λ is known but w is not, by defining

$$\dot{q} = \lambda q + y$$

then proceeding as described in part (a). What conditions do you need on the transfer function $c(sI - A)^{-1}$ in this case?

(c) The processes described in parts (a) and (b), where as build a replica of the system generating the disturbance inside the system, is called the *internal model principle*. Generalize this process to disturbances of the form

$$\sum_{i=1}^{b} w_i e^{\lambda_i t}$$

Problem 14. While proving the equivalence of statements about controllability, we showed that complete controllability on $[t_0, t_1]$ of a linear system is equivalent to positive definiteness of the controllability Gramian. Prove the analogous version for observability: complete observability on $[t_0, t_1]$ of a linear system is equivalent to positive (semi)definiteness of the observability Gramian.

Problem 15: Strong Observability [1]. Consider the zero-input response $\mathbf{y}(t)$ of $\dot{\mathbf{x}}(t) = A(t)\mathbf{x}(t) + B(t)\mathbf{u}(t)$ and $\mathbf{y}(t) = C(t)\mathbf{x}(t)$.

(a) Show that

$$I_0 \triangleq \int_{t_0}^{t_1} \|\mathbf{y}(t)\| \, dt = \mathbf{x}_0^* W_o(t_0, t_1)\mathbf{x}_0$$

(b) Determine the states of the unit sphere that are the most observable (i.e., with the longest I_0).

Minimal Realizations, Kalman Decomposition

Problem 16: Balanced Realizations. Write the SVDs of W_c and W_o as $U_c\Sigma_c^2U_c^\top$ and $U_o\Sigma_o^2U_o^\top$.

(a) Define $H = \Sigma_o U_o^\top U_c \Sigma_c^2 U_c^\top U_o \Sigma_o$. Show that H is positive-definite if the linear system is completely controllable and completely observable.
(b) Let the SVD of H be given by $U_h \Sigma_h^2 U_h^\top$. Verify that the entries of Σ_h are the square roots of the eigenvalues of $W_c W_o$.
(c) Derive the controllability and the observability Gramians for the transformed system with $T = U_o \Sigma_o^{-1} U_h \Sigma_h^{1/2}$. What can you conclude from your results?

Problem 17: Zeros of a Multivariable Linear System [2]. Consider a square linear system (i.e., number of inputs is the same as the number of outputs, $n_i = n_o$) which is completely controllable and observable. Eigenvalue λ is said to be a *zero* of the system if there exist $\mathbf{u}_0 \in \mathbb{R}^{n_i}$ and $\mathbf{x}_0 \in \mathbb{R}^n$ such that the input $\mathbf{u}(t) = \mathbf{u}_0 e^{\lambda t}$ with initial condition $\mathbf{x}_0$ produces zero output. Further, assume that $\text{rank}(B) = \text{rank}(C) = n_i$. Show that

$$\det \begin{bmatrix} \lambda I - A & -B \\ C & D \end{bmatrix} = 0$$

Set $D = 0$ and define $F \in \mathbb{R}^{n_i \times n}$ to be such that $F\mathbf{x}_0 = \mathbf{u}_0$. Is the system $(A + BF, B, C, 0)$ completely controllable.

Problem 18: Cartpole Revisited. Compute the Kalman decomposition of the system

$$\dot{\mathbf{x}}(t) = \begin{bmatrix} 1\,1\,0\,0\,0 \\ 0\,1\,0\,0\,0 \\ 0\,0\,2\,0\,0 \\ 0\,0\,0\,3\,0 \\ 0\,0\,0\,0\,3 \end{bmatrix} \mathbf{x}(t) + \begin{bmatrix} 0 \\ 1 \\ 0 \\ 0 \\ 1 \end{bmatrix} u(t)$$

Try to do as much of the calculation by hand. You may use software to do basic computation (e.g., find matrix inverses). Do not use MATLAB commands that give you the answer in one line.

Problem 19. This problem concerns the Kalman decomposition of LTI systems.

(a) Compute the Kalman decomposition of the following system

$$A = \begin{bmatrix} 0\,1\,0\,0 \\ 0\,2\,0\,0 \\ 1\,0\,0\,1 \\ 0\,0\,0\,2 \end{bmatrix}, \quad B = \begin{bmatrix} 0\,1 \\ 0\,0 \\ 2\,0 \\ 0\,0 \end{bmatrix}, \quad C = \begin{bmatrix} 2\,1\,0\,0 \end{bmatrix}$$

Write down the matrix P (or P^{-1}) used to transform the system. Identify both the reachable and unobservable subspaces.

(b) What is the Kalman decomposition of a general LTI (A, B, C, D) when $D \neq 0$? Justify your answer.

Problem 20. This problem is concerned with minimal realizations.

(a) Show that the LTI system $\dot{\mathbf{x}}(t) = A\mathbf{x}(t) + B\mathbf{u}(t)$, $\mathbf{y}(t) = C\mathbf{x}(t)$ with $\mathbf{x}(t) \in \mathbb{R}^n$ and $\mathbf{u}(t), \mathbf{y}(t) \in \mathbb{R}^m$ is minimal if and only if $\dot{\mathbf{z}}(t) = (A + BC)\mathbf{z}(t) + B\mathbf{u}(t)$, $\mathbf{y}(t) = C\mathbf{z}(t)$ is minimal.
Hint: Check the controllability/observability matrices for each system.

(b) Show that the two realizations

$$\dot{\mathbf{x}}(t) = \begin{bmatrix} 2 & 1 \\ 0 & 1 \end{bmatrix} \mathbf{x}(t) + \begin{bmatrix} 1 \\ 0 \end{bmatrix} \mathbf{u}(t), \quad \mathbf{y}(t) = \begin{bmatrix} 2 & 2 \end{bmatrix} \mathbf{x}(t)$$

and

$$\dot{\mathbf{x}}(t) = \begin{bmatrix} 2 & 0 \\ -1 & -1 \end{bmatrix} \mathbf{x}(t) + \begin{bmatrix} 1 \\ 2 \end{bmatrix} \mathbf{u}(t), \quad \mathbf{y}(t) = \begin{bmatrix} 2 & 0 \end{bmatrix} \mathbf{x}(t)$$

are realizations of $H(s) \triangleq (2s+2)/(s^2-s-2)$. Are they minimal realizations? Are they equivalent?

Observer Design

Problem 21. This problem is concerned with state observers.

(a) Complete the derivation of the reduced-order observer discussed in this chapter: LTI system $\dot{\mathbf{x}}(t) = A\mathbf{x}(t) + B\mathbf{u}(t)$ with measurement equation $\mathbf{y}(t) = \begin{bmatrix} C_m & 0 \end{bmatrix} \mathbf{x}(t)$ has asymptotically stable error dynamics if we choose

$$\hat{A} = A_{22} - LC_m A_{12}, \quad G = (A_{21} - LC_m A_{11})C_m^{-1} + \hat{A}L, \quad H = B_2 - LC_m B_1$$

and L to make $\hat{A}$ Hurwitz.

(b) For the system

$$\dot{\mathbf{x}}(t) = \begin{bmatrix} 0 & -1 \\ 1 & -2 \end{bmatrix} \mathbf{x}(t) + \begin{bmatrix} 2 \\ 1 \end{bmatrix} \mathbf{u}(t), \quad \mathbf{y}(t) = \begin{bmatrix} 1 & 1 \end{bmatrix} \mathbf{x}(t)$$

compute a 2D *full*-order observer such that the error decays exponentially with a rate of $\lambda = 10$. Repeat the same problem to compute a *reduced*-order observer.

Problem 17: Echo Canceller [2]. This problem concerns with the design of the chip that is typically used to cancel echoing effects in telephone calls. The echo $y(t) \in \mathbb{R}$ is represented as a linear combination of delayed versions of your input (spoken) message signal $u(t)$ as follows.

$$y(t) = \sum_{i=1}^{N} a_i u(t-i)$$

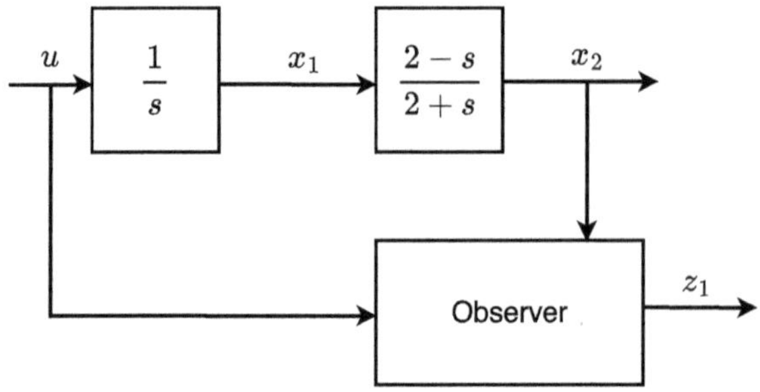

Fig. 16.1 The velocity system

where t is a discrete time variable dependent on the sampling rate of the voice signal. The coefficients $a_i \in \mathbb{R}$ model the characteristics of the telephone line, and are assumed unknown when you pick up the telephone at time $t = 0$. However, you can obtain estimates of them over time, denoted $\hat{a}_i(t)$. The aim of the echo canceller is to update the estimates using the measurement of the echo $y(t)$ and the prediction error

$$e(t) \triangleq y(t) - \sum_{i=1}^{N} \hat{a}_i u(t-i)$$

From the model, note that it will take a N-timestep delay after picking up the phone to be able to get all the $u(t-i)$. Thus, the echo canceller is initialized with $u(-1) = u(-2) = \cdots = u(-N+1) = 0$.

In this problem, we will set the echo canceller up as an observability problem, with the vector $\mathbf{a} \in \mathbb{R}^N$ representing the vector of the unknown coefficients a_i.

(a) Find $\hat{a}_1$ so that it is the vector closest to $\hat{a}_0$ in the norm and gives the correct value of $y(1)$.
(b) Turn your answer to part (a) into a recursive expression so that you can solve for $\hat{a}_{t+1}$ from $\hat{a}_t$ and $y(t)$.

Problem 22: Linearization by State-Feedback [2]. Consider the following single-input, single-output nonlinear system

$$\dot{\xi}_1 = \xi_2, \quad \dot{\xi}_2 = \xi_3, \quad \ldots, \dot{\xi}_r = \alpha(\boldsymbol{\xi}, \boldsymbol{\eta}) + \beta(\boldsymbol{\xi}, \boldsymbol{\eta})\mathbf{u}, \quad \dot{\eta} = q(\boldsymbol{\xi}, \boldsymbol{\eta})$$

where $\boldsymbol{\xi} \in \mathbb{R}^r$, $\boldsymbol{\eta} \in \mathbb{R}^{n-r}$, $\alpha, \beta : \mathbb{R}^r \times \mathbb{R}^{n-r} \to \mathbb{R}$, and $q : \mathbb{R}^r \times \mathbb{R}^{n-r} \to \mathbb{R}^{n-r}$. Let the output be $y = \xi_1$, and consider the system with the state-feedback law

$$u = -\frac{-\alpha(\boldsymbol{\xi}, \boldsymbol{\eta}) + v}{b(\boldsymbol{\xi}, \boldsymbol{\eta})}$$

Use this law to find a relationship between y and $\mathbf{u}$. Is the system with feedback observable? If not, what are the unobservable states, and why?

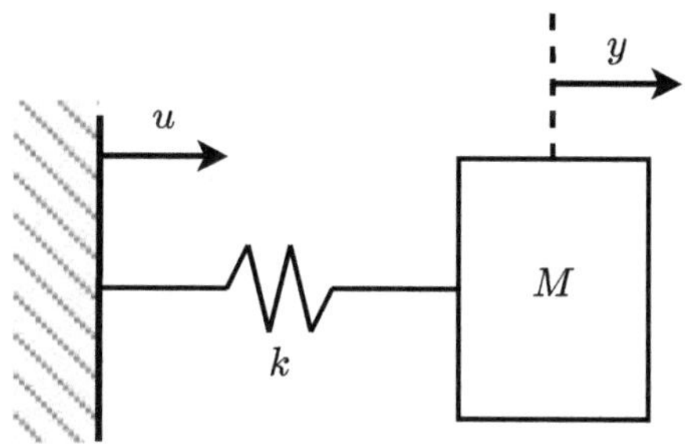

Fig. 16.2 A simplified model for the control of a robot arm

Problem 23: Observer Design [1]. Consider the a velocity observation system Fig. 16.1 where x_1 is the velocity to be observed. An observer is to be constructed to track x_1, using u and x_2 as inputs. The variable x_2 is obtained from x_1 through a sensor having the known transfer function $G(s) = (2 - s)/(2 + s)$ (as shown in Fig. 16.1).

(a) Derive a set of state-space equations for the system with state variables x_1 and x_2, input u, and output x_2.
(b) Design an observer with states z_1 and z_2 to track x_1 and x_2, respectively. Choose both observer eigenvalues to be at -5. What are the state-space equations for the observer.
(c) Derive the combined state equation for the system plus observer. Take, as state variables, x_1, x_2, and errors $e_i = x_i - z_i$ with $i = 1, 2$. Take u as input and z_1 as the output. Is this system controllable and/or observable?
(d) What is the transfer function relating u to z_1? Explain your result.

Problem 24: Simple Model of Robot Arm [1]. A simplified model for the control of a flexible robotic arm is shown in Fig. 16.2. Here is k is a spring constant which models the flexibility of the arm, M is the mass of the arm, y is the mass position, and u is the position of the end of the spring. Here, $k/M = 900\ \text{rad/sec}^2$.

(a) The equations of motion for this system are thus given by $M\ddot{y} + k(y - u) = 0$. Define state variables $x_1 = y$ and $x_2 = \dot{y}$.
(b) Design a full-state observer with observer eigenvalues at $s = -100 \pm 100j$.
(c) Could both state-variables of the system be estimated if only a measurement of $\dot{y}$.

(d) Design a state-feedback controller with gain matrix F giving the closed-loop system roots at $s = -20 \pm 20j$.
(e) Would it be reasonable to design a control law for the system with roots at $s = -200 \pm 200j$? Why or why not?

References

1. Claire Tomlin. *Lecture notes for EE221A: Linear Systems*. 2017.
2. Shankar Sastry. *Lecture notes for EE221A: Linear Systems*. 2013.

Part IV
Linear Optimal Control and Estimation

Chapter 17
Feedback Stabilization

17.1 Parametrizing Stabilizing Controllers

In the previous Part III, we addressed the question about how to design a stabilizing (state-)feedback controller for any given LTI system $\mathcal{H}$. In this chapter, we will address an important, related question: how can we characterize an entire set $S(\mathcal{H})$ of *all* controllers $\mathcal{K}$ which internally stabilizes $\mathcal{H}$?

As done before, we make the interconnection $(\mathcal{H}, \mathcal{K})$ more general by allowing $\mathcal{K}$ to have its own internal dynamics. In Fig. 17.1, we emphasize this with the argument (s), borrowing inspiration from the Laplace transform notation.

The LTI plant dynamics (with $\mathbf{d}_1, \mathbf{d}_2 \equiv 0$) on the left side of Fig. 17.1 follows the dynamical systems equation

$$\mathcal{H} \triangleq \begin{cases} \dot{\mathbf{x}}(t) = A\mathbf{x}(t) + \begin{bmatrix} B_1 & B_2 \end{bmatrix} \begin{bmatrix} \mathbf{w}(t) \\ \mathbf{u}(t) \end{bmatrix} \\ \begin{bmatrix} \bar{\mathbf{z}}(t) \\ \mathbf{y}(t) \end{bmatrix} = \begin{bmatrix} C_1 \\ C_2 \end{bmatrix} \mathbf{x}(t) + \begin{bmatrix} D_{11} & D_{12} \\ D_{21} & D_{22} \end{bmatrix} \begin{bmatrix} \mathbf{w}(t) \\ \mathbf{u}(t) \end{bmatrix} \end{cases}$$

Here, we require that (A, B_2, C_2) are stabilizable and detectable. A more compact notation for $\mathcal{H}$ which is commonly used is

$$H(s) = \left[\begin{array}{c|c} A & B \\ \hline C & D \end{array}\right] = \left[\begin{array}{c|cc} A & B_1 & B_2 \\ \hline C_1 & D_{11} & D_{12} \\ C_2 & D_{21} & D_{22} \end{array}\right] = \begin{bmatrix} H_{11}(s) & H_{12}(s) \\ H_{21}(s) & H_{22}(s) \end{bmatrix}$$

Just like $\mathcal{H}$, a general controller $\mathcal{K}$ can be written with an internal state $\boldsymbol{\xi}(t)$. In fact, we've seen this before with (15.5) in Sect. 15.2. We rewrite it here for self-containment.

$$\mathcal{K} \triangleq \begin{cases} \dot{\boldsymbol{\xi}}(t) = A_k \boldsymbol{\xi}(t) + B_k \mathbf{y}(t) \\ \mathbf{u}(t) = C_k \boldsymbol{\xi}(t) + D_k \mathbf{y}(t) \end{cases} \quad (A_k, B_k, C_k) \text{ stabilizable, detectable} \tag{17.1}$$

SJ. Han, *Linear Systems*, KAIST Research Series,
https://doi.org/10.1007/978-981-96-9774-8_17

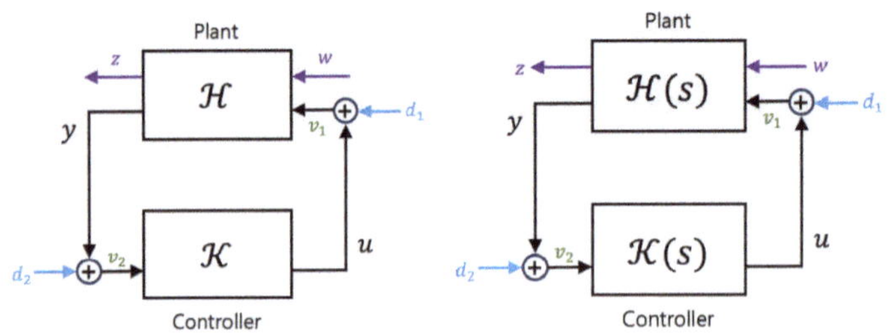

Fig. 17.1 A redrawing of Fig. 1.1 from Chap. 1, with additional signal notations and extra arguments (s) to emphasize all internal dynamics

There are two primary cases of the controller $\mathcal{K}$ to ensure internal stability, which we have discussed before. We will organize it more closely as follows.
Case 1. When $\mathcal{K} \triangleq K \in \mathbb{R}^{m\times n}$ is static gain and the control law is state-feedback, we have $\mathbf{u}(t) = K\mathbf{x}(t)$.

$$\left.\begin{array}{l} C_2 = I,\, D_{22} = 0 \\ A_k = 0,\, B_k = 0,\, C_k = 0,\, D_k = K \end{array}\right\} \Longrightarrow \dot{\mathbf{x}}(t) = \underbrace{(A + B_2 K)}_{\triangleq A_{\text{cl}}} \mathbf{x}(t)$$

Internal stability here is equivalent to ensuring A_{cl} is Hurwitz, which can be achieved using the pole placement techniques described in Chap. 15. To characterize all such K where A_{cl} is Hurwitz, we can use *linear matrix inequalities (LMIs)*. Recall that a LMI in the variable $\mathbf{x} \in \mathbb{R}^n$ is an expression of the form

$$F(x) = F_0 + x_1 F_1 + x_2 F_2 + \ldots + x_n F_n \succeq 0$$

where the $F_i \in \mathbb{R}^{m\times m}$ are symmetric.

Lemma 17.16 *A_{cl} is Hurwitz if $\exists$ symmetric $X \succ 0$ such that*

$$A_{cl}X + XA_{cl}^\top \prec 0 \Longrightarrow (A + B_2K)X + X(A + B_2K)^\top \prec 0$$

which can be rewritten as

$$AX + XA^\top + B_2KX + XK^\top B_2^\top \prec 0 \tag{17.2}$$

Note that (17.2) is not a LMI of K and X because of bilinear cross terms KX. Instead, let $Z \triangleq KX$ and rewrite (17.2) as a LMI in terms of Z and X. This is written in the following theorem.

Theorem 17.41 *Static state-feedback controller $\mathcal{K}$ (i.e., (17.1) with $D_k = K$) stabilizes interconnection $(\mathcal{H}, \mathcal{K})$ iff $\exists$ symmetric $X \prec 0$, Z s.t. $K = ZX^{-1}$ and*

$$\begin{bmatrix} A & B_2 \end{bmatrix} \begin{bmatrix} X \\ Z \end{bmatrix} + \begin{bmatrix} X & Z^\top \end{bmatrix} \begin{bmatrix} A^\top \\ B_2^\top \end{bmatrix} \prec 0 \tag{17.3}$$

Thus, the set of all stabilizing controllers of $\mathcal{H}$ is written as:

$$\mathcal{S}(\mathcal{H}) \triangleq \left\{ ZX^{-1} \mid X \succ 0 \textit{ symmetric, } Z \textit{ s.t. (17.3) holds} \right\}$$

__Case 2__. When $\mathcal{K}$ is a general (non-static) controller of the form (15.5), we first need to determine the following things before parametrizing anything:

1. the conditions for which $(\mathcal{H}, \mathcal{K})$ is *well-posed*, i.e., if unique solutions exist for $\mathbf{x}(t), \boldsymbol{\xi}(t), \mathbf{y}(t), \mathbf{u}(t)$
   $\forall$ initial $\mathbf{x}_0, \boldsymbol{\xi}_0$ and all disturbances $\mathbf{w}(t)$.
2. a definition of internal stability for generic $(\mathcal{H}, \mathcal{K})$.

To address point 1, note that combining $\mathcal{H}$ and $\mathcal{K}$ above together, we get

$$\underbrace{\begin{bmatrix} I & -D_k \\ -D_{22} & I \end{bmatrix}}_{(*)} \begin{bmatrix} \mathbf{u}(t) \\ \mathbf{y}(t) \end{bmatrix} = \begin{bmatrix} 0 & C_k \\ C_2 & 0 \end{bmatrix} \begin{bmatrix} \mathbf{x}(t) \\ \boldsymbol{\xi}(t) \end{bmatrix} + \begin{bmatrix} 0 \\ D_{21} \end{bmatrix} \mathbf{w}(t) \tag{17.4}$$

System is well-posed if the matrix $(*)$ is nonsingular (equivalently, if $I - D_{22}D_k$ nonsingular).

To address point 2, let $\mathbf{w} \equiv 0$ first. If the system is well-posed, we can write:

$$\begin{bmatrix} \mathbf{u}(t) \\ \mathbf{y}(t) \end{bmatrix} = \begin{bmatrix} I & -D_k \\ -D_{22} & I \end{bmatrix}^{-1} \begin{bmatrix} 0 & C_k \\ C_2 & 0 \end{bmatrix} \begin{bmatrix} \mathbf{x}(t) \\ \boldsymbol{\xi}(t) \end{bmatrix}$$

The closed-loop dynamics become

$$\begin{bmatrix} \dot{\mathbf{x}}(t) \\ \dot{\boldsymbol{\xi}}(t) \end{bmatrix} = \underbrace{\left(\begin{bmatrix} A & 0 \\ 0 & A_k \end{bmatrix} + \begin{bmatrix} B_2 & 0 \\ 0 & B_k \end{bmatrix} \begin{bmatrix} I & -D_k \\ -D_{22} & I \end{bmatrix}^{-1} \begin{bmatrix} 0 & C_k \\ C_2 & 0 \end{bmatrix} \right)}_{A_{\text{cl}}} \begin{bmatrix} \mathbf{x}(t) \\ \boldsymbol{\xi}(t) \end{bmatrix} \tag{17.5}$$

Note that if the matrix A_{cl} is Hurwitz, then $\mathbf{x}(t), \boldsymbol{\xi}(t) \to 0$ as $t \to \infty$.

Definition 17.96 When $\mathbf{w} \equiv 0$, interconnection $(\mathcal{H}, \mathcal{K})$ is *internally stable* if it is well-posed and $\mathbf{x}(t), \boldsymbol{\xi}(t) \to 0$ as $t \to \infty$ for all initial conditions $\mathbf{x}_0, \boldsymbol{\xi}_0$.

This addresses the two preliminary points from above. Now we are interested in the main question: how should we design internally stabilizing $\mathcal{K}$ for $\mathcal{H}$ when $\mathbf{w} \equiv 0$?

First, there exists such a $\mathcal{K}$ iff (A, B_2, C_2) is stabilizable and detectable. Directly from (17.4) and (17.5), we have

$$\mathcal{S}(\mathcal{H}) \triangleq \left\{ K \triangleq (A_k, B_k, C_k, D_k) \middle| \begin{bmatrix} A & 0 \\ 0 & A_k \end{bmatrix} + \begin{bmatrix} B_2 & 0 \\ 0 & B_k \end{bmatrix} \begin{bmatrix} I & -D_k \\ -D_{22} & I \end{bmatrix}^{-1} \begin{bmatrix} 0 & C_k \\ C_2 & 0 \end{bmatrix} \prec 0 \right\}$$

This is not a LMI in terms of (A_k, B_k, C_k, D_k), but we can impose a specific structure on (A_k, B_k, C_k, D_k) to turn it into a LMI. One choice to achieve this is as follows:

$$A_k = A + B_2K + LC_2 + LD_{22}K, \quad B_k = -L, \quad C_k = K, \quad D_k = 0 \tag{17.6}$$

where K and L are constant matrices of appropriate dimension. This way,

$$A_{\mathrm{cl}} = \begin{bmatrix} A & 0 \\ 0 & A + B_2K + LC_2 + LD_{22}K \end{bmatrix} + \begin{bmatrix} B_2 & 0 \\ 0 & -L \end{bmatrix} \underbrace{\begin{bmatrix} I & 0 \\ -D_{22} & I \end{bmatrix}^{-1}}_{=\begin{bmatrix} I & 0 \\ D_{22} & I \end{bmatrix}} \begin{bmatrix} 0 & K \\ C_2 & 0 \end{bmatrix}$$

$$= \begin{bmatrix} A & B_2K \\ -LC_2 & A + B_2K + LC_2 \end{bmatrix}$$

We've seen before in Chap. 15, that A_{cl} is Hurwitz iff we choose K and L s.t. $A + B_2K$ and $A + LC_2$ are Hurwitz. (Again, as in Remark 15.24, be mindful of the sign convention.) A LMI characterization of $\mathcal{S}(\mathcal{H})$ can then be derived as

$$\mathcal{S}(\mathcal{H}) \triangleq \{(A_k, B_k, C_k, D_k) |\ (17.6) \text{ holds, and } \exists L, K \text{ s.t. } A + B_2K \prec 0, A + LC_2 \prec 0\}$$

When $\mathbf{w} \neq 0$, this kind of LMI characterization based on the state-space matrices is generally difficult to do. In the following chapters of this part, we will discuss a few cases where it is possible.

Now, let us consider the *transfer function* (input-output) description instead of the state-space description we've been mostly using in the previous parts.

$$H(s) = \begin{bmatrix} H_{11}(s) & H_{12}(s) \\ H_{21}(s) & H_{22}(s) \end{bmatrix}$$

Then when $\mathbf{d}_1 = 0, \mathbf{d}_2 = 0$:

$$\begin{bmatrix} \mathbf{z} \\ \mathbf{y} \end{bmatrix} = \begin{bmatrix} H_{11} & H_{12} \\ H_{21} & H_{22} \end{bmatrix} \begin{bmatrix} \mathbf{w} \\ \mathbf{u} \end{bmatrix}, \quad \mathbf{u} = K(s)\mathbf{y} \tag{17.7}$$

Here, there is a slight abuse of notation, in that $\mathbf{w}, \mathbf{u}, \mathbf{y}$, etc. are used to denote the respective signal in both the time-domain and the frequency-domain. We write it this way with the understanding that the reader will be able to infer which domain is being used from the given context.

We have a definition of internal stability based on transfer functions, more suitable for cases when $\mathbf{w} \neq 0$.

Definition 17.97 (*Internal Stability: Transfer Function Version*) Interconnection $(\mathcal{H}, \mathcal{K})$ is *internally stable* if it is well-posed and for bounded exogenous inputs,

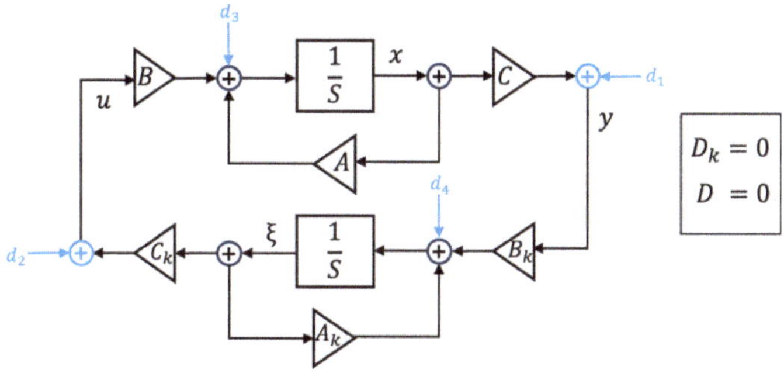

Fig. 17.2 An example block diagram of an internally-stable transfer function

every combination of transfer functions between every possible input-output pair of signals is stable.

One example of an internally-stable transfer function block diagram is shown in Fig. 17.2. Note that each of the 16 transfer functions from $(\mathbf{d}_1, \mathbf{d}_2, \mathbf{d}_3, \mathbf{d}_4)$ to $(\mathbf{x}, \mathbf{u}, \mathbf{y}, \boldsymbol{\xi})$ is stable (because they have negative real-part poles). By Definition 17.97, the entire system is internally stable.

Note that we don't need to check *literally every* possible transfer function from every possible input to every possible output because the stability of some transfer functions may automatically imply stability of others. Many schemes to parametrize the set of all internally stabilizing controllers $\mathcal{S}(\mathcal{H})$ rely on this fact. For example, the algebraic manipulation of (17.7) allows us to get the $\mathbf{w} \mapsto \mathbf{z}$ transfer function G_{zw} as follows:

1. $\mathbf{y} = H_{21}\mathbf{w} + H_{22}\mathbf{u} = H_{21}\mathbf{w} + H_{22}K\mathbf{y} \Longrightarrow \mathbf{y} = (I - H_{22}K)^{-1}H_{21}\mathbf{w}$
2. $\mathbf{z} = H_{11}\mathbf{w} + H_{12}K\mathbf{y} = \underbrace{(H_{11} + H_{12}K(I - H_{22}K)^{-1}H_{21})}_{\triangleq G_{zw}}\mathbf{w}$

Define $Q \triangleq K(I - H_{22}K)^{-1}$ so that G_{zw} is an affine function of Q:

$$G_{zw} = H_{11} + H_{12}QH_{21}$$

The transfer matrix Q is often called the *Youla parameter*. Once Q has been designed, K can be computed using

$$K = (I + QH_{22})^{-1}Q \tag{17.8}$$

Parametrization of the controller K is in the way of (17.8) is called the *Youla parametrization*. This parametrization is nice since it lets us characterize the set $\mathcal{S}(\mathcal{H})$ of all stabilizing controllers for a given plant $\mathcal{H}$ as (17.8) for any arbitrary stable transfer function Q.

17.2 Youla Parametrization

We defined what it meant for two rational transfer functions to be coprime in Remark 14.23: if we can express a proper rational transfer function $H(s)$ as $H(s) = \frac{N(s)}{D(s)}$ without any further simplifications, this means the polynomials $N(s)$ and $D(s)$ are coprime. We will use this concept in many parametrization methods for $\mathcal{S}(\mathcal{H})$, starting with the Youla parametrization.

Our goal remains the same: to stabilize (LTI) plant $H(s)$ via feedback interconnection with controller $K(s)$. We further assume H is strictly proper and K is proper. So far, we defined the Youla parameter Q s.t. $Q \triangleq K(I - H_{22}K)^{-1}$. We claimed that a necessary and sufficient condition to make the closed-loop response internally stable is to find such a Q which stabilizes G_{zw}.

Definition 17.98 ($\mathcal{RH}_\infty$) A transfer function $G(s)$ is an element of $\mathcal{RH}_\infty$ if it is real, rational, and proper.

Example 17.30 To explain the exact meaning of real, rational, and proper, as used in Definition 17.98, let's consider each of the terms individually.

- *Real*: The coefficients of the polynomial or transfer function are all real numbers. For example, $3s^2 + 2s + 1$.
- *Rational*: The transfer function can be expressed as the ratio of two polynomials. For example, $H(s) = \frac{1}{s+1}$ is the ratio of the constant polynomial 1 and the linear polynomial $s + 1$.
- *Proper*: The degree of the numerator polynomial is less than or equal to the degree of the denominator polynomial. For example, for $H(s) = \frac{1}{s+1}$, degree(num) (0) is less than degree(den) (1).

Note that $\forall Q, R \in \mathcal{RH}_\infty$, we have that $QR \in \mathcal{RH}_\infty$ and $Q + R \in \mathcal{RH}_\infty$. However, Q^{-1} may not be in $\mathcal{RH}_\infty$ since it might not be proper. For example, $Q(s) = \frac{s+1}{s^2+4s+4}$ $\implies Q^{-1}(s) = \frac{s^2+4s+4}{s+1}$ but degree 2 > degree 1.

Definition 17.99 ($\mathcal{Q}$) Let $\mathcal{Q} \subseteq \mathcal{RH}_\infty$ be the space of all stable, real rational proper transfer functions.

17.2.1 *Case 1: Stable Plant*

Suppose $H \in \mathcal{Q}$. Then the set of all stabilizing controllers is given by

$$\mathcal{S}(\mathcal{H}) = \{(I + QH_{22})^{-1} \mid Q \in \mathcal{Q}\} \tag{17.9}$$

Proof In the scalar case, (17.9) isn't too hard to prove.
Case $\subseteq$. Suppose $K \in \mathcal{S}(\mathcal{H})$ achieves internal stability. From the general block diagram Fig. 17.1:

$$\begin{bmatrix} \mathbf{z} \\ \mathbf{y} \end{bmatrix} = \begin{bmatrix} H_{11} & H_{12} \\ H_{21} & H_{22} \end{bmatrix} \begin{bmatrix} \mathbf{w} \\ \mathbf{u} + \mathbf{d}_2 \end{bmatrix}$$

where $\mathbf{u} = K(\mathbf{y} + \mathbf{d}_2)$ and therefore:

$$\mathbf{u} = K(H_{21}\mathbf{w} + H_{22}\mathbf{u} + H_{22}\mathbf{d}_2) + K\mathbf{d}_2$$

which simplifies to:

$$\mathbf{u} = \frac{K}{1 - H_{22}K}(H_{21}\mathbf{w} + H_{22}\mathbf{d}_1 + \mathbf{d}_2)$$

Define Q as:

$$Q \triangleq \frac{K}{1 - H_{22}K}$$

Note that $Q \subseteq \mathcal{Q}$ since it is the $\mathbf{d}_2 \to \mathbf{u}$ transfer function, which is stable because $K \in S(\mathcal{H})$ and $Q - QH_{22}K = K$ implies $K = \frac{Q}{1+QH_{22}}$.

Case $\supseteq$. Suppose $K = \frac{Q}{1+QH_{22}}$ with $Q \in \mathcal{Q}$.

Internal stability of the feedback system is achieved if all 9 transfer functions from $(\mathbf{w}, \mathbf{d}_1, \mathbf{d}_2)$ to $(\mathbf{z}, \mathbf{y}, \mathbf{u})$ are stable.

The $(\mathbf{w}, \mathbf{d}_1, \mathbf{d}_2) \to \mathbf{u}$ transfer functions are already written above as:

$$\mathbf{u} = -\frac{K}{1 - H_{22}K}(H_{21}\mathbf{w} + H_{22}\mathbf{d}_1 + \mathbf{d}_2)$$

and thus for $\mathbf{z}$:

$$\begin{aligned} \mathbf{z} &= H_{11}\mathbf{w} + H_{12}\mathbf{u} + H_{12}\mathbf{d}_1 \\ &= \left(H_{11} + \frac{KH_{21}}{1 - H_{22}K}\right)\mathbf{w} + \left(H_{12} + \frac{KH_{22}}{1 - H_{22}K}\right)\mathbf{d}_1 + \frac{KH_{12}}{1 - H_{22}K}\mathbf{d}_2 \end{aligned}$$

The $(\mathbf{w}, \mathbf{d}_1, \mathbf{d}_2) \to \mathbf{y}$ transfer functions are similar.

In particular, the shared $-\frac{K}{1-H_{22}K}$ term in all 9 transfer functions is exactly equal to Q after some algebra.

Together with the fact that $H \in \mathcal{Q}$, all 9 transfer functions are in $\mathcal{Q}$. ∎

Example 17.31 (*Reference Tracking Problem*) While the stabilization problem has the objective of driving all internal states $\mathbf{x}(t)$ and output $\mathbf{y}(t)$ to zero, the *reference-tracking* problem aims to design controller $K(s)$ to drive the output signal $\mathbf{y}(t)$ to some desired reference signal $\mathbf{r}(t)$.

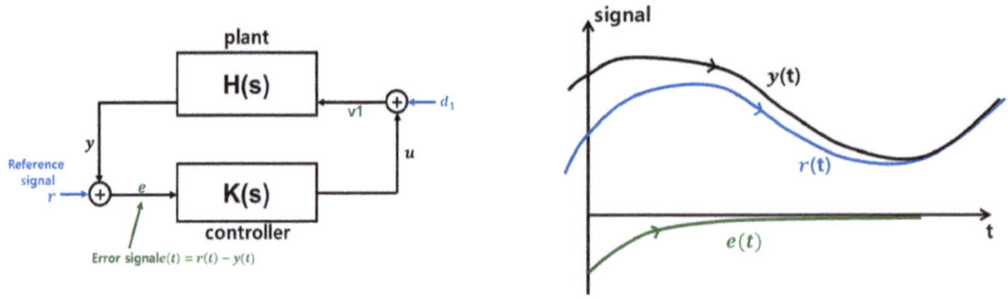

Reference-tracking is another very common application of control theory. For example, an autonomous vehicle should stay in its lane, and aircraft should maintain a desired angle of attack during takeoff.

In this example, let's try to find an internally stabilizing K for the plant $H(s) = \frac{1}{(s+1)(s+2)}$ so that scalar output $y(t)$ tracks a ramp signal, defined as $r(t) = t$ for $t \geq 0$ and $r(t) = 0$ otherwise. The ramp has a Laplace transform $\mathcal{L}\{r(t)\} = \frac{1}{s^2}$ with a region of convergence $\mathrm{Re}(s) > 0$.

Designing the Controller: It turns out that in order to track a ramp, the transfer function from $r(t)$ to $e(t)$ should have two zeros at $s = 0$. The error signal $e(t)$, which is $r(t) - y(t)$, is affected by the controller $K(s)$ through the feedback loop involving $H(s)$. We can express $e(s)$ as:

$$e(s) = r(s) - H(s)K(s)e(s) - H(s)d_1(s)$$

$$\implies e(s) = \frac{1}{1 + H(s)K(s)}r(s) - \frac{H(s)}{1 + H(s)K(s)}d_1(s)$$

This is simplified via Youla parametrization as:

$$e(s) = \frac{1}{1 + H(s)\frac{Q(s)}{1-Q(s)H(s)}}r(s) - \frac{H(s)}{1 + H(s)K(s)}d_1(s) \tag{17.10}$$

where $Q(s)$ is chosen as a function that will provide the necessary cancellation of poles/zeros.

Choosing $Q(s)$: Try choosing $Q(s) = \frac{as+b}{s+1}$ with parameters a and b to be determined. Then the closed-loop transfer function from r to e (which is the first term of (17.10)) becomes:

$$\frac{1}{1 + \frac{H(s)Q(s)}{1-H(s)Q(s)}} = 1 - HQ = \frac{s^3 + 4s^2 + (5-a)s + (2-b)}{(s+1)^2(s+2)}$$

Choosing $a = 5$ and $b = 2$ yields:

$$\frac{s^2(s+4)}{(s+1)^2(s+2)}$$

This setup not only provides the desired zeros at $s = 0$ but also cancels both poles of $H(s)$ at $s = -1$ and $s = -2$. The resulting $K(s)$, by incorporating $H(s)$, becomes:

$$K(s) = \frac{(5s+2)(s+1)(s+2)}{s^2(s+4)}$$ □

17.2.2 Case 2: General, Possibly Unstable Plant

First, we require a few further definitions.

Definition 17.100 (*Right and Left Coprime*) Two transfer functions (or transfer matrices) $M, N \in \mathcal{RH}_\infty$ are *right coprime* if there exist $X, Y \in \mathcal{RH}_\infty$ such that

$$XN + YM = I \tag{17.11}$$

Likewise, they are *left coprime* if there exist $\tilde{X}, \tilde{Y} \in \mathcal{RH}_\infty$ such that

$$N\tilde{X} + M\tilde{Y} = I \tag{17.12}$$

Equations (17.11) and (17.12) are often called the *Bezout Identities*.

Lemma 17.17 *A right coprime factorization of transfer function H is the factorization $H = NM^{-1}$ where $N, M \in \mathcal{RH}_\infty$, and M^{-1} is proper. Likewise, a left coprime factorization is $H = \tilde{M}^{-1}\tilde{N}$ where $\tilde{N}, \tilde{M} \in \mathcal{RH}_\infty$, and $\tilde{M}^{-1}$ is proper.*

Our general method for parametrizing $\mathcal{S}(\mathcal{H})$ is to invoke coprime factorization. We will first express the right coprime factorization $H = NM^{-1}$, then parameterize $K = XY^{-1}$ where X and Y satisfy $XN + YM = I$.

Before we proceed with this method, there are a few preliminary remarks to address.

1. Does there always *exist* such a factorization?
2. There may be problem cases such as $Y(s) = 0$ or $K(s) \notin \mathcal{RH}_\infty$ because it might not be proper. For example, for scalar transfer function $H(s) = \frac{s-1}{s-2}$, one could take $N(s) = 1, M(s) = \frac{s-2}{s-1}$. The easiest solution to $XN + YM = 1$ is $X(s) = 1$, $Y(s) = 0$ but $K(s) = X/Y$ is undefined.
3. To ensure internal stability, we will add an extra requirement and restrict N, M, $X, Y \in \mathcal{Q}$. How to construct such a coprime factorization under this new requirement?

To address point 1 above, we have the following result, whose proof we will defer.

Lemma 17.18 *For any transfer function H corresponding to a stabilizable and detectable system, there exist both a right and left coprime factorization $H = NM^{-1} = \tilde{M}^{-1}\tilde{N}$ s.t. $\exists \tilde{X}, \tilde{Y}, X, Y \in \mathcal{RH}_\infty$ with*

$$\begin{bmatrix} X & Y \\ \tilde{M} & \tilde{N} \end{bmatrix} \begin{bmatrix} N \\ \tilde{Y} \end{bmatrix} = I \quad (\text{"doubly coprime factorization"})$$

To address point 3 above, let's further divide our analysis into two subcases.

Case: Scalar Version. In the scalar version of the problem, there is *no distinction* between "left" or "right" coprime factorization. *Euclid's Algorithm* can be used to obtain X, Y given N, M with $\deg(M) \geq \deg(N)$.

1. Divide to get $M = NQ_1 + R_1$, where $\deg(R_1) < \deg(N)$.
2. Divide to get $N = R_1 Q_2 + R_2$, where $\deg(R_2) < \deg(R_1)$.
3. Divide to get $R_{k-1} = R_k Q_k + R_{k+1}$, where $\deg(R_{k+1}) < \deg(R_k) \forall k \geq 2$.
4. Stop until R_{k+1} is a constant number.
5. If $R_{k+1} = 0$, N and M are not coprime and GCD is R_k. Else, rewind steps 3 to 1 to get

$$R_{k+1} = WN + ZM \Longrightarrow 1 = \frac{W}{R_{k+1}} N + \frac{Z}{R_{k+1}} M \Longrightarrow 1 = XN + YM$$

Example 17.32 (*Using Euclid's Algorithm*) Suppose we have a transfer function with $n(\lambda) = \lambda^2$ and $m(\lambda) = 6\lambda^2 - 5\lambda + 1$. Applying Euclid's algorithm gives us

$$q_1(\lambda) = \frac{1}{6}, \quad r_1(\lambda) = \frac{5}{6}\lambda - \frac{1}{6}, \quad q_2(\lambda) = \frac{36}{5}\lambda - \frac{114}{25}, \quad r_2(\lambda) = \frac{6}{25}.$$

Since r_2 is a nonzero constant, we stop after Step 2. Then the equations are

$$n = mq_1 + r_1, m = r_1 q_2 + r_2,$$

yielding

$$r_2 = (1 + q_1 q_2)m - q_2 n.$$

So we should take

$$x = -\frac{q_2}{r_2}, \quad y = \frac{1 + q_1 q_2}{r_2},$$

that is,

$$x(\lambda) = -30\lambda + 19, \quad y(\lambda) = 5\lambda + 1.$$ □

Now to develop a full stabilization procedure with Euclid involved, the main idea is to transform variables, $s \to \lambda$, so that polynomials in λ yield transfer functions in terms of s.

1. If $H \in \mathbb{Q}$, set $N = H, M = 1, X = 0, Y = 1$. End.
2. If $H \notin \mathbb{Q}$, substitute $s \to \frac{1-\lambda}{\lambda}$ to get $H\left(\frac{1-\lambda}{\lambda}\right)$.
3. Write $H\left(\frac{1-\lambda}{\lambda}\right) = \frac{N(\lambda)}{M(\lambda)}$.
4. Apply Euclid's Algorithm to get $X(\lambda), Y(\lambda)$.
5. Transform back $\lambda \to \frac{1}{s+1}$ to get $N(s), M(s), X(s), Y(s)$. Essentially, we are placing poles at -1 with this transformation.

Example 17.33 (*Full Stabilization Procedure*) For $G(s) = \frac{1}{(s-1)(s-2)}$, the algorithm gives:

$$\tilde{G}(\lambda) = \frac{\lambda^2}{6\lambda^2 - 5\lambda + 1} \Longrightarrow n(\lambda) = \lambda^2, m(\lambda) = 6\lambda^2 - 5\lambda + 1$$

$$\Longrightarrow x(\lambda) = -30\lambda + 19, \quad y(\lambda) = 5\lambda + 1 \quad \text{(from Example 32)}.$$

Now, use the mapping $\lambda = \frac{1}{s+1}$.

$$N(s) = \frac{1}{(s+1)^2}, \quad M(s) = \frac{(s-1)(s-2)}{(s+1)^2},$$

$$X(s) = \frac{19s - 11}{s+1}, \quad Y(s) = \frac{s+6}{s+1}.$$

□

Theorem 17.42 (Youla-Kucera Parametrization: Scalar Case) *Suppose $H = \frac{N}{M}$ with $N, M \in \mathcal{Q}$ and suppose $\exists X, Y \in \mathcal{Q}$ such that N, M are coprime. Then the set of all internally stabilizing controllers $\mathcal{S}(\mathcal{H})$ is given by:*

$$\mathcal{S}(\mathcal{H}) = \left\{ \frac{X + MQ}{Y - NQ} \mid Q \in \mathcal{Q} \right\}$$

Proof We use the following result without proof: Let $K = \frac{W}{Z}$ be some coprime factorization over $\mathcal{Q}$. Then the feedback system is internally stable if and only if $(NW + MZ)^{-1} \in \mathcal{Q}$.

Suppose $K \in \mathcal{S}(\mathcal{H})$. We must find a $Q \in \mathcal{Q}$ such that $K = \frac{X+MQ}{Y-NQ}$. Choose W, Z such that $K = \frac{W}{Z}$, a coprime factorization over $\mathcal{Q}$, and define V such that $(NW + MZ)V = NWU + MZV = 1$.

Let Q be the solution of $2V = Y - NQ$ so that $NWU + M(Y - NQ) = 1$. Also, from $NX + MY = 1$, we get $NX + MY + NMQ - MNQ = 1$. Comparing yields $WV = X + MQ$, thus $K = \frac{W}{Z} = \frac{X+MQ}{Y-NQ}$.

Now, does $Q \in \mathcal{Q}$? Using the results, we get $ZUX = XY - NQX$ and $WVY = XY + MQY$, thus $(MY + NX)Q = WVY - ZUX$. Therefore, $Q \in \mathbb{Q}$ and the theorem is proved. ∎

Case: Matrix Version. The Youla parametrization described in Theorem 17.42 can be extended to the more general matrix-vector case.

Theorem 17.43 (Youla-Kucera Parametrization: Matrix Case) *Suppose there exists a doubly coprime factorization* $H = NM^{-1} = \tilde{M}^{-1}\tilde{N}$ *such that* $\exists X, Y, \tilde{X}, \tilde{Y} \in \mathcal{Q}$ *with*

$$\begin{bmatrix} X & Y \\ \tilde{M} & \tilde{N} \end{bmatrix} \begin{bmatrix} N \\ \tilde{Y} \end{bmatrix} = I$$

Then the set of all internally stabilizing controllers $\mathcal{S}(\mathcal{H})$ *is given by:*

$$\mathcal{S}(\mathcal{H}) = \left\{(X + MQ)(Y - NQ)^{-1} \mid Q \in \mathcal{Q}\right\} = \left\{(\tilde{Y} - \tilde{N}Q)^{-1}(\tilde{X} + \tilde{M}Q) \mid Q \in \mathcal{Q}\right\}$$

Remark 17.30 The transfer function $S \triangleq M(Y - NQ)$ is typically called the *sensitivity function*, and $T = N(X + MQ)$ is called the *complementary sensitivity function*. They are often used in control system engineering to measure how variations in the plant parameters affect the closed-loop response.

Remark 17.31 There are a lot of other interesting concepts which build upon what we discussed so far:

- *Strong stabilization*: if H can be stabilized with a $K \in \mathcal{Q}$.
- *Simultaneous stabilization*: if H_1 and H_2 can be stabilized with the same K.

Chapter 18
The Linear Quadratic Regulator

So far, we have approached the problem of feedback stabilization using eigenvalue (pole) placement techniques: design a state-feedback stabilizing controller K for LTI system (A, B) s.t. $\lambda_i(A - BK)$ for $i = 1, \ldots, n$ are placed at new pole locations $\{\tilde{\lambda}_i\}_{i=1}^n$. Feedback stabilization techniques like these assume that there already is a predetermined set of eigenvalues that we wish for the system to possess. In practice, it is more conventional to be given a certain performance index, which is a functional of the control and state trajectories (i.e., $J_{\mathbf{u}}(\mathbf{x}_0)$), and seek to find a control trajectory $\mathbf{u} : \mathbb{R}^{\geq 0} \to \mathbb{R}^m$ which *optimizes* it. This is the general framework of *optimal control*.

18.1 Dynamic Programming

Finding the optimal control trajectory by searching through all possible trajectories is computationally expensive and inefficient. In such cases, *dynamic programming (DP)* can be a powerful tool for solving complex optimization problems. Dynamic programming is a systematic approach that enables efficient computation by breaking a problem into smaller subproblems, solving them independently, and storing the results for reuse. It is particularly effective for problems that exhibit two characteristics: *optimal substructure* and *overlapping subproblems*.

Optimal Substructure. The key principle here is that the optimal solution to a problem can be constructed from the optimal solutions of its subproblems.

$$\texttt{OPT}(P) = \texttt{Combine}\,(\texttt{OPT}(P_1), \texttt{OPT}(P_2), \ldots, \texttt{OPT}(P_k))\,,$$

where $\texttt{OPT}(P)$ is the optimal solution to the overall problem P, $\texttt{OPT}(P_1), \ldots,$ $\texttt{OPT}(P_k)$ is the optimal solutions to subproblems $P_1, \ldots, P_k$, and $\texttt{Combine}$ is a function that combines the subproblem solutions to construct the overall solution.

SJ. Han, *Linear Systems*, KAIST Research Series,
https://doi.org/10.1007/978-981-96-9774-8_18

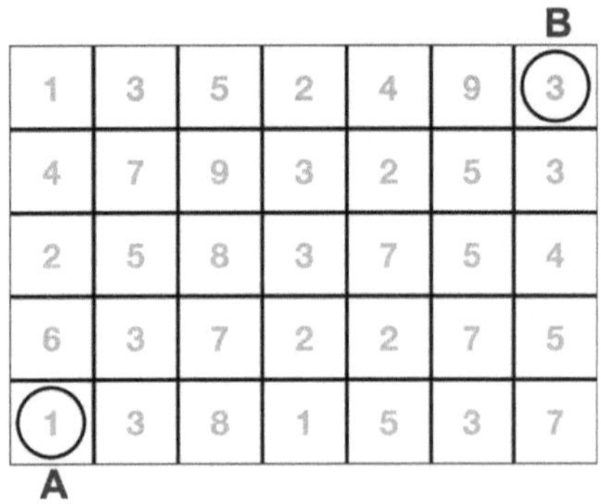

Fig. 18.1 The original grid G, upon which we are seeking to find the optimal path

Overlapping Subproblems. When the same subproblem is solved multiple times, its solution can be stored and reused, significantly reducing computational cost. This is implemented using *memoization*: if subproblem P_i has already been memoized, with $\texttt{Memo}[P_i]$ being the previously-computed solution to subproblem P_i, then $\texttt{OPT}(P_i) \triangleq \texttt{Memo}[P_i]$ is the optimal solution.

Example 18.34 (*Shortest Path DP*) Consider a rectangular grid $G \in \mathbb{R}^{H \times L}$ of numbers, as shown in Fig. 18.1. We are interested in determining the shortest path going from A to B.

We first prepare a memoization table $T \in \mathbb{R}^{H \times L}$ to keep track of all the optimal solutions to each subproblem. The optimal computation is performed backwards from the final goal B. Note that the edges are the easiest entries to fill in, because the optimal path starting from any location along the edge is just a direct line to B.

$$T[1, L] = 3, \quad T[1, \ell] = G[1, \ell] + T[1, \ell + 1], \quad T[h, L] = G[h, L] + T[h + 1, L]$$

To fill in the interior of the table T, first consider the $[H - 1, L - 1]$th entry. There are two possible paths: $[H - 1, L - 1] \rightarrow [H, L - 1] \rightarrow [H, L]$ versus $[H - 1, L - 1] \rightarrow [H - 1, L] \rightarrow [H, L]$. By DP memoization, to get the optimal path between from $[H - 1, L - 1]$, we only need to compare the two paths:

$$T[H - 1, L - 1] = G[H - 1, L - 1] + \min(T[H - 1, L], T[H, L - 1]) = 5 + 6 = 11$$

Now, we can fill in the $(H - 1)$th row and the $(L - 1)$th column.

In general, we can see that

$$T[h, \ell] = G[h, \ell] + \min(T[h, \ell + 1], T[h + 1, \ell])$$

and we obtain $T[H, 1] = 34$, as the shortest path from $A \rightarrow B$. All steps of the memoization process are shown in Fig. 18.2, with the final path highlighted in blue.

27	26	23	18	16	12	3
						6
						10
						15
						22

⇒

27	26	23	18	16	12	3
31	32	25	16	13	11	6
					15	10
					22	15
					25	22

⇒

27	26	23	18	16	12	3
31	32	25	16	13	11	6
33	32	27	19	20	15	10
37	31	28	21	22	22	15
34	33	30	22	27	25	22

Fig. 18.2 The memoization table T, in which we are saving all optimal solutions to each subproblem. Eventually, we reach the total optimal solution (shortest distance from $A \to B$) and the final optimal path $A \to B$

Essentially, for all DP problems, the computation of the optimal solutions to all subproblems is done backwards, while the final result is obtained as a forwards pass. □

18.2 Basic Derivations of the Linear Quadratic Regulator

Since the primary focus of this text is on *linear* systems, we mainly investigate the specific class of *linear optimal control* techniques. Among these techniques, the *linear quadratic regulator (LQR)* is one of the most common methods used in practice. In state-feedback LQR, the cost functional $J_{\mathbf{u}}(\mathbf{x}_0)$ is represented as a weighted sum of terms which are quadratic in $\mathbf{x}$ and $\mathbf{u}$. There is also an output-feedback version of LQR which represents $J_{\mathbf{u}}(\mathbf{x}_0)$ as a weighted sum of terms that are quadratic in $\mathbf{y}$ and $\mathbf{u}$. We will write and derive the LQR formulation for both CT and DT systems. As with III, we focus almost exclusively on *LTI* systems.

Remark 18.32 Just as with eigenvalue placement in Chap. 17, (A, B) is assumed to be stabilizable and (A, C) is assumed detectable.

18.2.1 Discrete-Time Dynamics

First, we recall the DT LTI dynamics $\mathbf{x}[t+1] = A\mathbf{x}[t] + B\mathbf{u}[t]$, with $\mathbf{y}[t] = \mathbf{x}[t]$ (full-state observable), and initial condition $\mathbf{x}[0] = \mathbf{x}_0$. There are two types of cost functionals depending on the *horizon* of the problem:

$$\text{Finite-Horizon.} \quad J_{\mathbf{u}}(\mathbf{x}_0) \triangleq \underbrace{\sum_{t=0}^{T-1} \begin{bmatrix}\mathbf{x}^\top[t] \; \mathbf{u}^\top[t]\end{bmatrix} \tilde{Q} \begin{bmatrix}\mathbf{x}[t] \\ \mathbf{u}[t]\end{bmatrix}}_{\text{“running cost”}} + \underbrace{\mathbf{x}^\top[T] Q_f \mathbf{x}[T]}_{\text{“terminal cost”}} \tag{18.1a}$$

$$\text{Infinite-Horizon.} \quad J_{\mathbf{u}}(\mathbf{x}_0) \triangleq \sum_{t=0}^{\infty} \left[\mathbf{x}^\top[t]\; \mathbf{u}^\top[t]\right] \tilde{Q} \begin{bmatrix} \mathbf{x}[t] \\ \mathbf{u}[t] \end{bmatrix} \tag{18.1b}$$

where $\tilde{Q} \triangleq \begin{bmatrix} Q & S \\ S^\top & R \end{bmatrix}$, $Q = Q^\top \succeq 0$, $R = R^\top \succ 0$, $Q_f = Q_f^\top \succeq 0$ and $u \in \mathcal{U} \subseteq \mathcal{L}_2([0, T]; \mathbb{R}^m)$. Here, $\tilde{Q}$ and Q_f can be thought of as *weights* which determine how much to penalize the deviation of $\mathbf{x}(t)$ from 0, as well as the control input $\mathbf{u}(t)$. Starting from any initial state $\mathbf{x}_0$, we essentially seek to control the system s.t. $\mathbf{x}(t) \to 0$ as quickly as possible without exerting too much *control effort* (i.e., $\|\mathbf{u}(t)\|$ should not be too large). Thus, the optimal control problem is posed as

$$\tilde{J}(\mathbf{x}_0) \triangleq \min_{u \in \mathcal{U}} J_{\mathbf{u}}(\mathbf{x}_0) \text{ s.t. } \mathbf{x}[t+1] = A\mathbf{x}[t] + B\mathbf{u}[t] \tag{18.2}$$

and the optimal control input $\mathbf{u}^*$ which solves (18.2) is given by a linear state-feedback form $\mathbf{u}[t] = K[t]\mathbf{x}[t]$, where the gain $K[t]$ is to be determined. Note that $K[t]$ is not necessarily static-gain, as indicated by the time index $[t]$.

Remark 18.33 Minimum-energy input is special case of LQR with $Q = 0$, $R = I$ (For DT finite horizon).

Because (18.2) is a convex optimization problem, we can use a variety of standard convex optimization techniques to solve it, including dynamic programming (DP) and via *Lagrange multipliers*. We will begin by applying DP to the case where $S = 0$ in the finite-horizon LQR cost (18.1a).

$$\text{Finite-Horizon.} \quad J_{\mathbf{u}}(\mathbf{x}_0) \triangleq \sum_{t=0}^{T-1} \mathbf{x}^\top[t]Q\mathbf{x}[t] + \mathbf{u}^\top[t]R\mathbf{u}[t] + \mathbf{x}^\top[T]Q_f\mathbf{x}[T] \tag{18.3a}$$

$$\text{Infinite-Horizon.} \quad J_{\mathbf{u}}(\mathbf{x}_0) \triangleq \sum_{t=0}^{\infty} \mathbf{x}^\top[t]Q\mathbf{x}[t] + \mathbf{u}^\top[t]R\mathbf{u}[t] \tag{18.3b}$$

This means that the cost functional does not penalize any cross-term dependencies between $\mathbf{x}$ and $\mathbf{u}$. (The case $S \neq 0$ is more natural in the output-feedback LQR case, which we will address in Sect. 18.2.5.)

Lemma 18.19 (Bellman's Principle of Optimality) *While the principle of optimality can be posed for general optimization problems, we pose it specifically in the context of our DT LTI LQR problem. Let $\mathbf{x}^* \equiv \{\mathbf{x}^*[t]\}_{t=0}^{T}$ be the optimal trajectory which results from applying optimal control $\mathbf{u}^*$ to the system $\mathbf{x}[t+1] = A\mathbf{x}[t] + B\mathbf{u}[t]$ starting from $\mathbf{x}_0$. If $\mathbf{x}^*[t]$ for any $t \in \{0, 1, \ldots, T\}$ is a state on this optimal path, then the "sub-path" $\{\mathbf{x}[t], \mathbf{x}[t+1], \ldots, \mathbf{x}[T]\}$ is also an optimal path.*

Lemma 18.19 allows us to express (18.3a) in terms of the following recursion

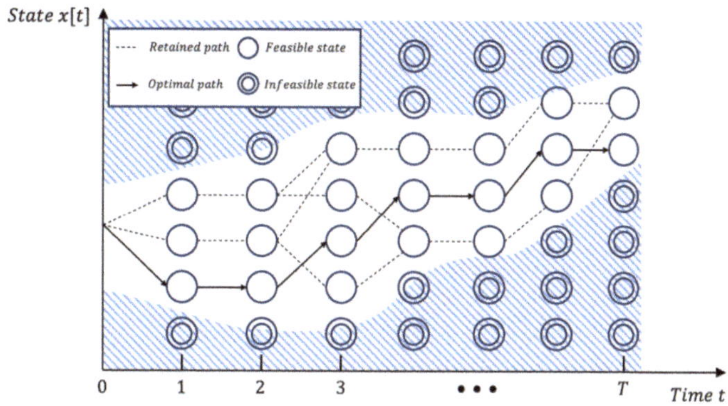

Fig. 18.3 The figure demonstrates how the *cost-to-go function* is optimized through backward recursion, evaluating feasible paths while minimizing the total cost. The optimal path (solid arrows) minimizes the cost function, while retained paths (dashed lines) indicate alternative path. The blue shaded regions denote infeasible state areas

$$\begin{aligned}\mathbf{V}_t(\mathbf{x}[t]) &\triangleq \min_{\mathbf{u}[t]\in\mathbb{R}^m}\left(\mathbf{x}^\top[t]Q\mathbf{x}[t]+\mathbf{u}^\top[t]R\mathbf{u}[t]+\mathbf{V}_{t+1}(\mathbf{x}_{t+1})\right) \text{ s.t. } \mathbf{x}[t+1]=A\mathbf{x}[t]+B\mathbf{u}[t]\\ &= \min_{\mathbf{u}[t]\in\mathbb{R}^m} \mathbf{x}^\top[t]Q\mathbf{x}[t]+\mathbf{u}^\top[t]R\mathbf{u}[t]+\mathbf{V}_{t+1}(A\mathbf{x}[t]+B\mathbf{u}[t]) \end{aligned} \tag{18.4}$$

Here, $\mathbf{V}_t(\mathbf{x}[t])$ is typically called the *cost-to-go function* or the *value function*. In the terminology of Sect. 18.1, V is the memoization table with an entry for each t and $\mathbf{x}$. The first two terms of (18.4) indicate the cost incurred at current time, while the last term is the cost-to-go starting from time $t+1$ and current state $\mathbf{x}[t]$. Since we are computing the memoization table backwards, $\mathbf{V}_{t+1}$ is a function we already have all the information about (Fig. 18.3).

Theorem 18.44 (Finite-Horizon DT LQR) *The optimal cost-to-go and optimal control at each time is given by*

$$V_t(\mathbf{x}) = \mathbf{x}^\top P_t\mathbf{x} \quad \forall\mathbf{x}\in\mathbb{R}^n, \quad \mathbf{u}[t] = K[t]\mathbf{x}[t], \quad \forall\, t\in\{0,1,\dots,T\}$$

where P_t satisfies the following final-value problem and control gain K can be computed from P:

$$P_t = \begin{cases} Q + K[t]^\top RK[t] + (A+BK[t])^\top P_{t+1}(A+BK[t]) & \text{if } t\in\{0,1,\dots T-1\}\\ Q_f & \text{if } t = T\end{cases} \tag{18.5a}$$

$$K[t] \triangleq -\left(R+B^\top P_{t+1}B\right)^{-1}B^\top P_{t+1}A \tag{18.5b}$$

One short remark before we prove the theorem: we use subscript notation for P_t in order to match the notation of the subscript in the value function $\mathbf{V}_t$. We do not do the same for the control gain K.

Proof The proof of Theorem 18.44 follows directly by induction. First, when $t = T$, the cost is independent of the control and so the optimal cost-to-go is

$$\mathbf{V}_T(\mathbf{x}) = \mathbf{x}^\top Q_f \mathbf{x} \quad \forall \mathbf{x} \in \mathbb{R}^n \implies P[T] = Q_f.$$

Now, suppose (18.5) holds for $t = \tau + 1$. From (18.4), we get

$$\begin{aligned}
\mathbf{V}_\tau(\mathbf{x}[\tau]) &\triangleq \min_{\mathbf{u}[\tau]\in\mathbb{R}^m} \mathbf{x}^\top[\tau]Q\mathbf{x}[\tau] + \mathbf{u}^\top[\tau]R\mathbf{u}[\tau] + \mathbf{V}_{\tau+1}(A\mathbf{x}[\tau] + B\mathbf{u}[\tau]) \\
&= \min_{\mathbf{u}[\tau]\in\mathbb{R}^m} \mathbf{x}^\top[\tau]Q\mathbf{x}[\tau] + \mathbf{u}^\top[\tau]R\mathbf{u}[\tau] + (A\mathbf{x}[\tau] + B\mathbf{u}[\tau])^\top P_{\tau+1}(A\mathbf{x}[\tau] + B\mathbf{u}[\tau]) \\
&= \min_{\mathbf{u}[\tau]\in\mathbb{R}^m} \mathbf{x}^\top[\tau]Q\mathbf{x}[\tau] + \mathbf{u}^\top[\tau]R\mathbf{u}[\tau] + \mathbf{x}^\top[\tau]A^\top P_{\tau+1}A\mathbf{x}[\tau] + \mathbf{u}^\top[\tau]B^\top P_{\tau+1}B\mathbf{u}[\tau] \\
&\qquad + \mathbf{x}^\top[\tau]A^\top P_{\tau+1}B\mathbf{u}[\tau] + \mathbf{u}^\top[\tau]B^\top P_{\tau+1}A\mathbf{x}[\tau]
\end{aligned}$$

To get the optimal variable $\mathbf{u}[\tau]$, simply differentiate the above with respect to $\mathbf{u}[\tau]$ and set the expression equal to zero:

$$\begin{aligned}
0 &= 2\mathbf{u}^\top[\tau]R + 2\mathbf{x}^\top[\tau]A^\top P_{\tau+1}B + 2\mathbf{u}^\top[\tau]B^\top P_{\tau+1}B \implies \mathbf{u}^*[\tau] \\
&= \underbrace{-\left(R + B^\top P_{\tau+1}B\right)^{-1} B^\top P_{\tau+1}A}_{\triangleq K[\tau] \text{ as defined by (18.5a)}}\mathbf{x}[\tau]
\end{aligned}$$

Substituting $\mathbf{u}^*[\tau]$ back into $\mathbf{V}_\tau(\mathbf{x}[\tau])$ yields:

$$\begin{aligned}
\mathbf{V}_\tau(\mathbf{x}[\tau]) &= \mathbf{x}^\top[\tau]Q\mathbf{x}[\tau] + \mathbf{u}[\tau]^{*\top}R\mathbf{u}^*[\tau] + \left(A\mathbf{x}[\tau] + B\mathbf{u}^*[\tau]\right)^\top P_{\tau+1}\left(A\mathbf{x}[\tau] + B\mathbf{u}^*[\tau]\right) \\
&= \mathbf{x}^\top[\tau]\underbrace{\left(Q + K[\tau]^\top RK[\tau] + (A + BK[\tau])^\top P_{\tau+1}(A + BK[\tau])\right)}_{\triangleq P_\tau \text{ as defined by (18.5b)}}\mathbf{x}[\tau]
\end{aligned}$$

Thus, the formula holds for the case $t = \tau$, and our induction is complete. ■

In the infinite-horizon case ($T \to \infty$), DP can still be applied.

Theorem 18.45 (Infinite-Horizon DT LQR) *As $T \to \infty$ in Theorem 18.44, the recursion in (18.5) reaches a steady-state solution.*

$$P = Q + K^\top RK + (A + BK)^\top P(A + BK) \tag{18.6a}$$

$$K = -(R + B^\top PB)^{-1}B^\top PA \tag{18.6b}$$

The optimal feedback control law becomes a static-gain law $\mathbf{u}(t) = K\mathbf{x}(t)$.

Remark 18.34 (18.6a) can alternatively be written as

$$P = Q + A^\top PA - A^\top PB\left(R + B^\top PB\right)^{-1}B^\top PA \tag{18.7}$$

This is often called *discrete algebraic Riccati equation (DARE)*.

18.2.2 Continuous-Time Dynamics

We recall the CT LTI dynamics $\dot{\mathbf{x}}(t) = A\mathbf{x}(t) + B\mathbf{u}(t)$ with initial condition $\mathbf{x}(0) = \mathbf{x}_0$. As with the DT case, there are two types of cost functionals depending on the horizon of the problem:

$$\text{Finite-Horizon.} \quad J_{\mathbf{u}}(\mathbf{x}_0) \triangleq \int_0^T \underbrace{\left[\mathbf{x}^\top(s)\ \mathbf{u}^\top(s)\right] \tilde{Q} \begin{bmatrix} \mathbf{x}(s) \\ \mathbf{u}(s) \end{bmatrix}}_{\text{"running cost"}} ds + \underbrace{\mathbf{x}^\top(T) Q_f \mathbf{x}(T)}_{\text{"terminal cost"}} \tag{18.8a}$$

$$\text{Infinite-Horizon.} \quad J_{\mathbf{u}}(\mathbf{x}_0) \triangleq \int_0^\infty \left[\mathbf{x}^\top(s)\ \mathbf{u}^\top(s)\right] \tilde{Q} \begin{bmatrix} \mathbf{x}(s) \\ \mathbf{u}(s) \end{bmatrix} ds \tag{18.8b}$$

where the weighting matrices $\tilde{Q}$, Q, R, Q_f have same definitions and conditions as in the DT case. The optimal cost is determined by solving

$$\tilde{J}(\mathbf{x}_0) \triangleq \inf_{u \in \mathcal{U}} J_{\mathbf{u}}(\mathbf{x}_0) \text{ s.t. } \dot{\mathbf{x}}(t) = A\mathbf{x}(t) + B\mathbf{u}(t)$$

Much like the DT case, we can invoke Lemma 18.19 to solve this problem. We will again consider the case where the cross-term weight $S = 0$.

The continuous version of the cost-to-go recursion (18.4), which we will represent as $V(t, \mathbf{x}(t))$, can be written using the *Hamilton-Jacobi-Bellman (HJB) equation*. Although the HJB equation is used for more general optimal control problems, we focus specifically on its application to our LQR problem.

First, we assume the cost-to-go function $V(t, \mathbf{x})$ is C^1 in both t and $\mathbf{x}$. The terminal condition is given by

$$\mathbf{V}(T, \mathbf{x}) = \mathbf{x}^\top(T) Q_f \mathbf{x}(T)$$

For $t < T$, we take an infinitesimally small timestep Δt from time t to time $t + \Delta t$ to get, via the first-order Taylor expansion

$$\begin{aligned} &\mathbf{V}(t+\Delta t, \mathbf{x}(t+\Delta t)) = \mathbf{V}(t, \mathbf{x}(t)) + \partial_t \mathbf{V}(t, \mathbf{x}(t))\Delta t + \nabla_{\mathbf{x}} \mathbf{V}(t, \mathbf{x}(t))\dot{\mathbf{x}}(t)\Delta t + \text{h.o.t.} \\ \Longrightarrow\ &\frac{1}{\Delta t}(\mathbf{V}(t+\Delta t, \mathbf{x}(t+\Delta t)) - \mathbf{V}(t, \mathbf{x}(t))) = \partial_t \mathbf{V}(t, \mathbf{x}(t)) + \nabla_{\mathbf{x}} \mathbf{V}(t, \mathbf{x}(t))\dot{\mathbf{x}}(t) + \text{h.o.t.} \end{aligned}$$

where h.o.t. denotes higher order terms. Integrate both sides, use the fact that $\int_t^{t+\Delta t} l(s)\, ds = l(t)\Delta t + \text{h.o.t.}$ for some placeholder function l. Then rearranging terms and taking $\Delta t \to 0$ yields exactly the HJB equation we need. Substituting in the running cost from (18.8) and the system dynamics yields

$$\begin{aligned} 0 &= \min_{u \in \mathcal{U}} \left\{ \mathbf{x}^\top(t) Q \mathbf{x}(t) + \mathbf{u}^\top R \mathbf{u}(t) + \partial_t \mathbf{V}(t, \mathbf{x}) + \nabla_{\mathbf{x}} \mathbf{V}(t, \mathbf{x}) \dot{\mathbf{x}}(t) \right\} \\ &= \min_{u \in \mathcal{U}} \left\{ \mathbf{x}^\top(t) Q \mathbf{x}(t) + \mathbf{u}^\top R \mathbf{u}(t) + \partial_t \mathbf{V}(t, \mathbf{x}) + \nabla_{\mathbf{x}} \mathbf{V}(t, \mathbf{x}) \left(A\mathbf{x}(t) + B\mathbf{u}(t) \right) \right\} \end{aligned} \tag{18.9}$$

Theorem 18.46 (Finite-Horizon CT LQR) *The optimal cost-to-go and optimal control at each time is given by*

$$V(t, \mathbf{x}) = \mathbf{x}^\top P(t) \mathbf{x} \quad \forall \mathbf{x} \in \mathbb{R}^n, \quad \mathbf{u}(t) = K(t)\mathbf{x}(t), \quad \forall\, t \in \{0, 1, \ldots, T\}$$

where $P(t)$ satisfies the following final-value problem and the control gain K can be computed from P:

$$P(t) = \begin{cases} Q - P(t)BR^{-1}B^\top P(t) + P(t)A + A^\top P(t) + \dot{P}(t) = 0 & \textit{if}\ t < T \\ Q_f & \textit{if}\ t = T \end{cases} \tag{18.10a}$$

$$K(t) \triangleq -R^{-1}B^\top P(t) \tag{18.10b}$$

Proof We simply continue from the HJB equation (18.9). The minimizing $\mathbf{u}^*$ can be found by setting the gradient of (18.9) to 0.

$$0 = 2\mathbf{u}^{*\top}(t) R + \nabla_{\mathbf{x}} \mathbf{V}(t, \mathbf{x}) B \quad \Longrightarrow \quad \mathbf{u}^*(t) = -\frac{1}{2} R^{-1} B^\top \nabla_{\mathbf{x}} \mathbf{V}(t, \mathbf{x})^\top \tag{18.11}$$

Similar to the DT case, we try a solution of the form $\mathbf{V}(t, \mathbf{x}) = \mathbf{x}^\top P(t)\mathbf{x}$, where $P(t) = P^\top(t) \succ 0, \forall t$. Then $\partial_t \mathbf{V}(t, \mathbf{x}) = \mathbf{x}^\top \dot{P}(t)\mathbf{x}$ and $\nabla_{\mathbf{x}} V(t, \mathbf{x}) = 2\mathbf{x}^\top P(t)$. Substituting these back into the above $\mathbf{u}^*$ and the HJB equation (18.9), we get the desired control input (18.10b) and

$$0 = \mathbf{x}^\top \left\{ Q - P(t)BR^{-1}B^\top P(t) + 2P(t)A + \dot{P}(t) \right\} \mathbf{x}$$

which results in (18.10a). finally, substituting $\nabla_{\mathbf{x}} \mathbf{V}(t, \mathbf{x}) = 2\mathbf{x}^\top P(t)$ into (18.11) and using $\mathbf{u}^*(t) = -K(t)\mathbf{x}(t)$ gives us (18.10b). ∎

Theorem 18.47 (Infinite-Horizon CT LQR) *As $T \to \infty$ in Theorem 18.46, the recursion in (18.10a) reaches a steady-state solution, i.e., set $\dot{P}(t) = 0$.*

$$0 = Q - PBR^{-1}B^\top P + PA + A^\top P \tag{18.12}$$

As $P(t) \equiv P$ is constant, the optimal control law becomes a static-gain law, yielding $\mathbf{u}(t) = K\mathbf{x}(t) = -R^{-1}B^\top P\mathbf{x}(t)$.

This Eq. (18.12) is often referred to as the *continuous algebraic Riccati equation (CARE)*.

18.2.3 LQR via Lagrange Multipliers

Since the LQR is an optimization problem, it can also be solved directly via *Lagrange multipliers*. Consider the discrete-time (DT) infinite-horizon case:

$$\min_{\mathbf{u}} J_{\mathbf{u}}(\mathbf{x}_0) \triangleq \frac{1}{2}\sum_{t=0}^{\infty}\left(\mathbf{x}[t]^\top Q\mathbf{x}[t] + \mathbf{u}[t]^\top R\mathbf{u}[t]\right)$$
$$\text{s.t.} \quad \mathbf{x}[t+1] = A\mathbf{x}[t] + B\mathbf{u}[t], \quad \mathbf{x}(0) = \mathbf{x}_0$$

Define Lagrangian $\mathcal{L}(\boldsymbol{\lambda}, \mathbf{u}, \mathbf{x})$ with dual variable $\boldsymbol{\lambda} \triangleq (\boldsymbol{\lambda}_1, \boldsymbol{\lambda}_2, \ldots)$, $\boldsymbol{\lambda}_t \in \mathbb{R}^n$:

$$\mathcal{L}(\boldsymbol{\lambda}, \mathbf{u}, \mathbf{x}) \triangleq J_{\mathbf{u}}(\mathbf{x}_0) + \sum_{t=0}^{\infty}\boldsymbol{\lambda}^\top[t](A\mathbf{x}[t] + B\mathbf{u}[t] - \mathbf{x}[t+1])$$

For each t:

$$\nabla_{\mathbf{u}}\mathcal{L}(\boldsymbol{\lambda}, \mathbf{u}, \mathbf{x}) = \mathbf{u}^\top[t]R + \boldsymbol{\lambda}^\top[t]B \overset{\text{set}}{=} 0 \implies \mathbf{u}[t] = -R^{-1}B^\top\boldsymbol{\lambda}_t \tag{18.13}$$

$$\nabla_{\mathbf{x}}\mathcal{L}(\boldsymbol{\lambda}, \mathbf{u}, \mathbf{x}) = \mathbf{x}^\top[t]Q + \boldsymbol{\lambda}_{t+1}^\top A - \boldsymbol{\lambda}^\top[t] \overset{\text{set}}{=} 0 \implies \boldsymbol{\lambda}_t = A^\top\boldsymbol{\lambda}_{t+1} + Q\mathbf{x}[t] \tag{2}$$

Thus, the state equation runs forwards in t:

$$\mathbf{x}[t+1] = A\mathbf{x}[t] + B\mathbf{u}[t], \quad \mathbf{x}(0) = \mathbf{x}_0 \tag{18.14}$$

and the costate equation runs backwards in t:

$$\boldsymbol{\lambda}_{t-1} = A^\top\boldsymbol{\lambda}_t + Q\mathbf{x}[t] \tag{18.15}$$

The variable $\boldsymbol{\lambda}$ is often called the *costate*, and (18.15) is referred to as the *costate equation* (or adjoint system, similar to definition in Chap. 11). Substituting (18.13) into (18.14) and (18.15) yields:

$$\begin{bmatrix}\mathbf{x}[t+1]\\ \boldsymbol{\lambda}_t\end{bmatrix} = \begin{bmatrix}A + BR^{-1}B^\top A^\top Q & -BR^{-1}B^\top\\ -A^\top Q & A^\top\end{bmatrix}\begin{bmatrix}\mathbf{x}[t]\\ \boldsymbol{\lambda}_{t-1}\end{bmatrix}$$
$$\implies \begin{bmatrix}\mathbf{x}[t+1]\\ \boldsymbol{\lambda}_{t+1}\end{bmatrix} = \begin{bmatrix}A & -BR^{-1}B^\top\\ -A^\top QA & A^\top(I + QBR^{-1}B^\top)\end{bmatrix}\begin{bmatrix}\mathbf{x}[t]\\ \boldsymbol{\lambda}_t\end{bmatrix}$$

Define M_1, M_2 as:

$$M_1 \triangleq \begin{bmatrix}A + BR^{-1}B^\top A^\top Q & -BR^{-1}B^\top\\ -A^\top Q & A^\top\end{bmatrix}, \quad M_2 \triangleq \begin{bmatrix}A & -BR^{-1}B^\top\\ -A^\top QA & A^\top(I + QBR^{-1}B^\top)\end{bmatrix}$$

It turns out $\boldsymbol{\lambda}_t = P\mathbf{x}[t+1]$, where P is the solution to the DARE from Theorem 18.45. This can be shown by applying the *Schur decomposition* to M_1 and M_2, although we won't delve into that here. We will see other uses of the Schur complement in the next chapter.

Using $\boldsymbol{\lambda}_t = P\mathbf{x}[t+1]$:

$$\mathbf{u}[t] = -R^{-1}B^\top\boldsymbol{\lambda}_t = -R^{-1}B^\top P\mathbf{x}[t+1] = -R^{-1}B^\top P(A\mathbf{x}[t] + B\mathbf{u}[t])$$

$$\implies \mathbf{u}[t] = -(R + B^\top PB)^{-1}B^\top PA\mathbf{x}[t]$$

by multiplying R across and combining $\mathbf{u}[t]$ terms. This is exactly the form of $\mathbf{u}[t]$ we derived in Theorem 18.45.

The continuous-time (CT) case is similar. Let's consider the finite-horizon scenario this time:

$$\min_{\mathbf{u}(t)} J_{\mathbf{u}}(\mathbf{x}_0) \triangleq \frac{1}{2}\int_0^T x^\top(t)Q\mathbf{x}(t) + u^\top(t)R\mathbf{u}(t)dt + \frac{1}{2}x^\top(T)Q_f\mathbf{x}(T)$$
$$\text{s.t. } \dot{\mathbf{x}}(t) = A\mathbf{x}(t) + B\mathbf{u}(t)$$

The Lagrangian, with dual variable $\boldsymbol{\lambda} \in \mathcal{L}_2([0, T]; \mathbb{R}^n)$, given by

$$\begin{aligned}
\mathcal{L}(\boldsymbol{\lambda}, \mathbf{u}, \mathbf{x}) &\triangleq J_{\mathbf{u}}(\mathbf{x}_0) + \langle \boldsymbol{\lambda}, Ax + Bu - \dot{\mathbf{x}}\rangle_{\mathcal{L}_2} \\
&= J_{\mathbf{u}}(\mathbf{x}_0) + \int_0^T \boldsymbol{\lambda}^\top(s)\left(A\mathbf{x}(s) + B\mathbf{u}(s) - \dot{\mathbf{x}}(s)\right)ds \\
\nabla_{\mathbf{u}}\mathcal{L}(\boldsymbol{\lambda}, \mathbf{u}, \mathbf{x}) &= u^\top(t)R + \boldsymbol{\lambda}^\top(t)B \overset{\text{set}}{=} 0 \implies \mathbf{u}(t) = -R^{-1}B^\top\boldsymbol{\lambda}(t) \\
\nabla_{\mathbf{x}}\mathcal{L}(\boldsymbol{\lambda}, \mathbf{u}, \mathbf{x}) &= x^\top(t)Q + \boldsymbol{\lambda}(T)A - \nabla_{\mathbf{x}}\int_0^T \boldsymbol{\lambda}^\top(s)\dot{\mathbf{x}}(s)ds
\end{aligned}$$

and we use integration by parts:

$$\int_0^T \boldsymbol{\lambda}^\top(s)\dot{\mathbf{x}}(s)ds = \boldsymbol{\lambda}^\top(T)\mathbf{x}(T) - \boldsymbol{\lambda}^\top(0)\mathbf{x}_0 - \int_0^T \dot{\boldsymbol{\lambda}}^\top(s)\mathbf{x}(s)ds$$

$$\implies \nabla_{\mathbf{x}}\int_0^T \boldsymbol{\lambda}^\top(s)\dot{\mathbf{x}}(s)ds = -\dot{\boldsymbol{\lambda}}^\top(t)$$

$$\therefore \nabla_{\mathbf{x}}\mathcal{L}(\boldsymbol{\lambda}, \mathbf{u}, \mathbf{x}) \overset{\text{set}}{=} 0 \implies \dot{\boldsymbol{\lambda}}(t) = -A^\top\boldsymbol{\lambda}(t) - Q\mathbf{x}(t) \quad \forall t \in (0, T)$$

At $t = T$:

$$\nabla_{\mathbf{x}}\mathcal{L}(\boldsymbol{\lambda}, \mathbf{u}, \mathbf{x}) \overset{\text{set}}{=} 0 \implies \boldsymbol{\lambda}(T) = Q_f\mathbf{x}(T)$$

Really, we need *distribution theory* and *generalized functions* to make concepts such as "$\nabla_{\mathbf{x}}\mathcal{L}(f)$" more precise. However, we won't go through these details here, and use them informally.

Thus, the optimality conditions are

$$\begin{bmatrix} \dot{\mathbf{x}} \\ \dot{\boldsymbol{\lambda}} \end{bmatrix} = \begin{bmatrix} A & -BR^{-1}B^\top \\ -Q & -A^\top \end{bmatrix} \begin{bmatrix} \mathbf{x} \\ \boldsymbol{\lambda} \end{bmatrix} \quad \text{with } \mathbf{x}(0) = \mathbf{x}_0 \text{ and } \boldsymbol{\lambda}(T) = Q_f\mathbf{x}(T)$$

As in the DT case, we can show $\boldsymbol{\lambda}(t) = P(t)\mathbf{x}(t)$, where $P(t)$ satisfies the CARE from Theorem 18.47.

18.2.4 LQR is a Stabilizing Controller

So far, we have discussed the optimality of LQR control, but one might also wonder if the LQR actually *stabilizes* the plant (A, B)? To show this, we connect our discussion back to Lyapunov stability, namely, stability analysis using the Lyapunov inequality.

Recall Lyapunov equations of the form $A^\top P + PA + Q = 0$ where $Q = Q^\top$. Previous chapters on Lyapunov stability (see II) told us the following for the CT LTI case: Given any $Q \succ 0$, $\exists!\, P = P^\top \succ 0$ s.t. $A^\top P + PA + Q = 0$ iff $\dot{\mathbf{x}} = A\mathbf{x}$ is globally asymptotically stable. We also saw how the Lyapunov Direct Method with Lyapunov function $\mathbf{x}^\top P\mathbf{x}$ can verify this. The steady-state continuous-time Riccati equation can be rewritten as a Lyapunov equation:

$$\begin{aligned} & Q - PBR^{-1}B^\top P + PA + A^\top P = 0, \quad K \triangleq -R^{-1}B^\top P \\ \Longrightarrow\ & Q + P(A + BK) + A^\top P + K^\top \left(B^\top P + RK\right) = 0 \\ \Longrightarrow\ & (A + BK)^\top P + P(A + BK) + \left(Q + K^\top RK\right) = 0 \end{aligned}$$

where $Q + K^\top RK > 0$ since $R \succ 0$. Thus, by Lyapunov stability criteria above, $P = P^\top \succ 0$ if and only if $A + BK$ is Hurwitz.

So far, we saw several examples of matrix equations, including the Lyapunov equation $A^\top P + PA + Q = 0$, which is linear in the variable P (assuming known A, Q). Both Lyapunov inequalities and Riccati inequalities are LMIs. In particular, Riccati inequalities can be shown to be equivalent to a linear form as follows.

Definition 18.101 (*Schur Complement*) Riccati inequalities can be shown to be equivalent to a linear form as follows:

$$A^\top P + PA + PBR^{-1}B^\top P + Q \preceq 0, \tag{18.16a}$$

$$\Longleftrightarrow \begin{bmatrix} A^\top P + PA + \tilde{Q}^\top \tilde{Q} & P\tilde{B} \\ \tilde{B}^\top P & -I \end{bmatrix} \preceq 0. \tag{18.16b}$$

where $Q \triangleq \tilde{Q}^\top \tilde{Q}$ and $BR^{-1}B^\top \triangleq \tilde{B}^\top \tilde{B}$ can be defined because $Q = Q^\top \succeq 0$ and $R = \tilde{R}^\top \succ 0$. The form (18.16b) is often called the *Schur complement* of (18.16a).

Let's try an alternative derivation, using these matrix equations, that $\mathbf{u}(t) = -R^{-1}B^\top P\mathbf{x}(t)$ is the control input that minimizes $J_\mathbf{u}(\mathbf{x}_0)$ in the infinite-horizon CT LTI case.

Lemma 18.20 (Comparison of Riccati Solutions) *If $S \succ 0$ and $Q_2 \succeq Q_1 \succeq 0$, then the two solutions $P_1 \succ 0$, $P_2 \succ 0$ to the respective Riccati equations*

$$Q_1 - P_1SP_1 + A^\top P_1 + P_1A = 0, \quad Q_2 - P_2SP_2 + A^\top P_2 + P_2A = 0,$$

are such that $P_2 \succeq P_1$ if and only if $A - SP_2$ is Hurwitz.

Proof Let's add the terms $(P_1 - P_2)\,S\,(P_1 - P_2)$ to the first Riccati equation. First expand

$$(P_1 - P_2)\,S\,(P_1 - P_2) = P_1SP_1 - P_1SP_2 - P_2SP_1 + P_2SP_2$$

We rewrite as

$$\begin{aligned} 0 &= Q_1 - P_1SP_1 + A^\top P_1 + P_1A - P_1SP_2 - P_2SP_1 + P_2SP_2 + P_1SP_2 + P_2SP_1 - P_2SP_2 \\ &= Q_1 - (P_1 - P_2)\,S\,(P_1 - P_2) + (A - P_2S)^\top P_1 + P_1(A - SP_2) + P_2SP_2 \end{aligned} \tag{18.17}$$

For the second Riccati equation:

$$\begin{aligned} 0 &= Q_2 - P_2SP_2 + A^\top P_2 + P_2A + P_2SP_2 - P_2SP_2 \\ &= Q_2 + (A - SP_2)^\top P_2 + P_2(A - SP_2) + P_2SP_2 \end{aligned} \tag{18.18}$$

Subtracting (18.17) and (18.18) yields the Lyapunov equation

$$(A - SP_2)^\top (P_2 - P_1) + (P_2 - P_1)(A - SP_2) + \widetilde{Q} = 0$$

where $\tilde{Q} \triangleq (Q_2 - Q_1) + (P_1 - P_2)\,S\,(P_1 - P_2) \succeq 0$. Therefore, by the Lyapunov stability criteria above, $(P_1 - P_2) \succ 0$ if $A - SP_2$ is Hurwitz. ∎

We can apply Lemma 18.20 when $S \triangleq BR^{-1}B^\top \succ 0$ and $Q_1 \triangleq Q$. The Riccati equation becomes:

$$Q - P_1BR^{-1}B^\top P_1 + A^\top P_1 + P_1A = 0.$$

Considering:

$$\begin{aligned} &Q + K^\top K + P_2BK + K^\top B^\top P_2 + A^\top P_2 + P_2A = 0, \\ \Longrightarrow\ &\underbrace{Q + \left(P_2BR^{-1} + K^\top\right) R \left(R^{-1}B^\top P_2 + K\right)}_{\triangleq Q_2} - P_2BR^{-1}B^\top P_2 + A^\top P_2 + P_2A = 0. \end{aligned}$$

Note that $Q_2 \succeq Q_1$ since $R \succ 0$. Choose $K^* \triangleq -R^{-1}B^\top P_1$, which makes $A - SP_1 = A - BR^{-1}B^\top P_1$ Hurwitz. Then, for any other $K \triangleq -R^{-1}B^\top P_2 \neq K^*$ that stabilizes $A - SP_2$, Lemma 18.20 states $P_2 \succeq P_1$.

Thus, $\mathbf{x}^\top P_2\mathbf{x} \geq \mathbf{x}^\top P_1\mathbf{x}$ for all $\mathbf{x} \in \mathbb{R}^n$, i.e., the cost-to-go under $\mathbf{u}(t) = K\mathbf{x}(t)$ is always larger than the cost-to-go under $\mathbf{u}(t) = K^*\mathbf{x}(t)$. Hence, the minimum cost J_u is achieved with $\mathbf{u}(t) = -R^{-1}B^\top P_1\mathbf{x}(t)$.

18.2.5 Output-Feedback Case

LQR problems can also be posed for output-feedback control, i.e., when $C \neq I$ and $\mathbf{u} = K\mathbf{y}$. First, recall the general LTI system $\dot{\mathbf{x}}(t) = A\mathbf{x}(t) + B\mathbf{u}(t)$ ($\mathbf{x}[t+1] = A\mathbf{x}[t] + B\mathbf{u}[t]$ in DT) with measurement equation $\mathbf{y}(t) = C\mathbf{x}(t) + D\mathbf{u}(t)$ (i.e., $\mathbf{y}[t] = C\mathbf{x}[t] + D\mathbf{u}[t]$ in DT). The performance index becomes:

$$J_{\mathbf{u}}(\mathbf{x}_0) = \int_0^\infty (\mathbf{y}^\top(t)Q\mathbf{y}(t) + \mathbf{u}^\top(t)R\mathbf{u}(t))\, dt$$

and similarly for the finite-horizon and DT cases.

Note that substituting $\mathbf{y}(t) = C\mathbf{x}(t) + D\mathbf{u}(t)$ into $J_{\mathbf{u}}(\mathbf{x}_0)$ gives us the usual performance index in terms of $\mathbf{x}$ and $\mathbf{u}$:

$$\begin{aligned} J_{\mathbf{u}}(\mathbf{x}_0) &= \int_0^\infty \left(\mathbf{x}^\top(t)C^\top QC\mathbf{x}(t) + \mathbf{u}^\top(t)(R + D^\top QD)\mathbf{u}(t) + 2\mathbf{x}^\top(t)C^\top QD\mathbf{u}(t)\right) dt \\ &= \int_0^\infty \begin{bmatrix}\mathbf{x}^\top(t) & \mathbf{u}^\top(t)\end{bmatrix} \begin{bmatrix} C^\top QC & C^\top QD \\ D^\top QC & R + D^\top QD \end{bmatrix} \begin{bmatrix}\mathbf{x}(t) \\ \mathbf{u}(t)\end{bmatrix} dt \end{aligned}$$

In fact, output-feedback LQR is a clear application of the LQR problem (18.8) with $S \neq 0$.

A corresponding CARE can be derived.

$$A^\top P + PA + C^\top QC - (PB + C^\top QD)(R + D^\top QD)^{-1}(PB + C^\top QD)^\top = 0 \tag{18.19}$$

Lemma 18.21 *Suppose CARE (18.19) has a symmetric solution P. Then*

$$R + H^\top(s)QH(s) = (I + G(s))^\top(R + D^\top QD)(I + G(s))$$

where $H(s) \triangleq (C(sI - A)^{-1}B + D$ *and* $G(s) = K(sI - A)^{-1}B$ *with control gain* K *defined as:*

$$K \triangleq (R + D^\top QD)^{-1}(B^\top P + D^\top QC) \tag{18.20}$$

Proof Using the definition of K, (18.19) can be rewritten as

$$A^\top P + PA + C^\top QC - K^\top (R + D^\top QD)K = 0$$

$$\implies -(-sI - A)^\top P - P(sI - A) + C^\top QC - K^\top (R + D^\top QD)K = 0$$

Multiply on the left side by $B^\top(-sI - A)^{-\top}$ and on the right side by $(sI - A)^{-1}B$:

$$\begin{aligned} &\implies -B^\top P(sI - A)^{-1}B - B^\top(-sI - A)^{-\top}PB \\ &+ B^\top(-sI - A)^{-\top}C^\top QC - (K^\top(R + D^\top QD)K)(sI - A)^{-1}B = 0 \end{aligned}$$

Note that $B^\top P = (R + D^\top QD)K - D^\top QC$. Thus, we get

$$\begin{aligned} &(D^\top QC - (R + D^\top QD)K)(sI - A)^{-1}B - B^\top(-sI - A)^{-\top}(K^\top(R + D^\top QD) - C^\top QD) \\ &+ B^\top(-sI - A)^{-\top}C^\top QC - K^\top(R + D^\top QD)K(sI - A)^{-1}B = 0 \end{aligned}$$

Algebraic manipulation of this expression, using $C(sI - A)^{-1}B = H(s) - D$ and $K(sI - A)^{-1}B = J(s) - I$ yields the desired result. ■

Remark 18.35 Lemma 18.21 is sometimes called the *Kalman-Yakubovich-Popov (KYP) Lemma*. It essentially checks whether the quadratic cost functional $J_{\mathbf{u}}(\mathbf{x}_0)$ is nonnegative or nonpositive for all signals $\mathbf{x}(t)$ and $\mathbf{u}(t)$ satisfying the system dynamics. Several variations of the return-difference matrix and KYP lemma (and another result called the *KYP inequality*) have a lot of application to robust control and stochastic control. We will see different variations of the KYP lemma and KYP inequality in the following chapters.

18.3 Solving Riccati Equations

So far, in Sect. 18.2, we have expressed the LQR optimal control problem in terms of several types of *algebraic Riccati equations*. To complete our derivation, we need to actually solve these equations. For simplicity, we will focus our discussion on the continuous-time, infinite horizon case. The general CARE can be represented as

$$A^\top P + PA - PSP + Q = 0 \tag{18.21}$$

where $Q \succeq 0$ and $S \succ 0$. In the context of the LQR problem, S can be defined as $S = BR^{-1}B^\top$ in order to obtain the CARE (18.12). There are several ways to solve the general CARE (18.21).

18.3.1 With the Hamiltonian Matrix

We begin with a definition of a Hamiltonian matrix, which is a special type of matrix that is convenient for the analysis of Riccati equations.

Definition 18.102 (*Hamiltonian Matrix*) Define the matrix $M \triangleq \begin{bmatrix} M_{11} & M_{12} \\ M_{21} & M_{22} \end{bmatrix}$, $M_{ij} \in \mathbb{R}^{n\times n}$, $M \in \mathbb{R}^{2n\times 2n}$. Then M is *Hamiltonian* if it satisfies the property

$$JMJ = M^\top, \quad \text{where } J \triangleq \begin{bmatrix} 0 & I \\ -I & 0 \end{bmatrix} \tag{18.22}$$

Remark 18.36 The matrix J in Definition 18.102 satisfies several convenient properties, including $J^{-1} = J^\top = -J$. The condition (18.22) can be rewritten as $JM + M^\top J = 0$.

Based on (18.21), let's construct a Hamiltonian matrix

$$M \triangleq \begin{bmatrix} A & -BR^{-1}B^\top \\ -Q & -A^\top \end{bmatrix} \tag{18.23}$$

Our general method of solving the CARE from Theorem 18.47 is as follows. First, let $U, V \in \mathbb{R}^n$ be matrices such that

$$\begin{bmatrix} U \\ V \end{bmatrix} \equiv \begin{bmatrix} U \\ \text{---} \\ V \end{bmatrix} = \begin{bmatrix} u_{11} & \cdots & u_{1n} \\ \vdots & \ddots & \vdots \\ u_{n1} & \cdots & u_{nn} \\ \hline v_{11} & \cdots & v_{1n} \\ \vdots & \ddots & \vdots \\ v_{n1} & \cdots & v_{nn} \end{bmatrix}$$

have columns which span a M-invariant subspace of $\mathbb{R}^{2n}$, i.e.,

$$M\begin{bmatrix} U \\ V \end{bmatrix} = \begin{bmatrix} U \\ V \end{bmatrix} Z \tag{18.24}$$

where $Z \in \mathbb{R}^{n\times n}$ and the eigenvalues of Z are a subset of the eigenvalues of M. Now, if U is nonsingular, (18.24) yields

$$\begin{cases} AU - SV = UZ \\ -QU - A^\top V = VZ \end{cases} \implies 0 = Q + A^\top VU^{-1} + VU^{-1}A - VU^{-1}SVU^{-1}$$

which is precisely in the form of the general CARE (18.21) if $P \triangleq VU^{-1}$. Thus, VU^{-1} is a solution to (18.21).

There are several additional questions we must address to be able to use this approach. Namely, how to choose U, V

- which spans the M-invariant subspace?
- such that U is nonsingular?
- such that P stabilizes (A, B)?

To address these questions, we need a few more properties of the Hamiltonian matrix.

Lemma 18.22 (Eigenvalue Pairing) *If $\lambda \in \mathbb{C}$ is an eigenvalue of M, then $-\bar{\lambda}$ is also an eigenvalue.*

Lemma 18.23 (Eigenvector Relationship for Hamiltonian Matrices) *If $\mathbf{v}$ is an eigenvector of M, then $J\mathbf{v}$ is an eigenvector of $M^\top$.*

Lemma 18.24 (Characterization of Hamiltonian Matrices) *M is Hamiltonian if and only if $M_{22} = -M_{11}^\top$ and M_{12}, M_{21} are both symmetric.*

Proof Recall that $J = \begin{bmatrix} 0 & I \\ -I & 0 \end{bmatrix}$, and $J^2 = -\begin{bmatrix} I & 0 \\ 0 & I \end{bmatrix}$ (and this resembles the imaginary unit $j^2 = -1$, perhaps one of the reasons why this matrix is named J).

It is easy to compute that we have

$$JMJ = \begin{bmatrix} -M_{22} & M_{21} \\ M_{12} & -M_{11} \end{bmatrix}. \tag{18.25}$$

Therefore, $JMJ = M^\top$ if and only if M_{12}, M_{21} is symmetric and $M_{22} = -M_{11}^\top$. ■

Lemma 18.25 (Negative Eigenvalue Subspace) *Let $\mathcal{M}^-$ be the subspace of $\mathbb{R}^{2n}$ spanned by eigenvectors corresponding to $\mathrm{Re}(\lambda_i(M)) < 0$. Then $\mathcal{M}^-$ is an M-invariant subspace of $\mathbb{R}^{2n}$.*

Proof This follows directly from eigenvector-eigenvalue relationship. Define $\{\mathbf{v}_1^-, \ldots, \mathbf{v}_k^-\}$, $k \leq n$ to be the eigenvectors corresponding to $\mathrm{Re}(\lambda_i(M)) < 0$. Then $\mathbf{v} \in \mathcal{M}^-$ can be expressed as $\mathbf{v} = \alpha_1 \mathbf{v}_1^- + \ldots + \alpha_k \mathbf{v}_k^-$, which means $M\mathbf{v} = \alpha_1 M \mathbf{v}_1^- + \ldots + \alpha_k M \mathbf{v}_k^- = \alpha_1 \lambda_1 \mathbf{v}_1^- + \ldots + \alpha_k \lambda_k \mathbf{v}_k^- \in \mathrm{span}\{\mathbf{v}_1^-, \ldots, \mathbf{v}_k^-\} \triangleq \mathcal{M}^-$. ■

Remark 18.37 Suppose M as defined by (18.23) is such that $\mathrm{Re}(\lambda_i(M)) \neq 0$ for all $i = 1, \ldots, 2n$. This means the value of k in Lemma 18.25 is exactly equal to n, and we have the same number of negative eigenvalues as positive ones, by Lemma 18.22. This further means that M has at least two distinct Jordan blocks of the form

$$M \begin{bmatrix} V^+ & V^- \end{bmatrix} = \begin{bmatrix} V^+ & V^- \end{bmatrix} \begin{bmatrix} J_M^+ & \\ & J_M^- \end{bmatrix}, \tag{18.26}$$

where J_M^+, $J_M^- \in \mathbb{C}^{n \times n}$, V^+, $V^- \in \mathbb{C}^{2n \times n}$ and J_M^- and $-J_M^+$ are Hurwitz.

Definition 18.103 (*Sylvester Equation*) A *Sylvester equation*, like the Lyapunov and Riccati equations, is another type of matrix equation. It is given by

$$AX + XB = C \qquad A \in \mathbb{C}^{n\times n},\ B \in \mathbb{C}^{m\times m},\ C \in \mathbb{C}^{n\times m}$$

where X is the variable to be solved. The Sylvester equation has a unique solution $X \in \mathbb{C}^{n\times m}$ only when A and $-B$ have no common eigenvalues. Moreover, $AX + XB = 0$ admits only the solution $X = 0$.

Lemma 18.26 *Let* $V^- \triangleq \begin{bmatrix} X_1 \\ X_2 \end{bmatrix}$ *be a decomposition of the* V^- *from (18.26). If* X_1 *is nonsingular, then* $P \triangleq X_2 X_1^{-1}$ *solves (18.21).*

Proof Because $\mathcal{M}^-$ is M-invariant, we have that $MV^- = V^- J_M^-$ where J_M^- is Hurwitz (this fact is also observed in (18.26)). Rote calculation gives us

$$\begin{aligned}
MV^- = V^- J_M^- &\implies \begin{bmatrix} A & -S \\ -Q & -A^\top \end{bmatrix}\begin{bmatrix} X_1 \\ X_2 \end{bmatrix} = \begin{bmatrix} X_1 \\ X_2 \end{bmatrix} J_M^- \\
&\implies \begin{bmatrix} A & -S \\ -Q & -A^\top \end{bmatrix}\begin{bmatrix} I \\ P \end{bmatrix} = \begin{bmatrix} X_1 \\ X_2 \end{bmatrix} J_M^- X_1^{-1} = \begin{bmatrix} I \\ P \end{bmatrix} X_1 J_M^- X_1^{-1} \\
&\implies \begin{bmatrix} P & -I \end{bmatrix}\begin{bmatrix} A & -S \\ -Q & -A^\top \end{bmatrix}\begin{bmatrix} I \\ P \end{bmatrix} = \begin{bmatrix} P & -I \end{bmatrix}\begin{bmatrix} I \\ P \end{bmatrix} X_1 J_M^- X_1^{-1} = 0
\end{aligned}$$

Expanding the left side of the equation gives us $A^\top P + PA - PSP + Q = 0$ and indeed, P satisfies (18.21). ■

Lemma 18.27 *The* P *obtained from Lemma 18.26 is real, symmetric, and uniquely stabilizing.*

Proof (1) Since M is real, conjugate $\bar{v} \in \mathbb{C}^{2n}$ of complex eigenvector $v \in \mathbb{C}^{2n}$ is also an eigenvector. Thus, rearranging the existing eigenvectors via some permutation matrix Π will yield

$$\bar{V}^- = \begin{bmatrix} \bar{X}_1 \\ \bar{X}_2 \end{bmatrix} = \begin{bmatrix} X_1 \\ X_2 \end{bmatrix} \Pi = \begin{bmatrix} X_1 \Pi \\ X_2 \Pi \end{bmatrix}$$

$$\implies \bar{P} = \bar{X}_2 \bar{X}_1^{-1} = X_2 \Pi \Pi^{-1} X_1^{-1} = X_2 X_1^{-1} = P \quad \therefore P \text{ is real.}$$

(2) P is symmetric if $P = X_2 X_1^{-1} = X_1^{-\top} X_2^\top = P^\top \implies X_1^\top X_2 - X_2^\top X_1 = 0$, so we will prove this.

Let's call $T \triangleq X_1^\top X_2 - X_2^\top X_1$. Note

$$T = (V^-)^\top J V^- = \begin{bmatrix} X_1^\top & X_2^\top \end{bmatrix} \begin{bmatrix} 0 & I \\ -I & 0 \end{bmatrix} \begin{bmatrix} X_1 \\ X_2 \end{bmatrix}$$

Since M is Hamiltonian, $JM + M^\top J = 0$

$$\Longrightarrow (V^-)^\top JMV^- + (V^-)^\top M^\top JV^- = 0$$

Because $MV^- = V^- J_M^-$, $(V^-)^\top M^\top = (J_M^-)^\top (V^-)^\top$, $\mathcal{M}^-$ is M-invariant. Thus, $TJ_M^- + (J_M^-)^\top T = 0$. Because J_M^- is Hurwitz, the only solution to this Lyapunov equation is $T = 0$. (Alternatively, this equation can be viewed as a Sylvester equation, and its only solution is $T = 0$). Thus, P is symmetric.

(3) Similar to the proof of Lemma 18.26,

$$[I\ 0]\begin{bmatrix} A & -S \\ -Q & -A^\top \end{bmatrix}\begin{bmatrix} I \\ P \end{bmatrix} = [I\ 0]\begin{bmatrix} I \\ P \end{bmatrix} X_1 J_M^- X_1^{-1} \Longrightarrow A - SP = X_1 J_M^- X_1^{-1}$$

Therefore, $A - SP$ is stable because it is similar to J_M^-, which is a Hurwitz matrix. To show uniqueness, assume there is a different $\tilde{P}$ which is a stabilizing solution to (18.21).

$$\begin{aligned} 0 &= (A^\top P + PA - PSP + Q) - (A^\top \tilde{P} + \tilde{P}A - \tilde{P}S\tilde{P} + Q) \\ &= A^\top (P - \tilde{P}) + (P - \tilde{P})A - PSP + \tilde{P}S\tilde{P} - PS\tilde{P} + PS\tilde{P} \\ &= A^\top (P - \tilde{P}) + (P - \tilde{P})A - PS(P - \tilde{P}) - (P - \tilde{P})S\tilde{P} \\ &= (A - SP)^\top (P - \tilde{P}) + (P - \tilde{P})(A - SP) \end{aligned}$$

This is a Sylvester equation with matrix variable $(P - \tilde{P})$ because $A - SP$ and $A - SP$ are both Hurwitz, and so $A - SP$ has all negative eigenvalues while $-(A - S\tilde{P})$ has all positive eigenvalues. Clearly, there are no shared eigenvalues. Thus, the only solution is $P - \tilde{P} = 0$, which means $P = \tilde{P}$. Thus, P is unique. ∎

Now consider an example application of the Hamiltonian matrix method to fully solve a LQR problem.

Example 18.35 Consider $\dot{\mathbf{x}}(t) = A\mathbf{x}(t) + B\mathbf{u}(t)$ s.t. $A = \begin{bmatrix} 0 & 2 \\ 5 & 0 \end{bmatrix}$ and $B = \begin{bmatrix} 0 \\ 3 \end{bmatrix}$. Suppose we want to minimize

$$J_{\mathbf{u}}(\mathbf{x}_0) = \int_0^\infty x^\top(t) \begin{bmatrix} 1 & 0 \\ 0 & 1 \end{bmatrix} \mathbf{x}(t) + \mathbf{u}^\top(t)\mathbf{u}(t)dt$$

Our Hamiltonian matrix (18.23) is given by

$$M = \begin{bmatrix} A & -BR^{-1}B^\top \\ -Q & -A^\top \end{bmatrix} = \left[\begin{array}{cc|cc} 0 & 2 & 0 & 0 \\ 5 & 0 & 0 & -9 \\ \hline -1 & 0 & 0 & -5 \\ 0 & -1 & -2 & 0 \end{array}\right]$$

Get $\mathcal{M}^-$ via eigenvector decomposition (MATLAB)

$$J = \begin{bmatrix} 4.8080 & & & \\ & 2.4255 & & \\ & & -4.8080 & \\ & & & -2.4255 \end{bmatrix}, \quad V = \begin{bmatrix} V_1^+ \; V_2^+ \; V_1^- \; V_2^- \end{bmatrix}$$

Split $\begin{bmatrix} V_1^- \; V_2^- \end{bmatrix}$ into $\begin{bmatrix} X_1 \\ X_2 \end{bmatrix}$ s.t. X_1^{-1} exists. For this example:

$$\begin{bmatrix} V_1^- \; V_2^- \end{bmatrix} = \begin{bmatrix} -0.3624 & 0.5501 \\ 0.8713 & -0.6672 \\ 0.1993 & 0.4862 \\ 0.2641 & 0.1258 \end{bmatrix} = \begin{bmatrix} X_1 \\ X_2 \end{bmatrix}, \quad X_1 = \begin{bmatrix} -0.3624 & 0.5501 \\ 0.8713 & -0.6672 \end{bmatrix} \Longrightarrow X_1^{-1} \text{ exists.}$$

(In fact, for a controllable and observable system, X_1 can be shown to be always invertible.)

Thus, $P \triangleq X_2 X_1^{-1} = \begin{bmatrix} 0.1993 & 0.4862 \\ 0.2641 & 0.1258 \end{bmatrix} \begin{bmatrix} 2.8089 & 2.3161 \\ 3.6683 & 1.5259 \end{bmatrix} = \begin{bmatrix} 2.3432 & 1.2034 \\ 1.2034 & 0.8037 \end{bmatrix}$.
One can check $P = P^\top \succ 0$. Moreover, one can easily verify that the same result $K = -R^{-1} B^\top P = \mathtt{lqr}(A, B, Q, R, 0)$ comes from using MATLAB's built-in `lqr` command. □

Chapter 19
Linear Robust and Stochastic Control

The general optimization problem for LQR, as discussed in Chap. 18 can alternatively be written as follows:

$$\begin{cases} \min\limits_{\mathbf{u}\in\mathcal{U}} J_{\mathbf{u}}(\mathbf{x}_0) \\ \text{s.t. } \dot{\mathbf{x}}(t) = A\mathbf{x}(t) + B\mathbf{u}(t) \\ \quad (\text{ or } \mathbf{x}[t+1] = A\mathbf{x}[t] + B\mathbf{u}[t]) \end{cases}$$

$$\Longrightarrow \begin{cases} \min\limits_{K(t)(\text{or } K[t])} J_{\mathbf{u}}(\mathbf{x}_0) \\ \text{s.t. } \dot{\mathbf{x}}(t) = A\mathbf{x}(t) + B\mathbf{u}(t), \mathbf{u}(t) = K(t)\mathbf{x}(t) \\ \quad (\text{ or } \mathbf{x}[t+1] = A\mathbf{x}[t] + B\mathbf{u}[t], \mathbf{u}[t] = K[t]\mathbf{x}[t]) \end{cases}$$

and in the specific infinite-horizon case, this can be written generically as

$$\begin{cases} \min\limits_{K} \left\| \underline{\mathcal{S}}(H, K) \right\| \\ \text{s.t. algebraic Riccati equation} \end{cases} \tag{19.1}$$

where $\underline{\mathcal{S}}(H, K)$ is the closed-loop transfer function from $\mathbf{u} \to \mathbf{x}$ (or $\mathbf{u} \to \mathbf{y}$ in the output-feedback case). The norm $\left\| \underline{\mathcal{S}}(H, K) \right\|$ can be used to represent the weighted sum associated with the cost functional (18.8) (or (18.1) in the DT case). Thus, a natural extension to other optimal control problems arises depending on different choices of this norm. This chapter focuses on two particular extensions–*$\mathcal{H}_2$ and $\mathcal{H}_\infty$ optimal control*–as well as an introduction to stochastic optimal control via the *linear quadratic Gaussian (LQG)* framework.

Following the notation of (19.1), the $\mathcal{H}_2$ and $\mathcal{H}_\infty$ optimal control problems can generally be posed in the following way.

SJ. Han, *Linear Systems*, KAIST Research Series,
https://doi.org/10.1007/978-981-96-9774-8_19

$$\mathcal{H}_\infty \text{ control} \triangleq \min_{K \in \mathcal{H}_\infty} \left\| \underline{S}(H, K) \right\|_{\mathcal{H}_\infty}, \qquad \mathcal{H}_2 \text{ control} \triangleq \begin{cases} \min\limits_{K \in \mathcal{H}_\infty} \left\| \underline{S}(H, K) \right\|_{\mathcal{H}_2} \\ \text{s.t. } \underline{S}(H, K) \in \mathcal{H}_\infty \end{cases} \tag{19.2}$$

The remainder of this chapter is dedicated to defining each part of (19.2), such as the $\mathcal{H}_\infty$ space, as well as the $\mathcal{H}_\infty$ and $\mathcal{H}_2$ system norms. More importantly, we will rewrite both optimization problems so that they can be tractably solved using standard computational tools like CVX and YALMIP. Our main tool for these reformulations are *linear matrix inequalities (LMIs)* (which we've already seen in Chap. 17). Using LMIs enables more flexible design of controllers, including being able to handle robustness constraints, structural constraints (in the case of distributed/network control), and maintenance of performance guarantees. As such, $\mathcal{H}_2$ and $\mathcal{H}_\infty$ optimal control typically has wide applicability to the subfield of *robust control.*

19.1 Stabilization Using LMIs

We first investigate a LMI-based reformulation for the generic state-feedback stabilization problem studied in Chap. 17: find (static) control gain $K \in \mathbb{R}^{m\times n}$ s.t. $\mathbf{u}(t) = K\mathbf{x}(t)$ which makes the closed-loop system $\dot{\mathbf{x}}(t) = (A + BK)\mathbf{x}(t)$ asymptotically stable. As we've seen using the Lyapunov inequality, this is equivalent to finding a $P = P^\top \succ 0$ such that $(A + BK)^\top P + P(A + BK) \prec 0$. Note that $P \succ 0$ if and only if $P^{-1} \succ 0$, and so we can multiply across the entire inequality by P^{-1} without changing the order of the inequality:

$$\begin{aligned} &P^{-1}A^\top + P^{-1}K^\top B^\top + AP^{-1} + BKP^{-1} \prec 0 \\ &\implies XA^\top + Z^\top B^\top + AX + BZ \prec 0 \end{aligned} \tag{19.3}$$

Therefore, CT system $\dot{\mathbf{x}} = A\mathbf{x} + B\mathbf{u}$ is stabilizable by $\mathbf{u} = K\mathbf{x}$ iff $X = X^\top \succ 0$ and $Z \in \mathbb{R}^{m\times n}$ such that (19.3) holds. The stabilizing feedback gain is than given by $K \triangleq ZX^{-1}$.

Lemma 19.28 (Finsler) *Let $\mathbf{x} \in \mathbb{R}^n$, and $Z = Z^\top$, $Q = Q^\top \in \mathbb{R}^{n\times n}$. The following identities are equivalent.*

(a) $\mathbf{x}^\top Z\mathbf{x} = 0$ *and* $\mathbf{x} \neq 0 \implies \mathbf{x}^\top Q\mathbf{x} \prec 0$
(b) $\exists \mu > 0$ *s.t.* $Q - \mu Z \prec 0$

In addition, if $Z \succeq 0$ so that we can decompose $Z = Y^\top Y$ for some $Y \in \mathbb{R}^{m\times n}$:

(1) $\mathbf{x}^\top Q\mathbf{x} \prec 0\ \forall \mathbf{x} \in \mathbb{R}^n \setminus \{0\}$ *s.t.* $Y\mathbf{x} = 0$
(2) $\exists \mu > 0$ *s.t.* $Q - \mu Y^\top Y \prec 0$
(3) $\exists X \in \mathbb{R}^{m\times m}$ *s.t.* $Q + XY + Y^\top X^\top \prec 0$

Note that using the Lyapunov inequality to check closed-loop stability is the same form as condition (3) in Lemma 19.28: $A^\top P + PA + K^\top B^\top P + PBK \prec 0$. The equivalence between (2) and (3) lets us eliminate K:

$$\begin{aligned}
\exists\mu > 0 \text{ s.t. } \quad & A^\top P + PA - \mu PBB^\top P \prec 0 \\
\implies \quad & A^\top(\mu P) + (\mu P)A - (\mu P)BB^\top(\mu P) \prec 0 \\
\implies \quad & \tilde{A}^\top \tilde{P} + \tilde{P}\tilde{A} - \tilde{P}BB^\top \tilde{P} \prec 0.
\end{aligned}$$

where $\tilde{P} \triangleq \mu P$ is still positive definite because $\mu > 0$. This is not a LMI, but can be turned into one by simply multiplying across both sides via $\tilde{P}^{-1} \prec 0$ to get $\tilde{P}^{-1}A^\top + A\tilde{P}^{-1} - BB^\top \prec 0$. Later, we will see some common techniques to transform general matrix inequalities to LMIs, e.g., using the Schur complement.

Now let's consider again the state-feedback LQR studied in Chap. 18. Using transfer function/block-diagram form, the full loop gain from $\mathbf{u}$ to $\mathbf{v}$ is $G(s) \triangleq \frac{V(s)}{U(s)} = K(sI - A)^{-1}B$.

Adding and subtracting $-sI$ in the CARE (18.12) tells us:

$$-(sI - A)^\top P - P(sI - A) - PBR^{-1}B^\top P + Q = 0$$

Premultipling across by $B^\top\Phi(-s)^\top$ and postmultipling by $\Phi(s)B$ let's us simplify the above to

$$B^\top P\Phi(s)B + B^\top\Phi(-s)^\top PB + B^\top\Phi(-s)^\top PBR^{-1}B^\top P\Phi(s)B = +B^\top\Phi(-s)^\top Q\Phi(s)B + R$$

Note $G(s) = K(sI - A)^{-1}B = K\Phi B = R^{-1}B^\top P\Phi B$, and $H(s) = \Phi(s)B$. Further simplifying yields:

$$(I + G)^\sim R(I + G) = R + B^\top\Phi^\sim Q\Phi B \tag{19.4}$$

where we use notation $\sim$ for $G^\sim(s) \triangleq G^\top(-s)$ for any generic transfer function G. The transfer matrix $I + G(s)$ is often called the *return-difference matrix* because $G(s)\mathbf{u}$ is the difference between original signal $\mathbf{u}$ and the signal $\mathbf{v}$ "returning" from the closed loop. In fact:

$$\det(I + G(s)) = \det(I + K(sI - A)^{-1}B) = \det(I + BK(sI - A)^{-1}) \tag{19.5}$$

comes from the property $\det(I + AB) = \det(I + BA)$

$$(3.5) = \det(sI - A)^{-1}\det(sI - A + BK) \quad \text{since } \det(AB) = \det A \det B$$

but $\det(sI - A + BK)$ is the closed-loop characteristic polynomial, while $\det(sI - A)$ is the open-loop characteristic polynomial.

The relation (19.4) is another type of the *KYP equality* we've seen in Sect. 18.2. The KYP equality can be rewritten as an inequality condition by noticing that $B^\top \Phi^\sim Q \Phi B \succeq 0$ and $\tilde{\Phi} B \succeq 0$:

$$(I + G(s))^\top R(I + G(s)) \succeq R$$

This form of the KYP makes it easier to handle in the context of LMIs.

19.2 Hardy Spaces and System Norms

We address two main questions in this section. First, how can we define the norm of a *system*, i.e., what do the norms in (19.2) mean? And how do we compute norms in the *frequency domain*?

19.2.1 Signal Norms Review

Recall that for time-domain signals $\mathbf{x}$ we defined the $\mathcal{L}_2$ and $\mathcal{L}_\infty$ norms as:

$$\begin{aligned} \|\mathbf{x}\|_{\mathcal{L}_2} &\triangleq \left(\int_{-\infty}^{\infty} \|\mathbf{x}(\tau)\|_2^2 \, d\tau \right)^{\frac{1}{2}}, \\ \|\mathbf{x}\|_{\mathcal{L}_\infty} &\triangleq \operatorname*{ess\,sup}_t \|\mathbf{x}(t)\|_\infty \triangleq \inf\{B \mid \|\mathbf{x}(t)\|_\infty \leq B \text{ a.e.}\} \end{aligned} \tag{19.6}$$

where "a.e" stands for almost everywhere except on sets of measure 0. We also say that $\mathbf{x} \in \mathcal{L}_2(\mathbb{R}, \mathbb{R}^n)$ space or $\mathcal{L}_\infty(\mathbb{R}, \mathbb{R}^n)$ space if $\mathcal{L}_2$ and $\mathcal{L}_\infty$ norms are defined.

We can extend the above norm spaces for complex-valued and matrix-valued signals. A complex matrix function F is in $\mathcal{L}_2(j\mathbb{R})$ space if the $\mathcal{L}_2(j\mathbb{R})$ norm given below is bounded:

$$\|F\|_{\mathcal{L}_2} \triangleq \left(\int_{-\infty}^{\infty} \|F(\sigma + j\omega)\|_F^2 \, d\omega \right)^{\frac{1}{2}} \tag{19.7}$$

All real rational strictly proper transfer matrices with no poles on the imaginary axis form a (not closed) subspace of $\mathcal{L}_2(j\mathbb{R})$. The matrix norm inside the integral in (19.7) is the *Frobenius norm* of a matrix which is defined for a matrix A as: $\|A\|_F = \sqrt{\sum_{i=1}^n \sum_{j=1}^n |a_{ij}|^2}$.

Likewise, a complex matrix function F is in $\mathcal{L}_\infty(j\mathbb{R})$ space if the $\mathcal{L}_\infty(j\mathbb{R})$ norm given below is bounded:

$$\|F\|_{\mathcal{L}_\infty} \triangleq \operatorname*{ess\,sup}_{\sigma, \omega \in \mathbb{R}} \sigma_{\max}\left(F(\sigma + j\omega)\right). \tag{19.8}$$

where $\sigma_{\max}$ is the largest singular value. The space of all proper real rational transfer matrices with no poles on the imaginary axis is a subspace of $\mathcal{L}_\infty$. Note that while $\mathcal{L}_2$ is a Hilbert space (because it can be induced by an inner product), $\mathcal{L}_\infty$ is only a Banach space.

19.2.2 System Norms

We reintroduce external disturbances and auxiliary output signals $\bar{\mathbf{z}}$, $\mathbf{w}$ for our discussion, and focus on systems that are specifically in the configuration on the left of Fig. 17.1. As we've seen before, it is represented by the following set of equations (in the frequency domain):

$$\begin{bmatrix} \bar{\mathbf{z}} \\ \mathbf{y} \end{bmatrix} = \begin{bmatrix} H_{11} & H_{12} \\ H_{21} & H_{22} \end{bmatrix} \begin{bmatrix} \mathbf{w} \\ \mathbf{u} \end{bmatrix}, \quad \mathbf{u} = K\mathbf{y} \tag{19.9}$$

We would like to define a complex mapping with respect to $\bar{\mathbf{z}}$ to formulate the control problem.

Definition 19.104 (*Linear Fractional Transformation*) Let H be a complex matrix partitioned as Eq. (19.9). The *(lower) Linear Fractional Transformation (LFT)* $\underline{S}(H, K)$ is given by:

$$\underline{S}(H, K) \triangleq \frac{\bar{\mathbf{z}}}{\mathbf{w}} = H_{11} + H_{12}\,(I - K H_{22})^{-1}\, K H_{21} \tag{19.10}$$

provided the well-posedness conditions ($(I - K H_{22})^{-1}$ is invertible, etc., as in Chap. 17) are satisfied.

The LFT is useful because it shows how the nominal system behavior, represented by H_{11}, is influenced by control perturbations. This framework allows us to design a controller K that not only stabilizes the system but also maximizes some type of "system performance". This is similar to the LQR problem from Chap. 18, where the system performance was quantified by a weighted sum of terms that were quadratic in $\mathbf{x}$ and $\mathbf{u}$. By representing uncertainties and disturbances explicitly, the LFT helps in optimizing both the robustness and performance of the control system.

Definition 19.105 (*$\mathcal{H}_2$ Hardy Space*) The *$\mathcal{H}_2$ Hardy space* is a (closed) subspace of $\mathcal{L}_2(j\mathbb{R})$ consisting of all matrix functions $f : \mathbb{C} \to \mathbb{C}^{n\times m}$ which are analytic (more simply, C^∞ and bounded) on the open right-half plane $\mathbb{C}^+$, with norm $\|\cdot\|_{\mathcal{H}_2}$ defined as:

$$\|H\|_{\mathcal{H}_2} \triangleq \sup_{\sigma>0} \left\{ \left(\frac{1}{2\pi} \int_{-\infty}^{\infty} \|H(\sigma + j\omega)\|_F^2 \, d\omega \right)^{\frac{1}{2}} \right\} \tag{19.11}$$

Moreover, the orthogonal complement $\mathcal{H}_2^{\perp}$ of $\mathcal{H}_2$ is the subspace $\mathcal{L}_2$ that contains all matrix functions $f : \mathbb{C} \to \mathbb{C}^{n \times m}$ which are analytic on the open left-half plane $\mathbb{C}^-$, with norm $\| \cdot \|_{\mathcal{H}_2^{\perp}}$ defined as:

$$\|H\|_{\mathcal{H}_2^{\perp}} \triangleq \sup_{\sigma<0} \left\{ \left(\frac{1}{2\pi} \int_{-\infty}^{\infty} \|H(\sigma + j\omega)\|_F^2 \, d\omega \right)^{\frac{1}{2}} \right\} \tag{19.12}$$

Definition 19.106 *($\mathcal{H}_\infty$ Hardy Space)* The $\mathcal{H}_\infty$ *Hardy space* is a subspace of $\mathcal{L}_2(j\mathbb{R})$ which consists of all matrix functions $f : \mathbb{C} \to \mathbb{C}^{n \times m}$ that are analytic (more simply, C^∞ and bounded) on the open right-half plane $\mathbb{C}^+$, with norm $\| \cdot \|_{\mathcal{H}_\infty}$ defined as:

$$\|F\|_{\mathcal{H}_\infty} = \operatorname*{ess\,sup}_{\sigma>0} \sigma_{\max} \left(F(\sigma + j\omega) \right). \tag{19.13}$$

The equations for $\mathcal{H}_2$ and $\mathcal{H}_\infty$ can be further simplified using an important result from complex analysis called the *Maximum Modulus Theorem*, stated below without proof.

Theorem 19.48 (Maximum Modulus Theorem) *If scalar-valued $f(s)$ is defined and continuous on a closed-bounded set $\mathcal{S} \subset \mathbb{C}$ and analytic on the interior of $\mathcal{S}$, then $|f(s)|$ cannot attain the maximum in the interior of $\mathcal{S}$ unless $f(s)$ is a constant.*

Theorem 19.48 implies that $|f(s)|$ can only achieve its maximum on the boundary of $\mathcal{S}$. Applying this to the definition of $\mathcal{H}_2$ and $\mathcal{H}_\infty$ we get:

$$\|H\|_{\mathcal{H}_2} \triangleq \left(\frac{1}{2\pi} \int_{-\infty}^{\infty} \|H(j\omega)\|_F^2 \, d\omega \right)^{\frac{1}{2}} \tag{19.14a}$$

$$\|H\|_{\mathcal{H}_\infty} \triangleq \operatorname*{ess\,sup}_{\omega\in\mathbb{R}} \sigma_{\max} \left(H(j\omega) \right) \tag{19.14b}$$

since the boundary of $\{s \in \mathbb{C} \mid \text{Re}(s) > 0\}$ is $\{s \in \mathbb{C} \mid \text{Re}(s) = 0\}$. Moreover, in (19.14a), the Frobenius norm $\|H(j\omega)\|_F$ of the matrix $H(j\omega)$ sums the squares of its elements, or, equivalently, the squares of the singular values. The definition translates to the trace of the product of H and its conjugate transpose H^*. The expression (19.14a) integrates the squared Frobenius norm across all frequencies, effectively capturing the total energy of the system across the entire frequency spectrum.

The $\mathcal{H}_2$ norm defines the steady-state variance of output signal y when $\mathbf{x}_0 = 0$ and $\mathbf{u}(t) = \mathbf{w}(t)$. To intuitively understand what the $\mathcal{H}_2$ of a represents it's useful to interpret it as the measure of the system's response energy to stochastic inputs (e.g., Gaussian white noise). It effectively measures the expected energy of the system's output due to noisy input. We will discuss the physical interpretation of the $\mathcal{H}_2$ norm in more detail in Sect. 19.6 after establishing stochastic linear systems in Sect. 19.4.

The $\mathcal{H}_\infty$ norm of a system on the other hand measures the worst-case amplification of disturbances through the system. It considers the maximum gain (amplification) from the input to the output over all frequencies. Moreover, the $\mathcal{H}_\infty$ norm can be defined for more general distributions of noise $\mathbf{w}$ than Gaussian. The gain of a system $H = \frac{Y}{U}$ is defined by the $\mathcal{H}_\infty$ norm:

$$\max_{w \neq 0} \frac{\|y\|_{\mathcal{L}_2}}{\|u\|_{\mathcal{L}_2}} = \|H\|_{\mathcal{H}_\infty} \tag{19.15}$$

Remark 19.38 $\mathcal{RH}_\infty$, which is the space of real, rational proper transfer functions we defined in Definition 17.98, is a subspace of $\mathcal{H}_\infty$. In the same manner we can define other spaces for real-rational matrices:

- $\mathcal{RL}_2$: The space of real rational, strictly proper transfer matrices with no poles on imaginary axis.
- $\mathcal{RH}_2$: The space of real rational, strictly proper, stable transfer matrices.
- $\mathcal{RH}_2^\perp$: The space of real rational, strictly proper transfer matrices with no poles in $\mathrm{Re}\, s < 0$.
- $\mathcal{RL}_\infty$: The space of real rational, proper transfer matrices with no poles on imaginary axis.
- $\mathcal{RH}_\infty$: The space of real rational, proper, stable transfer matrices.

To compute these system norms in the frequency domain, we invoke the following *Payley-Wiener lemma* and *Parseval identity* to investigate the connection between Hilbert spaces in the time domain and frequency domain.

Lemma 19.29 (Paley-Wiener) *The Fourier transform is a Hilbert space isomorphism from $\mathcal{L}_2(-\infty, \infty)$ onto $\mathcal{L}_2$. It maps $\mathcal{L}_2[0, \infty)$ onto $\mathcal{H}_2$ and $\mathcal{L}_2(-\infty, 0]$ onto $\mathcal{H}_2^\perp$.*

Lemma 19.29 essentially says that $\mathcal{H}_2$ is just the set of Laplace transforms of signals in $\mathcal{L}_2[0, \infty)$, i.e., of signals on $t \geq 0$ of finite energy.

Lemma 19.30 (Parseval's Identity) *For $f, g \in \mathcal{L}_2(\mathbb{R}; \mathbb{R})$,*

$$\int_{-\infty}^{\infty} f(t)g(t)\, dt = \frac{1}{2\pi} \int_{-\infty}^{\infty} F(j\omega)G^*(j\omega)\, d\omega \tag{19.16}$$

where F, G are the Fourier transforms of f, g, i.e., $F(j\omega) \triangleq \mathcal{F}\{f(t)\} = \int_{-\infty}^{\infty} f(t)e^{-j\omega t}\, dt$, and $G^*(j\omega)$ is the complex conjugate transpose of $G(j\omega)$.

Remark 19.39 A caveat: neither $\mathcal{RH}_\infty$ nor $\mathcal{RH}_2$ are *complete* Banach spaces (i.e., the limit of a sequence of state-space realizations may itself not have a state-space realization).

Lemma 19.31 (Computation of the $\mathcal{H}_2$ Norm) *Consider the transfer matrix $H(s)$ in $\mathcal{RH}_\infty$ with realization $(A, B, C, 0)$ with A Hurwitz and $D = 0$ to make it strictly proper. Then*

$$\|H\|_{\mathcal{H}_2}^2 = \mathrm{tr}(B^\top W_o B) = \mathrm{tr}(C W_c C^\top) \tag{19.17}$$

where W_c, W_o are the controllability, observability Gramians of the system H.

Proof We've seen before that the impulse response of H can be computed as

$$h(t) = \begin{cases} Ce^{At}B & \text{if } t \geq 0 \\ 0 & \text{else} \end{cases}$$

Plugging this into the formula of $\mathcal{H}_2$-norm (and with Lemma 19.30):

$$\|H\|_{\mathcal{H}_2}^2 = \int_0^\infty \operatorname{tr}(h^\top(t)h(t))\,dt = \begin{cases} \int_0^\infty \operatorname{tr}(B^\top e^{A^\top t}C^\top Ce^{At}B)\,dt \\ \int_0^\infty \operatorname{tr}(Ce^{At}BB^\top e^{A^\top t}C^\top)\,dt \end{cases} \tag{19.18}$$

where the two cases are equivalent by the cyclic property of the trace. The top case of (19.18) is precisely $\operatorname{tr}(B^\top W_o B)$ while the bottom case is equal to $\operatorname{tr}(CW_cC^\top)$. ∎

Unlike the $\mathcal{H}_2$ norm, there is no straightforward way to compute the $\mathcal{H}_\infty$ norm by hand. Instead, a more common approach is to rewrite the norm $\|H\|_{\mathcal{H}_\infty}$ as an optimization problem $\min \gamma$ s.t. $\|H\|_{\mathcal{H}_\infty} \leq \gamma$ and to further convexify this constraint by using LMIs. In fact, this approach can be used for computing $\mathcal{H}_2$ norms too, and is more appealing to use than Lemma 19.31 in $\mathcal{H}_2$ optimal control problems where the Gramians of the LFT are difficult to compute.

Our general pipelines for $\mathcal{H}_\infty$ and $\mathcal{H}_2$ control are summarized as follows

$$\min_{K\in\mathcal{H}_\infty} \|\underline{S}(H,K)\|_{\mathcal{H}_\infty} \Longrightarrow \begin{cases} \min\limits_{K\in\mathcal{H}_\infty} \gamma \\ \text{s.t. } \|\underline{S}(H,K)\|_{\mathcal{H}_\infty} \leq \gamma \end{cases} \Longrightarrow \begin{cases} \min\limits_{K\in\mathcal{H}_\infty} \gamma \\ \text{s.t. linear matrix inequalities} \\ \text{equivalent to condition } \|\underline{S}(H,K)\|_{\mathcal{H}_\infty} \leq \gamma \end{cases}$$

$$\begin{cases} \min\limits_{K\in\mathcal{H}_\infty} \|\underline{S}(H,K)\|_{\mathcal{H}_2} \\ \text{s.t. } \underline{S}(H,K)\in\mathcal{H}_\infty \end{cases} \Longrightarrow \begin{cases} \min\limits_{K\in\mathcal{H}_\infty} \gamma \\ \text{s.t. } \|\underline{S}(H,K)\|_{\mathcal{H}_2} \\ \underline{S}(H,K)\in\mathcal{H}_\infty \end{cases} \Longrightarrow \begin{cases} \min\limits_{K\in\mathcal{H}_\infty} \gamma \\ \text{s.t. linear matrix inequalities} \\ \text{equivalent to condition } \|\underline{S}(H,K)\|_{\mathcal{H}_2} \leq \gamma \end{cases}$$

In the next section, we will describe what are the exact LMIs used in the pipeline for the $\mathcal{H}_\infty$ case. The $\mathcal{H}_2$ case is treated in Sect. 19.6.

Remark 19.40 One could use the MATLAB commands `h2norm(sys)` and `hinfnorm(sys)` to compute the $\mathcal{H}_2$ and $\mathcal{H}_\infty$ norms, respectively.

19.3 $\mathcal{H}_\infty$ Optimal Control

We begin by stating the analogous $\mathcal{H}_\infty$ norm computation version of Lemma 19.31.

Lemma 19.32 (Computation of the $\mathcal{H}_\infty$ Norm) *Let $\gamma > 0$ and let $H(s)$ be a transfer matrix in $\mathcal{RH}_\infty$ with realization (A, B, C, D) with A Hurwitz and no poles on the $j\omega$ axis.*

Then $\|H\|_{\mathcal{H}_\infty} < \gamma$ if and only if $\sigma_{max}(D) < \gamma$ and the Hamiltonian matrix

$$M \triangleq \begin{bmatrix} A + BR^{-1}D^\top C & BR^{-1}B^\top \\ -C^\top(I + DR^{-1}D^\top)C & -(A + BR^{-1}D^\top C)^\top \end{bmatrix}, \tag{19.19}$$

where $R \triangleq \gamma^2 I - D^\top D$, has no eigenvalues λ such that $\operatorname{Re}\lambda = 0$.

Remark 19.41 By dividing through by γ, we can rewrite $\|H\|_{\mathcal{H}_\infty} < \gamma$ as $\left\|\frac{1}{\gamma}H\right\|_{\mathcal{H}_\infty} < 1$. Thus, we can typically assume WLOG that $\gamma = 1$.

While we can use methods such as Sect. 18.3 to compute (19.19), a more common alternative approach to compute the $\mathcal{H}_\infty$ norm in general is to formulate the problem using LMIs.

Lemma 19.33 (Bounded Real KYP Lemma) *Let $H(s)$ be a transfer matrix defined the same way as in Lemma 19.32. Then the following are equivalent.*

1. $\|H\|_{\mathcal{H}_\infty} \leq \gamma$
2. $\exists P = P^\top \succ 0$ *such that*

$$\begin{bmatrix} A^\top P + PA & PB \\ B^\top P & -\gamma I \end{bmatrix} + \frac{1}{\gamma}\begin{bmatrix} C^\top \\ D^\top \end{bmatrix}\begin{bmatrix} C & D \end{bmatrix} \prec 0 \tag{19.20}$$

Proof The direction that (1) implies (2) can actually be proven in a similar fashion to Lemma 19.32. We will now prove the direction that (2) implies (1). Using (19.15), note that 1) is equivalent to the expression $\|\mathbf{y}\|_{\mathcal{L}_2} \leq \gamma\|\mathbf{u}\|_{\mathcal{L}_2}$. Because A is Hurwitz, $\exists P = P^\top \succ 0$ such that $A^\top P + PA \prec 0$. Because the inequality (19.20) is strict, there exists some $\epsilon > 0$ such that:

$$\begin{bmatrix} A^\top P + PA & PB \\ B^\top P & -(\gamma - \epsilon)I \end{bmatrix} + \frac{1}{\gamma}\begin{bmatrix} C^\top \\ D^\top \end{bmatrix}\begin{bmatrix} C & D \end{bmatrix} \preceq 0 \tag{19.21}$$

which comes from adding $\begin{bmatrix} 0 & 0 \\ 0 & -\epsilon I \end{bmatrix}$ to the left side of the expression. Then for any state and control trajectories $\mathbf{x}(t)$ and $\mathbf{u}(t)$ of the system modeled by $H(s)$, (19.21) is equivalent to:

$$0 \geq \begin{bmatrix} \mathbf{x}(t) \\ \mathbf{u}(t) \end{bmatrix}^\top \left(\begin{bmatrix} A^\top P + PA & PB \\ B^\top P & -(\gamma - \epsilon)I \end{bmatrix} \begin{bmatrix} \mathbf{x}(t) \\ \mathbf{u}(t) \end{bmatrix} + \frac{1}{\gamma} \begin{bmatrix} \mathbf{x}(t) \\ \mathbf{u}(t) \end{bmatrix}^\top \begin{bmatrix} C^\top \\ D^\top \end{bmatrix} \begin{bmatrix} C & D \end{bmatrix} \right) \begin{bmatrix} \mathbf{x}(t) \\ \mathbf{u}(t) \end{bmatrix}$$

$$= \begin{bmatrix} \mathbf{x}(t) \\ \mathbf{u}(t) \end{bmatrix}^\top \begin{bmatrix} A^\top P + PA & PB \\ B^\top P & -(\gamma - \epsilon)I \end{bmatrix} \begin{bmatrix} \mathbf{x}(t) \\ \mathbf{u}(t) \end{bmatrix} + \frac{1}{\gamma} \mathbf{y}^\top(t)\mathbf{y}(t)$$

$$= \mathbf{x}^\top (A^\top P + PA)\mathbf{x} + \mathbf{x}^\top PB\mathbf{u} + \mathbf{u}^\top B^\top P\mathbf{x} - (\gamma - \epsilon)\mathbf{u}^\top \mathbf{u} + \frac{1}{\gamma}\mathbf{y}^\top \mathbf{y}$$

$$= (A\mathbf{x} + B\mathbf{u})^\top P\mathbf{x} + \mathbf{x}^\top P(A\mathbf{x} + B\mathbf{u}) - (\gamma - \epsilon)\mathbf{u}^\top \mathbf{u} + \frac{1}{\gamma}\mathbf{y}^\top \mathbf{y} \tag{19.22}$$

Note that $\mathbf{u}^\top \mathbf{u} = \|\mathbf{u}\|_{\mathcal{L}_2}^2$ and likewise for $\mathbf{y}^\top \mathbf{y}$. Choose function $V(\mathbf{x}) \triangleq \mathbf{x}^\top P\mathbf{x}$. Then

$$(19.22) = \dot{V}(\mathbf{x}(t)) - (\gamma - \epsilon)\|\mathbf{u}(t)\|_{\mathcal{L}_2}^2 + \frac{1}{\gamma}\|\mathbf{y}(t)\|_{\mathcal{L}_2}^2$$

Integrating both sides of this inequality from 0 to some time T yields

$$0 \geq \int_0^T \left(\dot{V}(\mathbf{x}(t)) - (\gamma - \epsilon)\|\mathbf{u}(t)\|_{\mathcal{L}_2}^2 + \frac{1}{\gamma}\|\mathbf{y}(t)\|_{\mathcal{L}_2}^2 \right) dt.$$

Because A is Hurwitz, $\lim_{T\to\infty} \mathbf{x}(T) = 0 \implies \lim_{T\to\infty} V(\mathbf{x}(T)) = 0$. This implies

$$0 \geq -(\gamma - \epsilon) \int_0^\infty \|\mathbf{u}(t)\|_{\mathcal{L}_2}^2 \, dt + \frac{1}{\gamma} \int_0^\infty \|\mathbf{y}(t)\|_{\mathcal{L}_2}^2 \, dt$$

$$\implies 0 \geq -(\gamma - \epsilon)\|\mathbf{u}\|_{\mathcal{L}_2}^2 + \frac{1}{\gamma}\|\mathbf{y}\|_{\mathcal{L}_2}^2$$

Thus, $\|\mathbf{y}\|_{\mathcal{L}_2}^2 \leq (\gamma^2 - \gamma\epsilon)\|\mathbf{u}\|_{\mathcal{L}_2}^2$. By the equivalence from (19.15), $\|H\|_{\mathcal{H}_\infty}^2 \leq (\gamma^2 - \gamma\epsilon) \leq \gamma^2$. Taking the square root of both sides yields the desired condition (1). ■

Now, we are ready to take the first step in rewriting the original $\mathcal{H}_\infty$ optimal control problem. Note that under state-feedback:

$$\underline{\mathcal{S}}(H, K) = \left[\begin{array}{c|c} A + B_2K & B_1 \\ \hline C_1 + D_{12}K & D_{11} \end{array}\right], \quad \text{where } H \triangleq \left[\begin{array}{c|cc} A & B_1 & B_2 \\ \hline C_1 & D_{11} & D_{12} \\ C_2 & D_{21} & D_{22} \end{array}\right]$$

By the Bounded Real KYP Lemma, $\|\underline{\mathcal{S}}(H, K)\|_{\mathcal{H}_\infty} \leq \gamma$ iff $\exists P = P^\top \succ 0$ such that:

$$\begin{bmatrix} (A + B_2K)^\top P + P(A + B_2K) & PB_1 \\ B_1^\top P & -\gamma I \end{bmatrix} + \frac{1}{\gamma} \begin{bmatrix} (C_1 + D_{12}K)^\top \\ D_{11}^\top \end{bmatrix} \begin{bmatrix} C_1 + D_{12}K & D_{11} \end{bmatrix} \prec 0 \tag{19.23}$$

Thus, we can write: $\min_{K\in\mathcal{H}_\infty} \|\mathcal{S}(H, K)\|_{\mathcal{H}_\infty} \iff \min_{\gamma,K,P} \gamma$ s.t. (19.23) holds. However, (19.23) is still nonlinear in its decision variables γ, P, K, which means it is not a LMI and thus still difficult to solve using numerical methods (e.g., CVX).

A very common tool that is often used to convert nonlinear matrix inequalities into LMIs is the *Schur complement.*

Lemma 19.34 (Schur Complement) *For any* $Q = Q^\top \in \mathbb{R}^{n\times m}$, $R = R^\top \in \mathbb{R}^{m\times n}$, *and* $S \in \mathbb{R}^{n\times m}$, *the following are equivalent:*

(1) $\begin{bmatrix} Q & S \\ S^\top & R \end{bmatrix} \succ 0$ *(or* $\begin{bmatrix} Q & S \\ S^\top & R \end{bmatrix} \prec 0$*).*
(2) $Q \succ 0$ *and* $Q - SR^{-1}S^\top \succ 0$ *(or* $Q \prec 0$ *and* $Q - SR^{-1}S^\top \prec 0$*)*

Now, invoking Lemma 19.34 to (19.23) with

$$Q = \begin{bmatrix} \bar{A}^\top P + P\bar{A} & P\bar{B} \\ \bar{B}^\top P & -\gamma I \end{bmatrix}, \quad S = \begin{bmatrix} \bar{C}^\top \\ \bar{D}^\top \end{bmatrix}, \quad R = -\frac{1}{\gamma} I$$

where $\bar{A} \triangleq A + B_2K$, $\bar{B} \triangleq B_1$, $\bar{C} \triangleq C_1 + D_{12}K$, $\bar{D} \triangleq D_{11}$, yields another matrix inequality:

$$\begin{bmatrix} \bar{A}^\top P + P\bar{A} & P\bar{B} & \bar{C}^\top \\ \bar{B}^\top P & -\gamma I & \bar{D}^\top \\ \bar{C} & \bar{D} & -\gamma I \end{bmatrix} \prec 0 \tag{19.24}$$

which is bi-linear in both P and K.

In Sect. 19.1, we demonstrated how to transform a Lyapunov inequality to a LMI with two new variables showed last class how to transform Lyapunov inequality $(A + BK)^\top P + P(A + BK) \prec 0$ into a LMI with two new variables $X \triangleq P^{-1}$, $Z \triangleq KX$. A similar variable substitution trick can be used to further convert (19.24) into a LMI

$$\begin{bmatrix} -X & 0 & 0 & 0 \\ 0 & XA^\top + AX + Z^\top B_2^\top + B_2Z & B_1 & XC_1^\top + Z^\top D_{12}^\top \\ 0 & B_1^\top & -\gamma I & D_{11}^\top \\ 0 & C_1X + D_{12}Z & D_{11} & -\gamma I \end{bmatrix} \prec 0 \tag{19.25}$$

Thus, we can formalize the $\mathcal{H}_\infty$ optimal control pipeline mentioned at the end of Sect. 19.2 as follows:

$$\min_{K\in\mathcal{H}_\infty} \|\underline{S}(H,K)\|_{\mathcal{H}_\infty} \Longrightarrow \begin{cases} \min_{K\in\mathcal{H}_\infty} \gamma \\ \text{s.t. } \|\underline{S}(H,K)\|_{\mathcal{H}_\infty} \le \gamma \end{cases} \Longrightarrow \begin{cases} \min_{\gamma,K,P} \gamma \\ \text{s.t. (3.23) holds.} \end{cases} \Longrightarrow \begin{cases} \min_{\gamma,X,Z} \gamma \\ \text{s.t. (3.25) holds.} \end{cases} \tag{19.26}$$

The static state-feedback gain control law $\mathbf{u}^*(t) = K\mathbf{x}(t)$ which achieves the optimal $\mathcal{H}_\infty$ performance can then be implemented with $K \triangleq ZX^{-1}$ after solving for X, Z from (19.26).

Remark 19.42 Output feedback $\mathcal{H}_\infty$ optimal control undergoes a similar transformation to a LMI form.

Some more advanced treatments on $\mathcal{H}_\infty$ control are in [1, 2].

19.4 Stochastic Linear Systems

In order to begin a proper discussion of the $\mathcal{H}_2$ and LQG optimal control frameworks, we need to first delve into the mathematical foundations of how stochastic disturbances influence system dynamics. This section establishes some of the basic foundations of *stochastic linear systems*.

19.4.1 System Dynamics

Following the general feedback interconnection studied throughout this chapter, we are concerned with general linear systems of the form

$$\mathcal{H} \triangleq \begin{cases} \dot{\mathbf{x}}(t) = A\mathbf{x}(t) + B_2\mathbf{u}(t) + B_1\mathbf{w}(t) \\ \mathbf{y}(t) = C_2\mathbf{x}(t) + D_{22}\mathbf{u}(t) + D_{12}\mathbf{v}(t) \end{cases} \quad \text{or} \quad \mathcal{H} \triangleq \begin{cases} \mathbf{x}[t+1] = A\mathbf{x}[t] + B_2\mathbf{u}[t] + B_1\mathbf{w}[t] \\ \mathbf{y}[t] = C_2\mathbf{x}[t] + D_{22}\mathbf{u}[t] + D_{11}\mathbf{v}[t] \end{cases} \tag{19.27}$$

Here, $\mathbf{w}(t), \mathbf{w}[t] \in \mathbb{R}^m$ is the disturbance/process noise, and $\mathbf{v}(t), \mathbf{v}[t] \in \mathbb{R}^k$ is the measurement noise. Typically, they are assumed zero-mean Gaussian with covariance matrices

$$\begin{aligned} \mathbb{E}[\mathbf{w}(t)\mathbf{w}^\top(\tau)] = \Sigma_w\delta(t-\tau), \quad \mathbb{E}[\mathbf{v}(t)\mathbf{v}^\top(\tau)] &= \Sigma_v\delta(t-\tau) \\ \mathbb{E}[\mathbf{w}(t)\mathbf{v}^\top(\tau)] = 0, \quad \mathbb{E}[\mathbf{v}(t)\mathbf{w}^\top(\tau)] &= 0 \end{aligned}$$

and likewise in the DT case. In the common literature, stochastic system dynamics are described using two related notations. A more mathematically precise notation comes from the field of *stochastic differential equations (SDEs)*:

$$\mathcal{H} \triangleq \begin{cases} d\mathbf{x}(t) = (A\mathbf{x}(t) + B_2\mathbf{u}(t))dt + B_1 d\mathbf{w}(t), \\ \mathbf{y}(t) = C_2\mathbf{x}(t) + D_{22}\mathbf{u}(t) + D_{21}\mathbf{v}(t) \end{cases} \tag{19.28}$$

A majority of our discussion will involve the "*control-theoretic notation*" of (19.27). THe one time we will use the SDE notation, as well as some well-known results from SDEs, is when we compare the $\mathcal{H}_2$ optimal control formulation versus the LQG formulation in Sect. 19.7.

19.4.2 Interpretation of the $\mathcal{H}_2$ System Norm

Recall from Sect. 19.2 that the $\mathcal{H}_2$ norm of a transfer function, noted as $\|H\|_{\mathcal{H}_2}$, quantifies the energy of a system's response to white noise inputs. We will describe in more detail what this means by going through a review of selected concepts from signals processing and random stochastic processes.

19.4.3 Signal Processing Background

Given a (scalar-valued) CT signal $x(t)$, we may be interested in analyzing the statistical average of some kind of "quantity" in its frequency domain representation (which relates it to the Fourier transform). This statistical average, which we'll call S_{xx} (so that $S_{xx}(f)$ is its value at frequency $f \in \mathbb{C}$), is typically called the *spectrum* of $x(t)$. There are two types of spectrum that are commonly used in practice.

Definition 19.107 (*Energy Spectral Density*) The *energy spectral density* of a signal $x(t)$ is defined by the squared magnitude of its Fourier transform:

$$S_{xx}(f) = |X(f)|^2$$

where $X(f) \triangleq \int_{-\infty}^{\infty} x(t)e^{-j2\pi ft}dt$, representing the Fourier transform of $x(t)$. Essentially, the energy spectral density describes how the total energy of the signal, $E = \int_{-\infty}^{\infty} |x(t)|^2 dt$, is distributed across different frequencies.

This relationship is formally expressed by Parseval's identity (seen before in Lemma 19.30), which equates the total energy in the time domain with the total energy in the frequency domain.

$$\int_{-\infty}^{\infty} |x(t)|^2 \, dt = \int_{-\infty}^{\infty} |X(f)|^2 \, df$$

Analyzing the energy spectral density is the most meaningful when x has finite total energy. See, for example, the pulse signal of Fig. 19.1.

Definition 19.108 (*Power Spectral Density*) The *power spectral density*, $S_{xx}(f)$, measures the distribution of power into frequency components composing the signal.

$$S_{xx}(f) = \lim_{T\to\infty} \frac{1}{T} |X_T(f)|^2$$

Essentially, the power spectral density describes how the average power $P = \lim_{T\to\infty} \frac{1}{T} \int_{-\frac{T}{2}}^{\frac{T}{2}} |x(t)|^2 \, dt$ is distributed across frequency.

To compute power spectral density practically, consider the truncated signal $x_T(t) = x(t) \cdot \mathbf{1}_T(t)$, where $\mathbf{1}_T(t)$ is the indicator function that is non-zero only over an interval of length T. This interval is chosen arbitrarily and depends on the physical properties of the signal itself; for example, if x is periodic with period T, then the choice of T is obvious. The Fourier transform of the truncated signal x_T is:

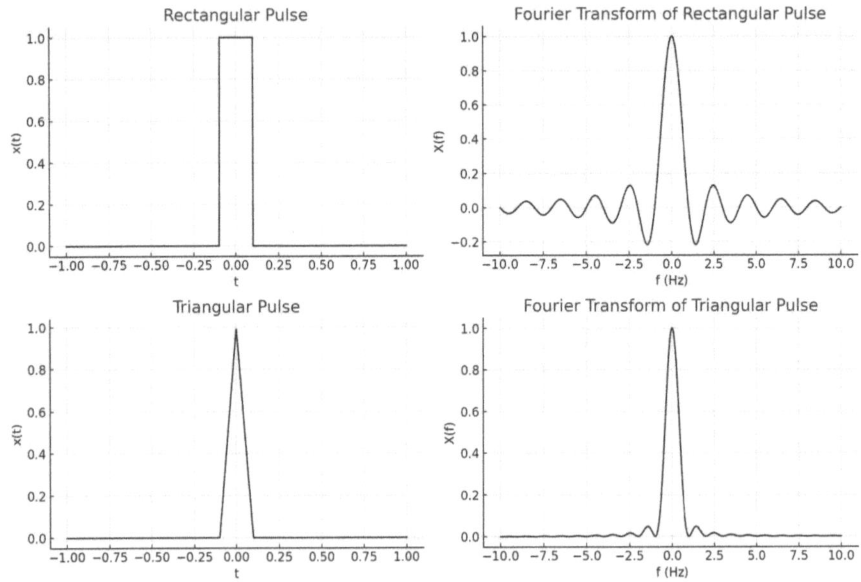

Fig. 19.1 Fourier transformation of rectangular and triangular pulses

$$X_T(f) = \int_{-\infty}^{\infty} x_T(t) e^{-j2\pi f t} dt$$

Using Parseval's identity, the total power in the truncated signal's time domain equals the total power in its frequency domain:

$$\lim_{T\to\infty} \int_{-\infty}^{\infty} |x_T(t)|^2 \, dt = \lim_{T\to\infty} \int_{-\infty}^{\infty} |X_T(f)|^2 \, df$$

19.4.4 Stochastic Processes Background

For the purposes of our optimal control problems, we will focus on the case where signal $x(t)$ considered throughout Sect. 19.4.3 represents a *stochastic process*. A few more definitions are needed.

Definition 19.109 (*(Strict-Sense) Stationary*) A stochastic process $\{X(t), t \geq 0\}$, is *strict-sense stationary* (or simply, *stationary*) if its statistical properties (e.g., mean, covariance, pdf) do not change over time t, i.e., the joint CDF of $\{X(t_1 + s), \ldots, X(t_n + s)\}$ is the same for all $s \in \mathbb{R}$:

$$P(X(t_1) \leq x_1, \ldots, X(t_n) \leq x_n) = P(X(t_1 + s) \leq x_1, \ldots, X(t_n + s) \leq x_n) \; \forall t_1, t_2, \ldots, t_n \in \mathbb{R}, n \in \mathbb{N}$$

(One may consider this as a "stochastic version" of the time-invariance property common in deterministic signals and functions).

Remark 19.43 The terminology "stationary" appears quite frequently in the stochastic processes literature. For example, the *Brownian motion process* has the following *stationary increments* property: for $t_1 > t_2 \geq 0$, $W(t_1 + s) - W(t_2 + s)$ has the same distribution as $W(t_1) - W(t_2)$ $\forall s \in \mathbb{R}$. This is not equivalent to Definition 19.109. Although all stationary processes clearly have stationary increments, not all processes with stationary increments are stationary. The Brownian motion itself is an example of this since $W(t) \sim N(0, t)$ has a variance which depends on time t.

In practice, stationarity of a stochastic process is too strict of a condition. *Wide-sense stationary* stochastic processes are much more common, and the power spectral density is usually defined for WSS processes.

Definition 19.110 (*Wide-Sense Stationary*) A stochastic process $\{X(t), t \geq 0\}$ is *wide-sense stationary* (WSS) if it satisfies:

1. $E[X(t_1)] = E[X(t_2)]$ for all $t_1, t_2 \in \mathbb{R}$.
2. $E[X(t_1)X(t_2)] = E[X(t_1 + s)X(t_2 + s)]$ $\forall t_1, t_2, s \in \mathbb{R}$.

i.e., the mean and variance of $X(t)$ are independent of t.

Definition 19.111 (*Autocorrelation*) The *autocorrelation function* of a stochastic process $\{X(t), t \geq 0\}$ is defined as $R_{xx}(t_1, t_2) \triangleq E[X(t_1)X(t_2)]$.

Note that "auto" refers to "self", and so R_{xx} essentially computes how correlated the process is with itself at two different times t_1 and t_2. For WSS processes, the autocorrelation function simplifies to $R_x(s) = E[X(t + s)X(t)]$; because mean and variance of $X(t)$ are independent of t, we only need to care about the *shift* s.

In the following definitions, we will revert back to lowercase notation $x(t)$ in describing our stochastic process. Although the capitalized X is common when discussing stochastic processes in the context of probability, we will henceforth use the signal notation and make them all lowercase x.

The power spectral density of WSS process $x(t)$ can be determined by its autocorrelation function R_x:

$$S_{xx}(f) = \mathcal{F}\{R_x(s)\} = \int_{-\infty}^{\infty} R_x(s) e^{-j2\pi f s}\, ds$$

$$R_x(s) = \mathcal{F}^{-1}\{S_{xx}(f)\} = \int_{-\infty}^{\infty} S_{xx}(f) e^{j2\pi f s}\, df$$

Note that when the shift is $s = 0$:

$$E[x^2(t)] = R_x(0) = \int_{-\infty}^{\infty} S_{xx}(f)\, df.$$

Thus, for WSS stochastic processes, $E[x^2(t)]$ is equal to the expected power of the signal x. Note that our entire discussion until now is independent of the fact that x is a scalar-valued signal, and is easily applicable to the case where $\mathbf{x}$ is vector-valued.

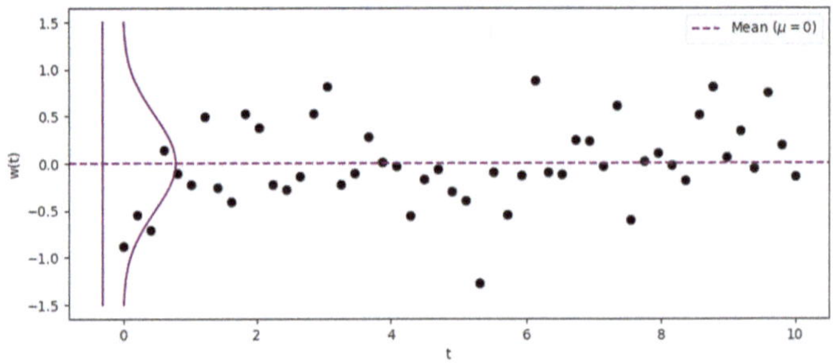

Fig. 19.2 A visualization of Gaussian noise

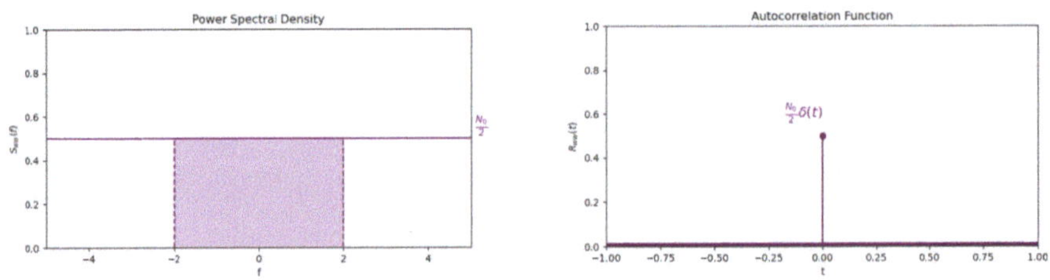

Fig. 19.3 A visualization of white noise. [Left] Its power spectral density. [Right] Its autocorrelation function. Without going into too much detail of the conventions from communications literature $\frac{N_0}{2}$ represents the noise power per unit bandwidth

19.4.5 The White Noise Terminology

We write a brief clarification remark about some common misconceptions when discussing "Gaussian white noise". First, not all Gaussian noise is white, and not all white noise is Gaussian. In fact, both terms are describing two distinct characteristics of a signal.

- "Gaussian" refers to the probability distribution of the signal's amplitude variations around its mean in the time domain. Specifically, the amplitudes follow a normal distribution with some mean μ and variance σ^2. See Fig. 19.2 for an illustration.
- "White" characterizes a signal that possesses a flat spectral density across all frequencies. In the frequency domain, this implies that the signal has equal power at all frequencies, commonly described as uniform noise power per unit bandwidth. Moreover, the autocorrelation function of white noise is the Dirac delta function $\delta(t)$, indicating that successive values of the noise signal $w(t)$ are uncorrelated with each other. See Fig. 19.3 for a visualization.

19.5 The Linear Quadratic Gaussian

We are now ready to discuss the optimal control methods common for stochastic linear systems of the form (19.27), starting with the linear quadratic Gaussian (LQG). We will focus exclusively on developing the optimal state-feedback law (i.e., $\mathbf{y}(t) = \mathbf{x}(t)$) for LTI systems, since the derivation of the output-feedback case follows similarly through the extensions discussed in Chap. 18.

The cost functional is defined very similarly to the LQR problem, but due to the additional randomness introduced by the noise process $\mathbf{w}$, the expectation needs to be taken. In the finite-horizon cases:

$$J_{\mathbf{u}}(x_0) = \mathbb{E}_{\mathbf{w}}\left[\int_0^T (\mathbf{x}^\top(t)Q\mathbf{x}(t) + \mathbf{u}^\top(t)R\mathbf{u}(t))dt + \mathbf{x}^\top(T)Q_f\mathbf{x}(T)\right] \text{ in CT} \tag{19.29a}$$

$$J_{\mathbf{u}}(x_0) = \mathbb{E}_{\mathbf{w}}\left[\sum_{t=0}^{T-1}(\mathbf{x}^\top[t]Q\mathbf{x}[t] + \mathbf{u}[t]^\top R\mathbf{u}[t]) + \mathbf{x}_T^\top Q_f \mathbf{x}_T\right] \text{ in DT} \tag{19.29b}$$

Moreover, in some cases, the initial condition $\mathbf{x}_0$ may be drawn from some initial distribution ρ. Then an additional expectation needs to be taken with respect to the initial condition $\mathrm{J}_{\mathbf{u}} \triangleq \mathbb{E}_{\mathbf{x}_0\sim\rho}\left[J_{\mathbf{u}}(\mathbf{x}_0)\right]$. In either case of (19.29), we seek the optimal $\mathbf{u}^*$ which attains $\min_{\mathbf{u}\in\mathcal{U}} J_{\mathbf{u}}(\mathbf{x}_0)$ (or $\min_{\mathbf{u}\in\mathcal{U}} \mathrm{J}_{\mathbf{u}}$).

Note that because of the additional uncertainty in the system, we need to distinguish two cases: (1) when $\mathbf{x}(t)$ is fully-known, and (2) when $\mathbf{x}(t)$ is not fully-known. The LQG approach is used in the first case, and the optimal control sequence $\{\mathbf{u}[t]\}$ can be attained via *stochastic* dynamic programming. The second case, like observer design (see Chap. 17), requires an additional component which develops an estimate $\hat{\mathbf{x}}$ of $\mathbf{x}$. This is usually achieved via *Kalman filters*, which can be thought of as the stochastic version of observers when $\mathbf{y}(t)$ is perturbed by noise $\mathbf{v}(t)$ too; we will discuss Kalman filtering in the following Chap. 20.

Theorem 19.49 (DT Finite-Horizon LQG) *For the DT finite-horizon LQG, define the cost-to-go function*

$$V_\tau(\mathbf{x}[\tau]) = \mathbb{E}_{\{\mathbf{w}[\tau],\mathbf{w}[\tau+1],\ldots\}}\left[\sum_{t=\tau}^{T-1}(\mathbf{x}^\top[t]Q\mathbf{x}[t] + \mathbf{u}[t]^\top R\mathbf{u}[t]) + V_{\tau+1}(\mathbf{x}[\tau+1]) \;\middle|\; \mathbf{x}[\tau]\right] \tag{19.30}$$

with the cost-to-go recursion:

$$V_\tau(\mathbf{x}) = \begin{cases} \min_{\mathbf{u}[\tau]\in\mathbb{R}^m}\left(\mathbf{x}^\top Q\mathbf{x} + \mathbf{u}[\tau]^\top R\mathbf{u}[\tau] + \mathbb{E}_{\mathbf{w}[\tau]}\left[V_{\tau+1}(\mathbf{x}[\tau+1])\right]\right) & \textit{for all } \tau \le T-1 \\ \mathbf{x}^\top Q_f\mathbf{x} & \textit{if } \tau = T \end{cases} \tag{19.31}$$

for all $\mathbf{x} \in \mathbb{R}^n$.

Proof (*Proof Sketch of Theorem* 19.49.) The derivation of the DT finite-horizon LQG follows very similarly to the DT finite-horizon LQR from Theorem 18.44. First, we assume a closed-form solution to recursion:

$$V_\tau(\mathbf{x}[\tau]) \triangleq \mathbf{x}^\top[\tau]P_\tau\mathbf{x}[\tau] + f_\tau$$

which is exactly the same original form as the LQR, with an extra term f_τ due to the noise. We proceed by induction. From the terminal condition, let $P_T = Q_f$, then for a fixed time $\tau \leq T$, suppose that the quadratic form of $V_{\tau+1}(\mathbf{x})$ holds. Calculations show that

$$\begin{aligned} V_\tau(\mathbf{x}[\tau]) &= \min_{\mathbf{u}[\tau]\in\mathbb{R}^m} \left(\mathbf{x}^\top[\tau]Q\mathbf{x}[\tau] + \mathbf{u}[\tau]^\top R\mathbf{u}[\tau] + \mathbb{E}_{\mathbf{w}[\tau]}\left[\mathbf{x}^\top[\tau+1]P_{\tau+1}\mathbf{x}[\tau+1] + f_{\tau+1}\right]\right) \\ &= \min_{\mathbf{u}[\tau]\in\mathbb{R}^m} \left(\mathbf{x}^\top[\tau](Q + A^\top P_{\tau+1}A)\mathbf{x}[\tau] + \mathbf{u}[\tau]^\top(R + B_2^\top P_{\tau+1}B_2)\mathbf{u}[\tau]\right. \\ &\quad \left. +2\mathbf{x}^\top[\tau]A^\top P_{\tau+1}B_2\mathbf{u}[\tau] + \operatorname{tr}(B_1^\top P_{\tau+1}B_1\Sigma_w) + f_{\tau+1}\right) \end{aligned} \tag{19.32}$$

where the other terms cancel since $\mathbb{E}[\mathbf{w}[\tau]] = 0$ for all $\tau \in \mathbb{N}$. Similar to the proof of Theorem 18.44, differentiate the argument inside the min operator above with respect to $\mathbf{u}$, and set it equal to 0 to get the optimal control law $\mathbf{u}[\tau]$. ■

Remark 19.44 (*Certainty-Equivalence*) An important result is observed: the noise process doesn't affect the optimal control law as derived in the LQR case via (18.5):

$$\mathbf{u}^*[\tau] = -(R + B_2^\top P_{\tau+1}B_2)^{-1}B_2^\top P_{\tau+1}A\mathbf{x}[\tau]$$

where P_τ needs to satisfy the same Riccati equation (18.5a). This is a property known as *certainty-equivalence*: the optimal control law for a system with and without noise are the same. Loosely-speaking, this property arises due to the noise process having mean zero.

Proof (*Proof Sketch of Theorem* 19.49, *continued.*) We can derive the f_τ extra term via recursion too.

$$f_\tau = \operatorname{tr}(B_1 P_{\tau+1}B_1\Sigma_w) + f_{\tau+1}, \quad f_T = 0,$$

which comes from comparing the original deterministic LQR cost-to-go against (19.32) and absorbing all extra terms into f_τ. This recursion is easy to solve iteratively, and we get

$$f_0 = \sum_{t=0}^{T-1} \operatorname{tr}(B_1 P_{t+1}B_1\Sigma_w) \tag{19.33}$$

The final optimum cost can be written as

$$J^* \triangleq \min_{u \in \mathcal{U}} \mathbb{E}_{\mathbf{x}_0}\left[J_{\mathbf{u}}(\mathbf{x}_0)\right] = \mathbb{E}_{\mathbf{x}_0}\left[V_0(\mathbf{x}_0)\right] = \mathbb{E}_{\mathbf{x}_0}\left[\mathbf{x}_0^\top P_0 \mathbf{x}_0\right] + f_0$$

■

19.6 $\mathcal{H}_2$ Optimal Control

Let $\mathbf{w}(t)$ be some input noise process with power spectral density $S_{ww}(j\omega)$, where $\omega = 2\pi f$ (representing frequencies as radians/sec instead of Hertz). This means:

$$\mathbb{E}[\mathbf{w}^\top \mathbf{w}(t)] = R_w(0) = \frac{1}{2\pi}\int_{-\infty}^{\infty} S_{ww}(j\omega)e^{j\omega 0}\,d\omega = \frac{1}{2\pi}\int_{-\infty}^{\infty} S_{ww}(j\omega)\,d\omega.$$

Lemma 19.35 (Power Spectral Density Transformation for LTI Systems) *For an LTI system with transfer function H, if the input signal $\mathbf{w}(t)$ has power spectral density $S_{ww}(j\omega)$, then the output signal $\overline{\mathbf{z}}(t)$, where $\overline{\mathbf{z}} = H\mathbf{w}$, has power spectral density:*

$$S_{zz}(j\omega) = H(j\omega)S_{ww}(j\omega)H^*(j\omega)$$

Using the above lemma, the expected power of the output signal can be written as

$$\begin{aligned}
\mathbb{E}[\overline{\mathbf{z}}^\top \overline{\mathbf{z}}(t)] &= \frac{1}{2\pi}\int_{-\infty}^{\infty} H(j\omega)S_{ww}(j\omega)H^*(j\omega)\,d\omega \\
&\leq \text{ess sup}_{\omega\in\mathbb{R}} S_{ww}(j\omega)\cdot \frac{1}{2\pi}\int_{-\infty}^{\infty} H(j\omega)H^*(j\omega)\,d\omega \\
&= \|S_{ww}\|_{H_\infty}\cdot \|H\|_{\mathcal{H}_2}^2
\end{aligned}$$

Thus, for general noise $\mathbf{w}(t)$ with power spectral density S_{ww}, the average power of the output signal $\overline{\mathbf{z}} = \underline{S}(H, K)\mathbf{w}$ is bounded as:

$$\mathbb{E}[\overline{\mathbf{z}}^\top \overline{\mathbf{z}}] \leq \|S_{ww}\|_{\mathcal{H}_\infty}\cdot \|\underline{S}(H, K)\|_{\mathcal{H}_2}^2$$

This gives as a nice physical interpretation of $\mathcal{H}_2$ optimal control. Namely, it aims to minimize the average power of the output signal, given the system is perturbed by some noise signal with a specified power spectral density.

We are now ready to proceed with formalizing the $\mathcal{H}_2$ optimal control pipeline described at the end of Sect. 19.2. The original optimal control problem (19.2) should be rewritten as a tractable convex program that can be processed using numerical tools like `CVX` or `YALMIP`. We begin with a lemma similar to Lemma 19.33.

Lemma 19.36 (The $\mathcal{H}_2$ Norm) *Suppose H has the realization $(A, B, C, 0)$ (i.e., $H(s) = C(sI - A)^{-1}B$) with A being Hurwitz. Then the following are equivalent:*

- $\|H\|_{\mathcal{H}_2} \leq \gamma$
- *there exists $X = X^\top \succ 0$ such that*

$$\mathrm{tr}(CXC^\top) < \gamma \text{ and } AX + XA^\top + BB^\top \prec 0 \tag{19.34}$$

The proof of Lemma 19.36 is related to Lemma 19.31, and we won't discuss it here. Under full-state feedback, i.e., $\mathbf{u}(t) = K\mathbf{x}(t)$ for some constant gain K for plant H, the $\mathcal{H}_2$ optimal control problem can be formulated with the following result.

Lemma 19.37 (The $\mathcal{H}_2$ Norm: Alternative) *Under the same conditions as Lemma 19.36, $\|\underline{S}(H, K)\|_{H_2} < \gamma$ is equivalent to saying there exists $X = X^\top \succ 0$ such that*

$$\begin{cases} \begin{bmatrix} A & B_2 \end{bmatrix} \begin{bmatrix} X \\ Z \end{bmatrix} + \begin{bmatrix} X & Z^\top \end{bmatrix} \begin{bmatrix} A^\top \\ B_2^\top \end{bmatrix} + B_1 B_1^\top \prec 0, \\ \mathrm{tr}\left\{ (C_1 X + D_{12} Z) X^{-1} (C_1 X + D_{12} Z)^\top \right\} < \gamma \end{cases} \tag{19.35}$$

The state-feedback gain is given by $K = ZX^{-1}$.

This follows directly from applying Lemma 19.36 to the closed-loop system.

$$\dot{\mathbf{x}}(t) = (A + B_2 K)\mathbf{x}(t) + B_1 \mathbf{w}(t), \quad \overline{\mathbf{z}}(t) = (C_1 + D_{12} K)\mathbf{x}(t)$$

Then use a change of variables $Z = KX$.

So far, the $\mathcal{H}_2$ optimal control problem posed generically at the end of Sect. 19.2 can be rewritten as $\min_{\gamma, X, Z} \gamma$ s.t. (19.35) holds. However, this matrix inequality constraint contains terms that are nonlinear in our decision variables. To make this optimization problem easier to solve numerically, we convert them into LMIs using tricks like the Schur complement, similar to how we did for the $\mathcal{H}_\infty$ case in Sect. 19.3.

Applying the Schur complement to (19.34) yields the following result:

$$\exists X = X^\top \succ 0, \quad W = W^\top \text{ such that}$$
$$AX + XA^\top + BB^\top \prec 0, \quad \begin{bmatrix} X & C^\top \\ C & W \end{bmatrix} \succ 0, \quad \mathrm{tr}(W) < \gamma$$

and likewise, for the closed-loop system, applying the Schur complement to (19.35) yields

$$
\exists X = X^\top \succ 0, \quad W = W^\top, \text{ and } Z \in \mathbb{R}^{m\times n} \text{ such that}
$$

$$
\begin{cases}
\begin{bmatrix} A & B_2 \end{bmatrix}\begin{bmatrix} X \\ Z \end{bmatrix} + \begin{bmatrix} X & Z^\top \end{bmatrix}\begin{bmatrix} A^\top \\ B_2^\top \end{bmatrix} + B_1 B_1^\top \prec 0, \\
\begin{bmatrix} X & (C_1 X + D_{12} Z)^\top \\ C_1 X + D_{12} Z & W \end{bmatrix} \succ 0, \\
\operatorname{tr}(W) < \gamma
\end{cases}
\tag{19.36}
$$

Finally, we can formalize the $\mathcal{H}_2$ optimal control pipeline as follows:

$$
\begin{cases}
\min\limits_{K\in\mathcal{H}_\infty} \|\underline{S}(H,K)\|_{\mathcal{H}_2} \\
\text{s.t. } \underline{S}(H,K) \in \mathcal{H}_\infty
\end{cases}
\Longrightarrow
\begin{cases}
\min\limits_{K\in\mathcal{H}_\infty} \gamma \\
\text{s.t. } \|\underline{S}(H,K)\|_{\mathcal{H}_2} \le \gamma
\end{cases}
\Longrightarrow
\begin{cases}
\min\limits_{\gamma,X,Z} \gamma \\
\text{s.t. (3.35) holds.}
\end{cases}
\Longrightarrow
\begin{cases}
\min\limits_{\gamma,X,Z} \gamma \\
\text{s.t. (3.6) holds.}
\end{cases}
\tag{19.37}
$$

The static state-feedback gain control law $\mathbf{u}^*(t) = K\mathbf{x}(t)$ which achieves the optimal $\mathcal{H}_2$ performance can then be implemented with $K \triangleq ZX^{-1}$ after solving for X, Z from (19.37).

19.7 Relationship Between $\mathcal{H}_2$ and LQG

To conclude this chapter, we provide a brief discussion about the relationship between the LQG studied in Sect. 19.5 and the $\mathcal{H}_2$ problem. We require the infinite-horizon CT formulation of the LQG problem, which aims to solve:

$$
\begin{aligned}
&\min_{\mathbf{u}\in\mathcal{U}} \mathbb{E}_{(\mathbf{x}_0,\mathbf{w})}\left[\int_0^\infty (\mathbf{x}^\top(t) Q\mathbf{x}(t) + \mathbf{u}(t)^\top R\mathbf{u}(t))\, dt\right] \\
&\text{s.t. } \dot{\mathbf{x}}(t) = A\mathbf{x}(t) + B_2\mathbf{u}(t) + B_1\mathbf{w}(t),
\end{aligned}
\tag{19.38}
$$

where $\mathbf{w}(t)$ is a Gaussian white noise signal and $\mathbf{u}(t) = K\mathbf{x}(t)$. For simplicity, we'll assume $\mathbf{x}_0$ is deterministic and fixed (although in many LQR problems, $\mathbf{x}_0 \sim \rho$ is assumed to be drawn randomly). Under state-feedback control law, the closed-loop system is given by

$$
\dot{\mathbf{x}}(t) = (A + B_2 K)\mathbf{x}(t) + B_1\mathbf{w}(t),
$$

where $A_{\text{cl}} = A + B_2 K$. In the SDE notation (see (19.28)):

$$
d\mathbf{x}(t) = A_{\text{cl}}\mathbf{x}(t)dt + B_1 d\mathbf{w}(t),
$$

which is a vector version of the Ornstein-Uhlenbeck process. The solution is given by:

$$\mathbf{x}(t) = e^{A_{cl}t}\mathbf{x}_0 + \int_0^t e^{A_{cl}(t-s)} B_1 d\mathbf{w}(s).$$

Substituting this solution into the cost functional, we get:

$$\begin{aligned}
\mathrm{J_u} &= \mathbb{E}\left[\int_0^\infty \mathbf{x}^\top(t)(Q + K^\top RK)\mathbf{x}(t)\, dt\right] \\
&= \int_0^\infty \left(\mathbf{x}_0^\top e^{A_{cl}^\top t}(Q + K^\top RK)e^{A_{cl}t}\mathbf{x}_0\right) dt \\
&\quad + 2\int_0^\infty \mathbb{E}\left[\mathbf{x}_0^\top e^{A_{cl}^\top t}(Q + K^\top RK)\int_0^t e^{A_{cl}(t-s)} B_1\, d\mathbf{w}(s)\right] dt \\
&\quad + \int_0^\infty \mathbb{E}\left[\left(\int_0^t e^{A_{cl}(t-s)} B_1\, d\mathbf{w}(s)\right)^\top (Q + K^\top RK)\left(\int_0^t e^{A_{cl}(t-s)} B_1\, d\mathbf{w}(s)\right)\right] dt.
\end{aligned} \tag{19.39}$$

Note that the second term is of the form $\mathbb{E}\left[\int_0^t g(s)d\mathbf{w}(s)\right]$ with matrix-valued function g. By first property of the (Itô or Payley-Wiener-Zygmund) stochastic integral, this is equal to zero.

Let $C_1 \triangleq \begin{bmatrix} Q^{\frac{1}{2}} \\ 0 \end{bmatrix}$, $D_{12} \triangleq \begin{bmatrix} 0 \\ R^{\frac{1}{2}} \end{bmatrix}$. Then

$$(C_1 + D_{12}K)^\top(C_1 + D_{12}K) = \begin{bmatrix} Q^{\frac{1}{2}\top} & K^\top R^{\frac{1}{2}\top} \end{bmatrix}\begin{bmatrix} Q^{\frac{1}{2}} \\ R^{\frac{1}{2}}K \end{bmatrix} = \begin{bmatrix} Q & Q^{\frac{1}{2}}(R^{\frac{1}{2}}K)^\top \\ R^{\frac{1}{2}}KQ^{\frac{1}{2}} & K^\top RK \end{bmatrix}$$

and the matrices on the block diagonals are Q and $K^\top RK$. Thus we can modify (19.39) as:

$$\begin{aligned}
J_u(x_0) &= \int_0^\infty \mathrm{tr}\left((C_1 + D_{12}K)e^{A_{cl}t}x_0\right)^\top \left((C_1 + D_{12}K)e^{A_{cl}t}x_0\right) dt \\
&\quad + \int_0^\infty \mathrm{tr}\left[\mathbb{E}\left[\int_0^t \left((C_1 + D_{12}K)e^{A_{cl}(t-s)} B_1\, dW(s)\right)^\top \int_0^t \left((C_1 + D_{12}K)e^{A_{cl}(t-s)} B_1\, dW(s)\right)\right]\right] dt
\end{aligned} \tag{19.40}$$

We can apply the cyclic property of the trace to simplify. Note that the last term of (19.40) is of the form:

$$\mathbb{E}\left[\left(\int_0^t g(s)d\mathbf{w}(s)\right)^\top \left(\int_0^t g(s)d\mathbf{w}(s)\right)\right]$$

Using the property of stochastic integrals (It's Isometry), this is equal to $\int_0^t \mathbb{E}[g(s)^2]\, ds$ and thus:

$$\begin{aligned}
(3.40) &= \int_0^\infty \mathrm{tr}\left(((C_1 + D_{12}K)e^{A_{cl}t}\mathbf{x}_0)^\top((C_1 + D_{12}K)e^{A_{cl}t}\mathbf{x}_0)\right) dt \\
&\quad + \int_0^\infty \mathrm{tr}\left(\mathbb{E}\left[((C_1 + D_{12}K)e^{A_{cl}(t-s)} B_1\, d\mathbf{w}(s))^\top((C_1 + D_{12}K)e^{A_{cl}(t-s)} B_1\, d\mathbf{w}(s))\right]\right) dt.
\end{aligned} \tag{19.41}$$

Now, let's derive the LFT $\underline{\mathcal{S}}(H, K)$. Considering the system dynamics:

$$\begin{cases} \dot{\mathbf{x}}(t) = A_{\text{cl}}\mathbf{x}(t) + B_1\mathbf{w}(t) \\ \overline{\mathbf{z}}(t) = (C_1 + D_{12})\mathbf{x}(t) \end{cases}$$

and transforming into the Laplace domain gives:

$$X(s) = (sI - A_{\text{cl}})^{-1}\mathbf{x}_0 + (sI - A_{\text{cl}})^{-1}B_1W(s) \tag{19.42a}$$

$$\overline{Z}(s) = (C_1 + D_{12})(sI - A_{\text{cl}})^{-1}\mathbf{x}_0 + (C_1 + D_{12})(sI - A_{\text{cl}})^{-1}B_1W(s) \tag{19.42b}$$

Taking the $\mathcal{H}_2$ norm and using Parseval's identity on Eq. (19.42b) gives us exactly Eq. (19.41) (recall that $(sI - A)^{-1}$ is the Laplace transform of e^{At}).

In conclusion, we can see that the LQG problem is a special case of the $\mathcal{H}_2$ problem when $\mathbf{w}(t)$ is Gaussian white noise, with:

$$A_e = A, \quad B_2 = B, \quad C_1 = \begin{bmatrix} Q^{1/2} \\ 0 \end{bmatrix}, \quad D_{12} = \begin{bmatrix} 0 \\ R^{1/2} \end{bmatrix}.$$

This formulates a direct relationship between the LQG and $\mathcal{H}_2$ control frameworks, illustrating how LQG is an $\mathcal{H}_2$ optimization under specific noise conditions.

References

1. Kemin Zhou, John C. Doyle, and Keith Glover. *Robust and Optimal Control*. Upper Saddle River, NJ, USA: Prentice-Hall, Inc., 1996.
2. Geir E. Dullerud and Fernando Paganini. *Robust and Optimal Control*. Upper Saddle River, NJ, USA: Prentice-Hall, Inc., 2000.

Chapter 20
Linear State Estimation

In Sect. 19.5, we discussed the need to consider two cases of optimal control for stochastic systems: (1) when the state $\mathbf{x}(t)$ is fully known, and (2) when the $\mathbf{x}(t)$ is unknown. Case (1) was addressed using stochastic dynamic programming and a variant of the LQR problem called the LQG, for when the external disturbance $\mathbf{w}(t)$ was an uncorrelated Gaussian white noise process. Case (2), however, requires an extra step which produces an estimate $\hat{\mathbf{x}}(t)$ of the true state $\mathbf{x}(t)$ from the measurements $\{\mathbf{y}(s) : s \leq t\}$. This problem, of optimal state estimation, is commonly known as *filtering*, especially *Kalman filtering* for linear stochastic systems, which is the focus of this chapter.

We remark that we will first separate the filtering problem from the optimal control problem, and consider mostly *uncontrolled* systems. It turns out, that similar to the observer-based feedback system we analyzed in Chap. 17, the *separation principle* still applies to stochastic systems, i.e., we can design an optimal filter separately from the state-feedback stochastic controller.

20.1 Preliminaries: Minimum Mean-Squared Estimator

At heart, all filtering problems are a type of *linear least-squares problem*, which we've seen before when discussing the minimum-energy input in Chap. 11. We will begin by investigating the filtering problem in DT, although we will later study both the DT and the CT cases in this chapter. Moreover, in contrast to the previous discussions on optimal control (Chaps. 18 and 19), we look at the linear *time-varying* case. (Note that the optimal control methods discussed in Chaps. 18 and 19 can also be easily extended to the LTV setting).

SJ. Han, *Linear Systems*, KAIST Research Series,
https://doi.org/10.1007/978-981-96-9774-8_20

Our (DT) state-space system model of consideration is

$$\begin{cases} \mathbf{x}[t+1] = A_t\mathbf{x}[t] + \mathbf{w}[t] \\ \mathbf{y}[t] = C_t\mathbf{x}[t] + \mathbf{v}[t] \end{cases} \tag{20.1}$$

where $\{\mathbf{w}[t]\}_{t\in\mathbb{N}}$, $\{\mathbf{v}[t]\}_{t\in\mathbb{N}}$ are independent Gaussian white noise processes with zero mean and covariances $\mathbb{E}[\mathbf{w}[t]\mathbf{w}^\top[s]] = \Sigma_w\delta(t-s)$ and $\mathbb{E}[\mathbf{v}[t]\mathbf{v}^\top[s]] = \Sigma_v\delta(t-s)$ for any $s, t \in \mathbb{N}$, much like the setting of the LQG in Sect. 19.5.

We further assume a *Markovian setting*, i.e.

1. Given $\mathbf{x}[t]$, the next state $\mathbf{x}[t+1]$ is independent of past states $\mathbf{x}[0], \ldots, \mathbf{x}[t-1]$ and all past measurements $\mathbf{y}[1], \ldots, \mathbf{y}[t]$.
2. Given $\mathbf{x}[t]$, the current measurement $\mathbf{y}[t]$ is independent of all past states $\mathbf{x}[0], \ldots, \mathbf{x}[t-1]$ and all past measurements $\mathbf{y}[1], \ldots, \mathbf{y}[t-1]$.

The goal of a filtering problem is to estimate $\hat{\mathbf{x}}[t]$ of $\mathbf{x}[t]$ from measurements $\mathscr{A}_t \equiv \{\mathbf{y}[s] : s \leq t\}$.

$$\forall t \in \mathbb{N}, \quad \hat{\mathbf{x}}[t] \equiv \underset{\mathbf{z}\in\mathbb{R}^n}{\operatorname{argmin}}\, \mathbb{E}\left[\|\mathbf{x}[t] - \mathbf{z}\|_2^2 \mid \mathscr{A}_t\right] \tag{20.2}$$

Remark 20.45 In relation to Chap. 15, filtering can be thought of as *stochastic* observer design.

20.1.1 Stochastic Linear Least-Squares

For the purpose of relating the filtering problem to the linear least squares problem we've seen before, let's (for the moment) forget about the time indices in (20.2). We consider static vectors $\mathbf{x} \equiv \mathbf{x}[t]$, $\mathbf{y} \equiv \mathbf{y}[t]$, etc., and keep the set of measurements equal to $\mathscr{A}_t$ with the understanding that it is just representing a set of vectors. We can expand the expression in (20.2) to get

$$\mathbb{E}[\|\mathbf{x} - \mathbf{z}\|_2^2 \mid \mathscr{A}_t] = \mathbb{E}[\mathbf{x}^\top\mathbf{x} \mid \mathscr{A}_t] - 2\mathbf{z}^\top\mathbb{E}\left[\mathbf{x} \mid \mathscr{A}_t\right] + \mathbf{z}^\top\mathbf{z}$$

Take the gradient of the argument with respect to $\mathbf{z}$ and set it equal to 0.

$$\mathbf{z} = \mathbb{E}\left[\mathbf{x} \mid \mathscr{A}_t\right] \implies \hat{\mathbf{x}} \equiv \mathbb{E}\left[\mathbf{x} \mid \mathscr{A}_t\right] \tag{20.3}$$

The linear least squares method can be applied to the case when we have some prior estimate $\tilde{\mathbf{x}}$, with covariance $\tilde{P}$, and measurement $\mathbf{y}$ of the current state $\mathbf{x}$, abiding by the linear measurement equation $\mathbf{y} = C\mathbf{x} + \mathbf{v}$, where $\mathbf{v}$ is a Gaussian white noise process with covariance Σ_v.

$$\hat{\mathbf{x}} \equiv \arg\min_{\mathbf{z}} \mathbb{E}\left[(\mathbf{z} - \tilde{\mathbf{x}})^\top\tilde{P}^{-1}(\mathbf{z} - \tilde{\mathbf{x}}) + (\mathbf{y} - C\mathbf{z})^\top\Sigma_v^{-1}(\mathbf{y} - C\mathbf{z})\right] \tag{20.4}$$

Note that the two terms in (20.4) represent a tradeoff in the two sources of information that are used to estimate $\mathbf{x}$. Here, $\tilde{P}^{-1}$ represents the amount of "confidence" we have in our prior estimate $\tilde{\mathbf{x}}$ and is used as a weighting factor: the less confidence we have in $\tilde{\mathbf{x}}$ being correct, the less we should use it to adjust our current estimate $\mathbf{z}$. Similarly, the noisy measurement's precision matrix, Σ_v^{-1}, represents the amount of confidence we have in the measurement being correct; a noisier measurement indicates a larger covariance and thus, less precision. In this case, we should not rely on $\mathbf{y}$ too much to get our current estimate $\mathbf{z}$.

Similar to (20.2), taking the gradient of (20.4) with respect to $\mathbf{z}$ and setting it equal to 0 yields:

$$\begin{aligned} &2\mathbf{z}^\top \tilde{P}^{-1} - 2\tilde{\mathbf{x}}^\top \tilde{P}^{-1} - 2\mathbf{y}^\top \Sigma_v^{-1} C + 2\mathbf{z}^\top C^\top \Sigma_v^{-1} C = 0 \\ \Longrightarrow\ &\hat{\mathbf{x}} = \tilde{\mathbf{x}} + \tilde{P} C^\top \Sigma_v^{-1} (\mathbf{y} - C\tilde{\mathbf{x}}) \quad \text{where} \quad \tilde{P}^{-1} \equiv P^{-1} + C^\top \Sigma_v^{-1} C \end{aligned} \tag{20.5}$$

We will see similar variations of expression (20.5) throughout this chapter.

20.1.2 The Orthogonality Principle

Define the space $\mathcal{L}_2(\Omega; \mathbb{R}^n)$ of random vectors in $\mathbb{R}^n$ over probability space $(\Omega, \mathcal{F}, \mathbb{P})$, equipped with the covariance $\text{Cov}(\mathbf{x}, \mathbf{y}) \equiv \mathbb{E}[(\mathbf{x} - \mathbb{E}(\mathbf{x}))(\mathbf{y} - \mathbb{E}(\mathbf{y}))^\top]$ as the inner product. While the notation is similar, this $\mathcal{L}_2$ can be viewed as like a stochastic version of the $\mathcal{L}_2$ space we've seen before for square-integrable deterministic functions. Let $\mathcal{V} \subset \mathcal{L}_2$ be a closed subspace.

Theorem 20.50 (MMSE) *The estimate* $\hat{\mathbf{x}} \in \mathcal{V}$ *which achieves the minimum MSE (i.e.,* $\mathbb{E}\left[\|\mathbf{x} - \hat{\mathbf{x}}\|^2\right] \leq \mathbb{E}\left[\|\mathbf{x} - \mathbf{z}\|^2\right]$ *for all* $\mathbf{z} \in \mathcal{V}$*) is unique and satisfies* $(\mathbf{x} - \hat{\mathbf{x}}) \perp \mathbf{z}\ \forall \mathbf{z} \in \mathcal{V}$ *(i.e.,* $Cov(\mathbf{x} - \hat{\mathbf{x}}, \mathbf{z}) = 0$*). Here,* $\hat{\mathbf{x}}$ *is called the minimum mean-squared estimator (MMSE) of* $\mathbf{x}$*, i.e., the projection of* $\mathbf{x}$ *onto* $\mathcal{V}$ *(Fig. 20.1).*

There are several useful properties of the MMSE:

1. $\forall t \in \mathbb{N},\ \mathbb{E}\left[\|\mathbf{x} - \hat{\mathbf{x}}\|_2^2\right] = \mathbb{E}[(\mathbf{x} - \hat{\mathbf{x}})^\top (\mathbf{x} - \hat{\mathbf{x}})]$

$$= \mathbb{E}[\mathbf{x}^\top \mathbf{x} - 2(\hat{\mathbf{x}}^\top \mathbf{x}) + \hat{\mathbf{x}}^\top \hat{\mathbf{x}}] = \mathbb{E}[\mathbf{x}^\top \mathbf{x}] - \mathbb{E}[\hat{\mathbf{x}}^\top \mathbf{x}]$$

2. The MMSE is an *unbiased* estimator, i.e., $\mathbb{E}\left[\mathbf{x} - \hat{\mathbf{x}}\right] = 0$. This can be seen from Theorem 20.50.

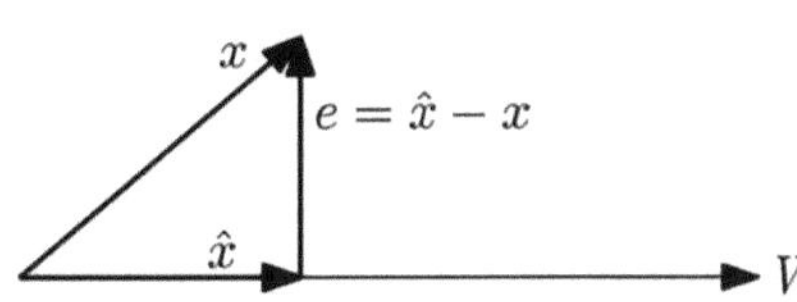

Fig. 20.1 A geometric meaning of orthogonality

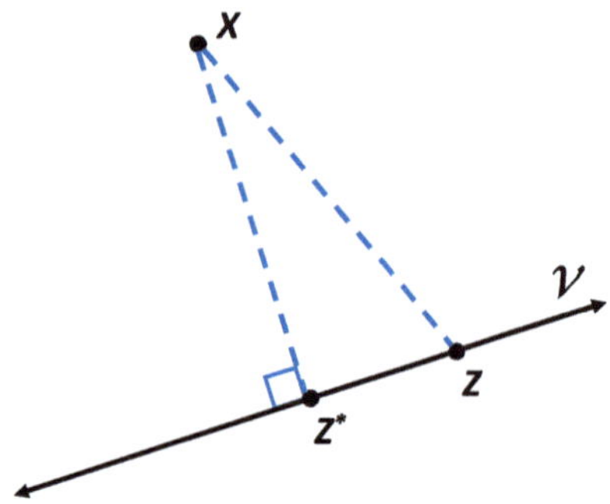

Fig. 20.2 A visualization of the projection mechanism where $\mathcal{V}$ is a line and $\mathbf{x} \in \mathbb{R}^n$. Note that any other $\mathbf{z} \in \mathcal{V}$ does not achieve the minimum error

Example 20.36 (*MMSE for Linear Gaussian Case*) Of particular interest is the form of the MMSE in the following *linear Gaussian case*:

1. $\mathbf{y} = C\mathbf{x} + \mathbf{v}$ (a linear measurement equation) with $\mathbf{v} \sim \mathcal{N}(0, \Sigma_v)$.
2. $\mathcal{V} \equiv \{A\mathbf{y} + \mathbf{b} \mid A \in \mathbb{R}^{n\times k}, b \in \mathbb{R}^n\}$.

We will return to this case later. □

With this setup, the estimation problem posed to us in the beginning of Sect. 20.1 can be restated as follows: determine the estimator of $\mathbf{x} \in \mathcal{L}^2$ over all $\mathcal{V}$ with the least mean-squared error, i.e., we want to find the MMSE $\mathbf{z}^* \in \mathcal{V}$ such that $\mathbb{E}[\|\mathbf{x} - \mathbf{z}^*\|^2] \leq \mathbb{E}[\|\mathbf{x} - \mathbf{z}\|^2]$ for all $z \in \mathcal{V}$. Correspondingly, $\mathbb{E}[\|\mathbf{x} - \mathbf{z}^*\|^2]$ is known as the *minimum mean squared error*. See Fig. 20.2 for visualization in the case where $\mathcal{V}$ is a line (e.g., as in Example 20.36).

Theorem 20.51 (Orthogonality Principle) *The following conditions are equivalent:*

1. *There exists a unique element $\mathbf{z}^* \in \mathcal{V}$ which achieves the MMSE.*
2. *Let $\mathbf{y} \in \mathcal{L}^2$. Then $\mathbf{y} = \mathbf{z}^*$ iff the following two conditions hold:*

   a. $\mathbf{y} \in \mathcal{V}$
   b. $(\mathbf{x} - \mathbf{y}) \perp \mathbf{z}$ *for all* $\mathbf{z} \in \mathcal{V}$.

3. *As a consequence of the above two conditions, the MMSE has a nice simplification:*

$$\mathbb{E}[(\mathbf{x} - \mathbf{z}^*)^\top(\mathbf{x} - \mathbf{z}^*)] = \mathbb{E}[\mathbf{x}^\top\mathbf{x}] - \mathbb{E}[(\mathbf{z}^*)^\top\mathbf{z}^*] \quad \textit{since } (\mathbf{x} - \mathbf{z}^*) \perp \mathbf{z}^*$$

Furthermore, the MMSE is an unbiased estimator, meaning $\mathbb{E}[\mathbf{x} - \mathbf{z}^*] = 0$.

Example 20.37 (*Orthogonality Principle for Linear Subspaces*) We continue our discussion from Example 20.36. We specialize the analysis to the affine subspace $\mathcal{V} = \{c_0 + c_1 y_1 + \cdots + c_m y_m;\ c_i \in \mathbb{R}^n\}$ for a specific value $\mathbf{y} \triangleq (y_1, \ldots, y_m)$. The projection of $\mathbf{x}$ onto $\mathcal{V}$ is of the form $\mathbf{z}^* = A\mathbf{y} + \mathbf{b}$. The estimation error becomes $\mathbf{e} = \mathbf{x} - (A\mathbf{y} + \mathbf{b})$. Let's use the Orthogonality Principle to determine what the coefficients A and $\mathbf{b}$ should be. Namely, in order to have $\mathbf{e} \perp \mathbf{z}$ for all $\mathbf{z} \in \mathcal{V}$, we need:

- $\mathbb{E}[\mathbf{e}] = 0$, which implies that $\mathbf{b} = \mathbb{E}[\mathbf{x}] - A\mathbb{E}[\mathbf{y}]$. Thus, $\mathbf{z}^* = \mathbb{E}[\mathbf{x}] + A(\mathbf{y} - \mathbb{E}[\mathbf{y}])$.
- $\text{Cov}(\mathbf{e}, \mathbf{y}) = 0$. Combined with the previous expression, we get:

$$\begin{aligned}\text{Cov}(\mathbf{x} - \mathbb{E}[\mathbf{x}] - A(\mathbf{y} - \mathbb{E}[\mathbf{y}]), \mathbf{y}) = 0 &\implies \text{Cov}(\mathbf{x}, \mathbf{y}) - A\text{Cov}(\mathbf{y}) = 0 \\ &\implies A = \text{Cov}(\mathbf{x}, \mathbf{y})\text{Cov}(\mathbf{y})^{-1}\end{aligned}$$

Combined together, we have the final expression for the MMSE. To distinguish the notation between the general and the linear cases, we denote the conditional expectation for the linear case with $\hat{E}$ instead of the usual $\mathbb{E}$.

$$\hat{\mathbf{x}} \triangleq \hat{E}[\mathbf{x}|\mathbf{y}] \triangleq \mathbb{E}[\mathbf{x}] + \text{Cov}(\mathbf{x}, \mathbf{y})\text{Cov}(\mathbf{y})^{-1}(\mathbf{y} - \mathbb{E}[\mathbf{y}]) \tag{20.6}$$

Moreover, we find that $\text{Cov}(\mathbf{e})$ satisfies:

$$\text{Cov}(\mathbf{e}) = \text{Cov}(\mathbf{x}) - \text{Cov}(\hat{\mathbf{x}}) = \text{Cov}(\mathbf{x}) - \text{Cov}(\mathbf{x}, \mathbf{y})\text{Cov}(\mathbf{y})^{-1}\text{Cov}(\mathbf{y}, \mathbf{x}).$$

□

Remark 20.46 Note that $\hat{\mathbf{x}}$ in (20.5) is linear in $\mathbf{y}$, and would belong to $\mathcal{V}$ when

$$A \equiv \tilde{P}C^\top\Sigma_v^{-1}, \quad b \equiv \tilde{\mathbf{x}} - \tilde{P}C^\top\Sigma_v^{-1}C\tilde{\mathbf{x}}$$

20.2 Sequential Estimation Over Time

Now, we are ready to bring back the time indices ($\mathbf{x} \mapsto \mathbf{x}[t], \mathbf{y} \mapsto \mathbf{y}[t]$, etc.) by using the dynamics and measurement Eq. (20.1). When there are successive measurements $\mathbf{y}[1], \ldots, \mathbf{y}[t]$ over time, it may be the case that a part of $\mathbf{y}[t]$ is already known (or can be inferred) from previous measurements $\mathbf{y}[1], \ldots, \mathbf{y}[t-1]$. This motivates the construction of an *innovations sequence* $\{\tilde{\mathbf{y}}[t]\}_{t\in\mathbb{N}}$, where:

$$\tilde{\mathbf{y}}[t] \equiv \mathbf{y}[t] - \mathbb{E}[\mathbf{y}[t] \mid \mathscr{A}_{t-1}] \tag{20.7}$$

We will touch more on (20.7) later in this section.

20.2.1 *The Linear Gaussian Case*

When the dynamics are linear and perturbed by an additive Gaussian noise process, as in Examples 20.36 and 20.37, two additional properties make the filtering approach more straightforward than the techniques we've seen previously. Namely,

- Affine combinations of Gaussian random vectors are still Gaussian-distributed. So, if our initial condition $\mathbf{x}(0)$ has a prior distribution which is Gaussian, all future state variables $\mathbf{x}(t)$ (or $\mathbf{x}[t]$, in the discrete-time case) will be Gaussian-distributed too.

- Gaussian distributions are fully characterized by its mean and covariance matrix. As a consequence of this property, determining the mean and the covariance of the true state is enough to know everything about the full distribution.

This gives rise to the *Kalman filtering algorithm*. Before we derive the Kalman filtering process, we first establish some necessary background in the next subsection about innovation processes, particularly in the case where the conditioned space is linear.

20.2.2 The Innovations Sequence

As mentioned previously, the motivation behind the construction of innovation sequences is as follows: it will often be the case that a new observation $\mathbf{y}[t]$ at time $t \in \mathbb{Z}$ is not totally new if we've already observed previous values $\mathbf{y}[1], \ldots, \mathbf{y}[t-1]$. The only innovation (new information about the system) will come from the component of $\mathbf{y}[t]$ that is *orthogonal* to the linear span of all previous observations $\mathbf{y}[1], \ldots, \mathbf{y}[t-1]$. Thus, we can rewrite the above equation as

$$\tilde{\mathbf{y}}[t] = \mathbf{y}[t] - \hat{E}[\mathbf{y}[t] | \mathscr{A}_{t-1}] \tag{20.8}$$

where $\hat{E}$ is the notation derived from (20.6), $\mathscr{A}_l$ represents the sigma algebra $\sigma(\mathbf{y}[1], \ldots, \mathbf{y}[l])$ spanned by the measurement vectors $\mathbf{y}[1], \ldots, \mathbf{y}[l]$ for any $l \in \mathbb{Z}$, and

$$\mathbb{E}[\mathbf{z} \mid \tilde{\mathscr{A}_t}] \equiv \bar{\mathbf{z}} + \sum_{s=1}^{t} \mathbb{E}\left[\mathbf{z} - \bar{\mathbf{z}} \mid \tilde{\mathbf{y}}[s]\right], \quad \bar{\mathbf{z}} \equiv \mathbb{E}[\mathbf{z}]$$

for any placeholder $\mathbf{z}$. Furthermore, with this definition, $\mathbb{E}[\tilde{\mathbf{y}}[k]] = 0$.

Remark 20.47 Both the innovation and observation sequences span the same space, i.e., span $\mathscr{A}_t =$ span $\tilde{\mathscr{A}_t}$. One can think of $\{\tilde{\mathbf{y}}_0, \ldots, \tilde{\mathbf{y}}[t]\}$ being the orthogonalized basis created from $\{\mathbf{y}_0, \ldots, \mathbf{y}[t]\}$. This consequently notes the similarity of this construction process to the *Gram-Schmidt orthonormalization* process used in the context of linear algebra. Thus, from this construction, $\mathbb{E}[\tilde{\mathbf{y}}[t]] = 0$ and $\tilde{\mathbf{y}}[s] \perp \tilde{\mathbf{y}}[t]\, \forall s \neq t$.

For this simplicity, we will often condition our estimate $\hat{\mathbf{x}}$ around this *linear innovations sequence* $\tilde{\mathbf{y}}[1], \tilde{\mathbf{y}}[2], \ldots$ as opposed to the original observation sequence $\mathbf{y}[1], \mathbf{y}[2], \ldots$. We write the notation as $\hat{E}[\mathbf{x}[t] \mid \tilde{\mathscr{A}_t}]$.

Theorem 20.52 *The estimate of* $\mathbf{x}$ *based on the sequence of orthogonal observations* $\tilde{\mathbf{y}}[1], \tilde{\mathbf{y}}[2], \ldots, \tilde{\mathbf{y}}[n]$ *is given by the joint projection:*

$$\hat{E}[\mathbf{x}|\tilde{\mathscr{A}_n}] = \overline{\mathbf{x}} + \sum_{i=1}^{n} \hat{E}[\mathbf{x} - \overline{\mathbf{x}}|\tilde{\mathbf{y}}[i]] \tag{20.9}$$

where we denote $\tilde{\mathscr{A}_n} = \sigma(\tilde{\mathbf{y}}[1], \ldots, \tilde{\mathbf{y}}[n])$ *and* $\overline{\mathbf{x}} \triangleq \mathbb{E}[\mathbf{x}]$.

Remark 20.48 We can simplify the expression (20.9) as follows:

$$\begin{aligned}\hat{E}[\mathbf{x}|\tilde{\mathscr{A}_n}] &= \overline{\mathbf{x}} + \sum_{i=1}^{n} \hat{E}[\mathbf{x} - \overline{\mathbf{x}}|\tilde{\mathbf{y}}[i]] = \overline{\mathbf{x}} + \sum_{i=1}^{n} \mathrm{Cov}(\mathbf{x}, \tilde{\mathbf{y}}[i])\mathrm{Cov}(\tilde{\mathbf{y}}[i])^{-1}\tilde{\mathbf{y}}[i] \\ &= \hat{E}[\mathbf{x}|\tilde{\mathscr{A}}_{n-1}] + \hat{E}[\mathbf{x} - \overline{\mathbf{x}}|\tilde{\mathbf{y}}[n]] \end{aligned} \tag{20.10}$$

This gives us a recursive formula in terms of each new observation $\tilde{\mathbf{y}}[n]$.

20.3 The General Bayesian Filtering Problem

We are now ready to continue the setup of the filtering problem started in the beginning of Sect. 20.1. Our dynamics are given by (20.1) and we will again assume the Markovian setting.

to restate the goal of a filtering problem more concretely beyond the form introduced by (20.2): we seek to estimate the full state $\mathbf{x}[t]$ given data observations $\mathscr{A}_t \triangleq \sigma(\mathbf{y}[1], \ldots, \mathbf{y}[t])$. Taking a *Bayesian inference* approach, we construct the pdf $p(\mathbf{x}[t]|\mathscr{A}_t)$ given $p(\mathbf{x}[t-1]|\mathscr{A}_{t-1})$ at each timestep k through a two-step iterative process: prediction and measurement update.

1. **Prediction**: given $p(\mathbf{x}[t-1] \mid \mathscr{A}_{t-1})$, predict $p(\mathbf{x}[t] \mid \mathscr{A}_{t-1})$. This uses the Chapman-Kolmogorov equation.

$$\begin{aligned} p(\mathbf{x}[t] \mid \mathscr{A}_{t-1}) &= \int p(\mathbf{x}[t] \mid \mathbf{x}[t-1], \mathscr{A}_{t-1}) p(\mathbf{x}[t-1] \mid \mathscr{A}_{t-1}) d\mathbf{x}[t-1] \\ &= \int p(\mathbf{x}[t] \mid \mathbf{x}[t-1]) p(\mathbf{x}[t-1] \mid \mathscr{A}_{t-1}) d\mathbf{x}[t-1] \end{aligned} \tag{20.11}$$

   where the second equality follows from Markovian assumptions. In this form, $p(\mathbf{x}[t]|\mathbf{x}[t-1])$ can be directly determined from the system dynamics.
2. **Measurement Update**: given observation $\mathbf{y}[t]$, update $p(\mathbf{x}[t] \mid \mathscr{A}_{t-1})$ to $p(\mathbf{x}[t] \mid \mathscr{A}_t)$. This uses Bayes' Rule.

$$
\begin{aligned}
p(\mathbf{x}[t] \mid \mathscr{A}_t) = p(\mathbf{x}[t] \mid \mathbf{y}[t], \mathscr{A}_{t-1}) &= \frac{p(\mathbf{x}[t], \mathbf{y}[t] \mid \mathscr{A}_{t-1})}{p(\mathbf{y}[t] \mid \mathscr{A}_{t-1})} \\
&= \frac{p(\mathbf{y}[t] \mid \mathbf{x}[t]) p(\mathbf{x}[t] \mid \mathscr{A}_{t-1})}{p(\mathbf{y}[t] \mid \mathscr{A}_{t-1})} \\
&\equiv \frac{p(\mathbf{y}[t] \mid \mathbf{x}[t]) p(\mathbf{x}[t] \mid \mathscr{A}_{t-1})}{\int p(\mathbf{y}[t] \mid \mathbf{x}[t]) p(\mathbf{x}[t] \mid \mathscr{A}_{t-1}) d\mathbf{x}[t]}
\end{aligned}
\tag{20.12}
$$

where the second equality follows from Markovian assumptions, $p(\mathbf{y}[t]|\mathbf{x}[t])$ can be determined from the observation equation, and

$$
p(\mathbf{y}[t] \mid \mathscr{A}_{t-1}) = \int p(\mathbf{y}[t] \mid \mathbf{x}[t]) p(\mathbf{x}[t] \mid \mathscr{A}_{t-1}) d\mathbf{x}[t]
$$

It is based off of this two-step construction that we will build the DTKF in the next section.

20.4 The Discrete-Time Kalman Filter

We introduce some further notations and assumptions: $\mathbf{x}_0, \mathbf{v}[0], \mathbf{v}[1], \ldots, \mathbf{w}[0], \mathbf{w}[1], \ldots$ are all pairwise uncorrelated Gaussian-distributed random variables, and $\mathbb{E}[\mathbf{w}[t]] = 0, \mathrm{Cov}(\mathbf{w}[t]) = Q_t, \mathbb{E}[\mathbf{v}[t]] = 0, \mathrm{Cov}(\mathbf{v}[t]) = R_t$ are known constant matrices. Further assume that $\mathbf{x}_0$ comes from a known Gaussian distribution with $\mathbb{E}[\mathbf{x}_0] = \bar{x}_0, \mathrm{Cov}(\mathbf{x}_0) = P_0$, and that it is pairwise uncorrelated with $\mathbf{w}[t]$ and $\mathbf{v}[t]$ for all t. Denote $\overline{\mathbf{x}}[t] = \mathbb{E}[\mathbf{x}[t]]$ and $P_t = \mathrm{Cov}(\mathbf{x}[t])$. These quantities are recursively determined for $t \geq 1$ by:

$$
\begin{aligned}
\mathbb{E}[\mathbf{x}[t+1]] = A_t \mathbb{E}[\mathbf{x}[t]] &\Longrightarrow \overline{\mathbf{x}}[k+1] = A_t \overline{\mathbf{x}}[t] \\
P_{t+1} &= A_t P_t A_t^\top + Q_t
\end{aligned}
$$

As before, we will define $\mathscr{A}_t = \sigma(\mathbf{y}[0], \mathbf{y}[1], \ldots, \mathbf{y}[t])$ to be the sigma algebra of observations up until time t. Then we define $\hat{\mathbf{x}}[t|s] \triangleq \hat{E}[\mathbf{x}[t]|\mathscr{A}_s]$ for nonnegative integers s, t, where $\hat{E}[\mathbf{x}[t]|\mathscr{A}_s]$ is the linear MMSE given by

$$
\begin{aligned}
&\hat{E}[\mathbf{x}[t]|\mathbf{y}[s]] = \mathbb{E}[\mathbf{x}[t]] + \mathrm{Cov}(\mathbf{x}[t], \mathbf{y}[s])\mathrm{Cov}(\mathbf{y}[s])^{-1}(\mathbf{y}[s] - \mathbb{E}[\mathbf{y}[s]]) \\
\Longrightarrow\ &\hat{E}[\mathbf{x}[t]|\mathscr{A}_s] = \hat{E}[\mathbf{x}[t]|\mathscr{A}_{s-1}] + \mathrm{Cov}(\mathbf{x}[t], \mathbf{y}[s])\mathrm{Cov}(\mathbf{y}[s])^{-1}(\mathbf{y}[s] - \mathbb{E}[\mathbf{y}[s]])
\end{aligned}
$$

Finally, denote the associated covariance of error matrices $\Sigma_{t|s} \triangleq \mathrm{Cov}(\mathbf{x}[t] - \hat{\mathbf{x}}[t|s])$ for nonnegative integers s, t.

The goal is to compute an estimate of $\mathbf{x}[t]$ at each timestep t. We will do this by deriving a recursive relationship between successive state estimates $\hat{\mathbf{x}}[t-1|t-1]$ and $\hat{\mathbf{x}}[t|t]$.

The filtering process takes the same two steps as in the Bayesian framework from Sect. 20.3:

1. **Prediction**: we predict the value of $\hat{\mathbf{x}}[t|t-1]$ given $\hat{\mathbf{x}}[t-1|t-1]$. To do this, directly use the dynamics (20.1) and the fact that the noise random variables are uncorrelated from all the system variables. Equations yield:

$$\hat{\mathbf{x}}[t|t-1] = A_{t-1}\hat{\mathbf{x}}[t-1|t-1] \tag{20.13a}$$

$$\Sigma_{t|t-1} = A_{t-1}\Sigma_{t-1|t-1}A_{t-1}^{\top} + Q_{t-1} \tag{20.13b}$$

2. **Measurement Update**: we modify our prediction from $\hat{\mathbf{x}}[t|t-1]$ to $\hat{\mathbf{x}}[t|t]$ in order to take into account a new observation $\mathbf{y}[t]$. Because we are able to predict a part of $\mathbf{y}[t]$ through the linear MMSE, the only innovation comes from the orthogonal component $\tilde{\mathbf{y}}[t] = \mathbf{y}[t] - \hat{E}[\mathbf{y}[t]|\mathscr{A}_{t-1}]$. Alternatively written, this is:

$$\tilde{\mathbf{y}}[t] = \mathbf{y}[t] - C_t\hat{\mathbf{x}}[t|t-1] \tag{20.14}$$

Derive $\hat{\mathbf{x}}[t|t]$ from $\hat{\mathbf{x}}[t|t-1]$:

$$\hat{\mathbf{x}}[t|t] = \hat{\mathbf{x}}[t|t-1] + \mathrm{Cov}(\mathbf{x}[t], \tilde{\mathbf{y}}[t])\mathrm{Cov}(\tilde{\mathbf{y}}[t])^{-1}\tilde{\mathbf{y}}[t] \tag{20.15}$$

Furthermore, the covariance of error is updated:

$$\Sigma_{t|t} = \Sigma_{t|t-1} - \mathrm{Cov}(\mathbf{x}[t], \tilde{\mathbf{y}}[t])\mathrm{Cov}(\tilde{\mathbf{y}}[t])^{-1}\mathrm{Cov}(\tilde{\mathbf{y}}[t], \mathbf{x}[t]) \tag{20.16}$$

Intuitively, the use of the new observation $\tilde{\mathbf{y}}[t]$ reduces the covariance of error for predicting $\mathbf{x}[t]$ from $\Sigma_{t|t-1}$ by the covariance matrix of the innovative part of the estimator.

Let us define the gain $L_t \triangleq \mathrm{Cov}(\mathbf{x}[t], \tilde{\mathbf{y}}^k)\mathrm{Cov}(\tilde{\mathbf{y}}[t])^{-1}$ so that we can simplify the information update equations as:

$$\hat{\mathbf{x}}[t|t] = \hat{\mathbf{x}}[t|t-1] + L_t\tilde{\mathbf{y}}[t] \tag{20.17a}$$

$$\Sigma_{t|t} = \Sigma_{t|t-1} - L_t\mathrm{Cov}(\tilde{\mathbf{y}}[t], \mathbf{x}[t]) \tag{20.17b}$$

We can further calculate

$$\mathrm{Cov}(\mathbf{x}[t], \tilde{\mathbf{y}}[t]) = \Sigma_{t|t-1}C_t^{\top}, \quad \mathrm{Cov}(\tilde{\mathbf{y}}[t]) = C_t\Sigma_{t|t-1}C_t^{\top} + R_t$$

so that $L_t = \Sigma_{t|t-1}C_t^{\top}(C_t\Sigma_{t|t-1}C_t^{\top} + R_t)^{-1}$.

Combining the results of the two steps above, we obtain our final equations.

Theorem 20.53 (The Discrete-Time Kalman Filtering Equations) *The DTKF equations are given as follows:*

$$\hat{\mathbf{x}}[t|t] = A_{t-1}\hat{\mathbf{x}}[t-1|t-1] + L_t\tilde{\mathbf{y}}[t] \tag{20.18a}$$

$$\Sigma_{t|t} = \Sigma_{t|t-1} - L_t C_t \Sigma_{t|t-1} = (I - L_t C_t)(A_{t-1}\Sigma_{t-1|t-1}A_{t-1}^\top + Q_{t-1}) \tag{20.18b}$$

where

$$L_t = \Sigma_{t|t-1}C_t^\top (C_t \Sigma_{t|t-1} C_t^\top + R_t)^{-1} \tag{20.19}$$

Remark 20.49 The alternative, more common form of the Kalman filter equations is the update process from $\mathbf{x}[t|t-1]$ to $\mathbf{x}[t+1|t]$:

$$\hat{\mathbf{x}}_{k+1|k} = A_t(\hat{\mathbf{x}}[t|t-1] + L_t\tilde{\mathbf{y}}[t])$$
$$\Sigma_{k+1|k} = A_t(\Sigma_{t|t-1} - L_t\mathrm{Cov}(\tilde{\mathbf{y}}^k, \mathbf{x}[t]))A_t^\top + Q_t$$

20.5 The Continuous-Time Kalman Filter

Now consider the continuous-time dynamics

$$\dot{\mathbf{x}}(t) = A\mathbf{x}(t) + B_w\mathbf{w}(t) \tag{20.20a}$$

$$\mathbf{y}(t) = C\mathbf{x}(t) + \mathbf{v}(t) \tag{20.20b}$$

where $\mathbf{x}(t), \mathbf{w}(t) \in \mathbb{R}^n$, $\mathbf{y}(t), \mathbf{v}(t) \in \mathbb{R}^m$, and $A, B_w \in \mathbb{R}^{n\times n}$, $C \in \mathbb{R}^{m\times n}$ are known for all $t > 0$.

As in the discrete-time case, we make the following assumptions: $\mathbf{x}(0) \triangleq \mathbf{x}_0$, $\{\mathbf{w}(t)\}$, and $\{\mathbf{v}(t)\}$ are pairwise uncorrelated for all $t > 0$, and $\mathbb{E}[\mathbf{w}(t)] = 0$, $\mathrm{Cov}(\mathbf{w}(s), \mathbf{w}(t)) = \mathbb{E}[\mathbf{w}(s)\mathbf{w}(t)] = Q\delta(t-s)$, $\mathbb{E}[\mathbf{v}(t)] = 0$, $\mathrm{Cov}(\mathbf{v}(s), \mathbf{v}(t)) = R\delta(t-s)$, where Q and R are known constant matrices. Further assume that $\mathbf{x}_0$ comes from a known Gaussian distribution with $\mathbb{E}[\mathbf{x}_0] = 0$ (assumed for simplicity), $\mathrm{Cov}(\mathbf{x}_0) = \Sigma_0$. We will define $\mathscr{A}(t) = \sigma\{\mathbf{y}(s) : 0 \le s < t\}$ to represent the observations made until time t.

The goal is to estimate $\hat{\mathbf{x}}(t)$ of $\mathbf{x}(t)$ given the observations $\mathscr{A}(t)$ such that the MSE:

$$J \triangleq \mathbb{E}\left[\mathrm{tr}((\mathbf{x}(t) - \hat{\mathbf{x}}(t))(\mathbf{x}(t) - \hat{\mathbf{x}}(t))^\top)\right] \tag{20.21}$$

is minimized. In analogue to the DTKF, the MMSE $\hat{\mathbf{x}}(t)$ is given by $\mathbb{E}[\mathbf{x}(t) \mid \mathscr{A}(t)]$. This implies that $\hat{\mathbf{x}}(0) = \mathbb{E}[\mathbf{x}_0] = 0$. Additionally, one can derive the dynamics:

$$\dot{\hat{\mathbf{x}}}(t) = A\hat{\mathbf{x}}(t) + L(\mathbf{y} - C\hat{\mathbf{x}}(t))$$

For the moment, we will take this as given, and use it to derive the covariance equation and the optimal Kalman filter gain.

Define the error vector $\mathbf{e} = \mathbf{x} - \hat{\mathbf{x}}$. Its dynamics are given by:

$$\dot{\mathbf{e}} = \dot{\mathbf{x}} - \dot{\hat{\mathbf{x}}} = A\mathbf{x} + B_w\mathbf{w} - A\hat{\mathbf{x}} - L(\mathbf{y} - C\hat{\mathbf{x}}) = (A - LC)\mathbf{e} + B_w\mathbf{w} - L\underbrace{(\mathbf{y} - C\mathbf{x})}_{\mathbf{v}}$$

Define the error covariance $\Sigma \triangleq \mathbb{E}[\mathbf{e}\mathbf{e}^\top]$. Then note that the MSE at time t is exactly $\mathrm{tr}(P)$. Derive the dynamics of Σ as follows:

$$\begin{aligned}
\dot{\Sigma} &= \mathbb{E}[\dot{\mathbf{e}}\mathbf{e}^\top + \mathbf{e}\dot{\mathbf{e}}^\top] \\
&= \mathbb{E}[(A - LC)\mathbf{e}\mathbf{e}^\top + \mathbf{e}\mathbf{e}^\top(A^\top - C^\top L^\top)] + \mathbb{E}[B_w\mathbf{w}\mathbf{e}^\top + \mathbf{e}\mathbf{w}^\top B_w^\top] + \mathbb{E}[-L\mathbf{v}\mathbf{e}^\top - \mathbf{e}\mathbf{v}^\top L] \\
&= (A - LC)\Sigma + \Sigma(A^\top - C^\top L^\top) + \mathbb{E}[B_w\mathbf{w}\mathbf{e}^\top + \mathbf{e}\mathbf{w}^\top B_w^\top] + \mathbb{E}[-L\mathbf{v}\mathbf{e}^\top - \mathbf{e}\mathbf{v}^\top L^\top] \quad (20.22)
\end{aligned}$$

Denote the state transition matrix $\Phi(t_0, t) \triangleq e^{(A-LC)(t-t_0)}$. Then we can write $\mathbf{e}(t)$ as:

$$\mathbf{e}(t) = \Phi(0, t)\mathbf{e}_0 + \int_0^t \Phi(s, t)B_w\mathbf{w}(s)ds - \int_0^t \Phi(s, t)L\mathbf{v}(s)ds \tag{20.23}$$

which implies that

$$\begin{aligned}
\mathbb{E}[\mathbf{e}\mathbf{w}^\top(t)B_w^\top] &= \Phi(0, t)\mathbb{E}[\mathbf{e}_0\mathbf{w}^\top]B_w^\top + \int_0^t \Phi(s, t)B_w\mathbb{E}[\mathbf{w}(s)\mathbf{w}^\top(t)]B_w^\top ds \\
&\quad - \int_0^t \Phi(s, t)L\mathbb{E}[\mathbf{w}(s)\mathbf{v}^\top(t)]B_w^\top ds \\
&= \int_0^t \Phi(s, t)B_w Q\delta(t - s)B_w^\top ds \\
&= \frac{1}{2}B_w Q B_w^\top \text{ since } \Phi(t, t) = I
\end{aligned}$$

One can make a symmetric argument for $\mathbb{E}[B_w\mathbf{w}\mathbf{e}^\top] = \frac{1}{2}B_w Q B_w^\top$.

For the last line, we also have the following formula:

$$\int_a^b f(x)\delta(b - x)dx = \int_{b-a}^0 f(b - u)\delta(u)du = \frac{1}{2}f(b)$$

Informally speaking, since the delta function occurs at one of the endpoints of the integral, only half the total weight gets integrated over.

Similarly, (20.23) implies that $\mathbb{E}[-\mathbf{e}\mathbf{v}^\top(t)L^\top] = \mathbb{E}[-L\mathbf{v}\mathbf{e}^\top] = (1/2)LRL^\top$ since the minus signs cancel.

Substituting everything back int (20.22) yields:

$$\begin{aligned}\dot{\Sigma} &= (A - LC)\Sigma + \Sigma(A^\top - C^\top L^\top) + B_w Q B_w^\top + LRL^\top \\ &= A\Sigma + \Sigma A^\top + B_w Q B_w^\top - LC\Sigma - \Sigma C^\top L^\top + LRL^\top \\ &= A\Sigma + \Sigma A^\top + B_w Q B_w^\top + (LR - \Sigma C^\top)R^{-1}(LR - \Sigma C^\top)^\top - \Sigma C^\top RC\Sigma\end{aligned} \tag{20.24}$$

where the last line follows from completing the square. To minimize $J = \mathrm{tr}(\Sigma(t))$, we can minimize $\Sigma(t)$ by choosing L so that $\dot{P}(t)$ decreases by the maximum amount possible at time t. This happens when the term $(LR - \Sigma C^\top)R^{-1}(LR + \Sigma C^\top)^\top$ is equal to 0, since it can never be negative (like a square term).

This implies:

$$L(t)R - \Sigma(t)C^\top = 0 \Longrightarrow L(t) = \Sigma(t)C^\top R^{-1}$$

Substituting this back into (20.24) yields the final covariance equation.

Remark 20.50 More mathematically rigorously, we can determine the optimal gain L by solving the following optimal control problem:

$$\begin{aligned} &\min_L \mathrm{tr}(\Sigma(t)) \\ \text{s.t. } &\dot{\Sigma} = A\Sigma + \Sigma A^\top + B_w Q B_w^\top - LC\Sigma - \Sigma C^\top L^\top + LRL^\top \\ &\Sigma(0) = \Sigma_0\end{aligned}$$

Other well-known modern Bayesian filtering techniques, such as the extended Kalman filter (EKF), the unscented Kalman filter (UKF), and the particle filter can now be derived using similar principles as the Kalman filter. Additional treatment on these filtering techniques can be found in [1–3].

References

1. Bruce Hajek. *Random Processes for Engineers*. 2014.
2. Paul Zarchan and Howard Musoff. *Fundamentals of Kalman Filtering*. American Institute of Aeronautics and Astronautics, 2000.
3. Yaakov Bar-Shalom, X.-Rong Li, and Thiagalingam Kirubarajan. *Estimation with Applications to Tracking and Navigation: Theory, Algorithms and Software*. John Wiley & Sons, Inc., 2002.

Chapter 21
Problems and Exercises

Feedback Stabilization

Problem 1: Coprime Factorization [1]. The *cascade compensation with unit feedback* system is typically configured as shown in Fig. 21.1. We will take no external inputs other than the reference signal $r(t)$ (i.e., set $d = 0$ and $n = 0$).

In this problem, we will compute a suitable $K(s)$ so that the configuration in Fig. 21.1 stabilizes plant $H(s) = 1/(s(s-2))$ such that the transfer function from r to y is given by

$$G_{yr}(s) = \frac{\beta(s)}{(s+1)^2}, \quad \deg\beta(s) \leq 2 \ \ \text{(which means } \beta(s) = \beta_0 + \beta_1 s + \beta_2 s^2 \text{ for some } \beta_i\text{'s)}$$

(a) Let $K(s) = N(s)/D(s)$. Write the polynomials $N(s)$ and $D(s)$ in terms of the coefficients of β.
(b) Choose β_2, β_1 such that $K(s) \in \mathcal{RH}_\infty$. Write the transfer function G_{yr} in terms of β_0.
(c) Choose three different values of β_0. For each value, identify whether G_{yr} is stable or not.

What you observe in part c is a consequence of requiring $G_{yr}(s)$ to have the same denominator degree as the original plant $H(s)$. This illustrates why the coprime factorization approach is needed.

Problem 2: Euclid's Algorithm. Consider the plant

$$H(s) = \frac{5(s-2)}{(s+2)(s^2-9s+20)}$$

Find a coprime factorization $H(s) = N(s)/M(s)$ such that $N, M \in \mathcal{RH}_\infty$, N and M do not share any zeros on the open right-half plane, and the poles of M are at $s = -3$ and $s = -4$. Use Euclid's Algorithm to verify that your choices of N and

SJ. Han, *Linear Systems*, KAIST Research Series,
https://doi.org/10.1007/978-981-96-9774-8_21

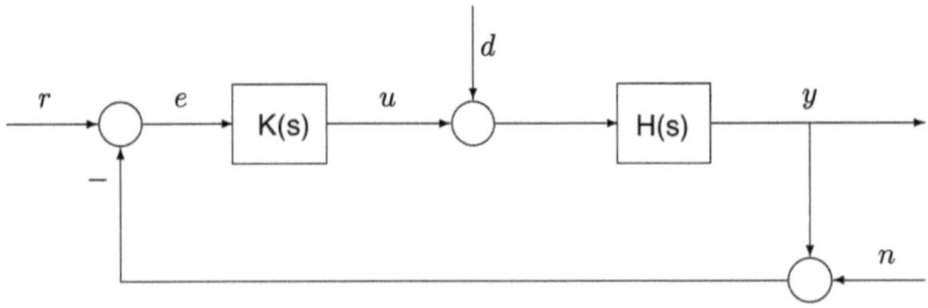

Fig. 21.1 The cascade compensation $K(s)$ of plant $H(s)$ with unit feedback. Modified from Doyle, Francis, Tannenbaum [2]. Here, we will take $d = 0$ and $n = 0$

M are coprime.
Hint: Try starting with $N(s) = 5(s-2)/p(s)$ and $M(s) = (s+2)(s^2 - 9s + 20)/p(s)$, where $p(s)$ is some polynomial of s which satisfies the given constraints.

Problem 3. Consider the plant as follows, with coprime factorization

$$H(s) = \frac{(s-3)(s-5)}{(s-p)^2(s+8)}, \quad N(s) = \frac{(s-3)(s-5)}{(s+2)^2(s+8)}, \quad M(s) = \frac{(s-p)^2}{(s+2)^2}$$

(a) Verify that the above N and M belong to the space $\mathcal{Q}$, defined in class.
(b) An internally stabilizing controller $K \in \mathcal{S}(H)$ is said to be *strongly stabilizing* if K itself is also a stable transfer function. If such a K exists for the plant H, then H is said to be *strongly stabilizable*.
Can you find a strongly stabilizing controller of H when $p = 4$? What about when $p = 3.1$?

Problem 4: Hamiltonian Matrices. A matrix

$$M = \begin{bmatrix} M_{11} & M_{12} \\ M_{21} & M_{22} \end{bmatrix}, \quad M_{ij} \in \mathbb{R}^{n \times n}$$

is said to be *Hamiltonian* if

$$\begin{bmatrix} 0 & I \\ -I & 0 \end{bmatrix} M \begin{bmatrix} 0 & I \\ -I & 0 \end{bmatrix} = M^\top.$$

Here, we denote $J : - \begin{bmatrix} 0 & I \\ -I & 0 \end{bmatrix}$.

(a) Prove that M is Hamiltonian if and only if $M_{22} = -M_{11}^\top$ and M_{12}, M_{21} are both symmetric.
(b) Show that if $\mathbf{v}$ is an eigenvector of M, then $J\mathbf{v}$ is an eigenvector of $M^\top$. Also, show that if $\lambda \in \mathbb{R}$ is an eigenvalue of M, then so is $-\lambda$.

Linear Quadratic Regulator

Problem 5: Cartpole Pendulum via LQR. We return to the inverted pendulum on a cart system in this problem. Consider again the linearized model about the upright pole position (around $(x, \dot{x}, \theta, \dot{\theta}) = (0, 0, \pi, 0)$):

$$\dot{\mathbf{x}}(t) = \begin{bmatrix} 0 & 1 & 0 & 0 \\ 0 & -\frac{b}{M} & -\frac{mg}{M} & 0 \\ 0 & 0 & 0 & 1 \\ 0 & -\frac{b}{ML} & -\frac{(M+m)g}{ML} & 0 \end{bmatrix} \mathbf{x}(t) + \begin{bmatrix} 0 \\ \frac{1}{M} \\ 0 \\ \frac{1}{ML} \end{bmatrix} F(t)$$

(a) Choose the following matrices for Q and R

$$Q = \begin{bmatrix} 1 & & & \\ & 1 & & \\ & & 10 & \\ & & & 100 \end{bmatrix}, \quad R = 0.01$$

and use MATLAB to design an infinite horizon LQR controller which stabilizes this system. Plot the position $x(t)$ versus time t, angle $\theta(t)$ versus time t, and control $F(t)$ versus time t on three separate (sub)plots. Be sure to choose at least 10 different initial conditions close to the equilibrium.

(b) Repeat part (a) for the following Q and R instead:

$$Q = 0.2I_4 \triangleq \begin{bmatrix} 0.2 & & & \\ & 0.2 & & \\ & & 0.2 & \\ & & & 0.2 \end{bmatrix}, \quad R = 20$$

What do you observe? How does the performance of this LQR controller compare to that of part (a)?

Problem 6: Infinite-Horizon LQR. Consider the following system described by the linear equations $\dot{\mathbf{x}} = A\mathbf{x} + Bu$ with $y = C\mathbf{x}$.

$$A = \begin{bmatrix} 0 & 1 \\ 0 & 0 \end{bmatrix}, \quad B = \begin{bmatrix} 0 \\ 1 \end{bmatrix}, \quad C = \begin{bmatrix} 1 & 0 \end{bmatrix}$$

(a) Determine the optimal control $\mathbf{u}^*(t) = F^*\mathbf{x}(t)$ with $t \geq 0$ which minimizes the performance index $J = \int_0^\infty (y^2(t) + \rho u^2(t))dt$, where ρ is positive and real.

(b) Observe how the eigenvalues of the dynamic matrix of the resulting closed-loop system change as a function ρ. What do the results tell you?

Problem 7. Consider an object of mass $m = 1$ moving along the x-axis in response to a force input $u(t)$. The object's dynamics can be described simply as $\ddot{x}(t) = u(t)$. Suppose you would like to design an input which will move the object from any initial position and velocity, then come to rest at the position $x_f = 5$. Formulate this problem as a LQR and solve it in a coding language of your choice (e.g., MATLAB). Plot the state and control input trajectories over time for multiple different Q and R weightings, as well as different initial conditions.

Problem 8: Receding-Horizon LQR. In this problem, we will consider a special version of the LQR called the *receding horizon LQR*. Let $\mathbf{x}_{t+1} = A\mathbf{x}_t + B\mathbf{u}_t$, $\mathbf{y}_t = C\mathbf{x}_t$, and choose costs Q, R, and a horizon of length $T \in \mathbb{N}$ for the LQR problem. For each horizon $T \in \mathbb{N}$, this controller takes a linear state-feedback form $\mathbf{u}_t = K_T\mathbf{x}_t$.

(a) What happens to the system if T is increased?
(b) What is the smallest value of T for which the closed-loop system becomes stable?
(c) Write down a Riccati equation for this receding horizon LQR problem. Express the feedback gain $K_T \in \mathbb{R}^{1\times 4}$ in terms of A, B, and P_i, the solution of the Riccati equation.
(d) Suppose we wanted to implement the receding horizon controller when

$$A = \begin{bmatrix} 1 & 0.4 & 0 & 0 \\ -0.7 & 1 & 0.5 & 0 \\ 0 & 0.5 & 1 & -0.7 \\ 0 & 0 & 0.4 & 1 \end{bmatrix}, \quad B = \begin{bmatrix} 1 \\ 0 \\ 0 \\ 0 \end{bmatrix}, \quad C = \begin{bmatrix} 0 & 0 & 0 & 1 \end{bmatrix}$$

with state cost $Q = C^\top C$, control cost $R = 1$. Plot the feedback gain K_T versus the horizon length T. Make sure each of the 4 entries of K_T are drawn as separate lines. Make sure that the range of T you choose includes your answer to part b).
(e) For the same values of A, B, C, Q, R, and the same range of T as part d), plot the spectral radius of the closed-loop $A_{\mathrm{cl},T} \triangleq A + BK_T$.

Problem 9: DT Finite-Horizon LQR. Prove that the optimal control input $\mathbf{u} \in \mathcal{U}$ for the discrete-time (DT) finite-horizon LQR cost functional

$$J_u(\mathbf{x}_0) = \sum_{t=0}^{T-1} \begin{bmatrix} \mathbf{x}_t^\top & \mathbf{u}_t^\top \end{bmatrix} \begin{bmatrix} Q & S \\ S^\top & R \end{bmatrix} \begin{bmatrix} \mathbf{x}_t & \mathbf{u}_t \end{bmatrix} + \mathbf{x}_t^\top Q_f \mathbf{x}_t,$$

with $Q, Q_f \geq 0$ and $R > 0$, is given by

$$\mathbf{u}_t^* = K_t\mathbf{x}_t, \text{ where } K_t \triangleq -(R + B^\top P_{t+1}B)^{-1}(S^\top + B^\top P_{t+1}A), \text{ where } P_{t+1} \text{ is defined by the recursion}$$

$$P_t = \begin{cases} Q + A^\top P_{t+1}A - (S^\top + B^\top P_{t+1}A)^\top (R + B^\top P_{t+1}B)^{-1}(S^\top + B^\top P_{t+1}A) & \text{if } 0 \leq t \leq T-1 \\ Q_f & \text{if } t = T \end{cases}$$

We did the $S = 0$ case earlier in this chapter.

Problem 10. Solve the following LQR problems without using any computer program. Note that you are asked to compute the **minimum value** of the optimization problem.

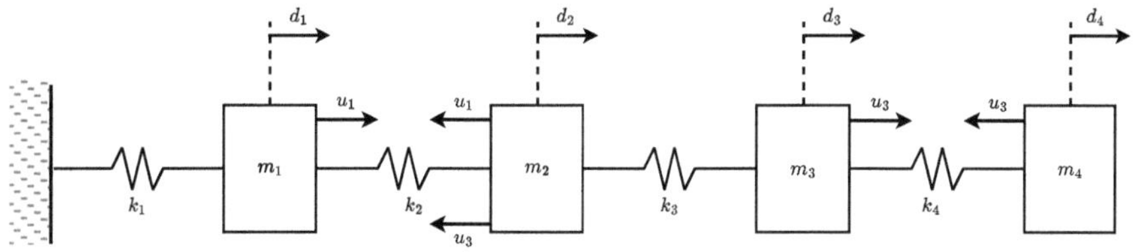

Fig. 21.2 The mass-spring system of Problem 11

Hint: The Riccati equations can be solved by hand.

(a) $\min_{u \in \mathcal{U}} \int_{t_0}^{t_f} (x^2(t) + u^2(t))dt$ s.t. $\dot{x}(t) = u(t)$

(b) $\min_{u \in \mathcal{U}} \int_{0}^{\infty} (x_1^2(t) + u^2(t))dt$ s.t. $\begin{cases} \dot{x}_1(t) = x_2(t), \ x_1(0) = x_{10} \\ \dot{x}_2(t) = u(t), \ x_2(0) = x_{20} \end{cases}$

Problem 11: Mass-Spring System. We will design an LQR controller for the mass-spring interconnection shown in Fig. 21.2.

For $i = 1, \ldots, 4$, block i has mass $m_i \in \mathbb{R}^+$ and its state is represented by is displacement $d_i \in \mathbb{R}$ from some designated equilibrium point. For simplicity, we will take $m_i = 1$ for all i. We assume there are three main forces acting on these blocks, denoted u_1, u_2, u_3.

The dynamics of such a system are given by

$$\ddot{d}_1 = -k_1 d_1 + k_2(d_2 - d_1) + u_1 \qquad \ddot{d}_2 = -k_2(d_2 - d_1) + k_3(d_3 - d_2) - u_1 - u_3$$
$$\ddot{d}_3 = -k_3(d_3 - d_2) + k_4(d_4 - d_3) + u_2 \qquad \ddot{d}_4 = -k_4(d_4 - d_3) - u_2$$

(a) Define an appropriate choice of state $\mathbf{x}(t)$, then write this system in LTI state-space form $\dot{\mathbf{x}}(t) = A\mathbf{x}(t) + B\mathbf{u}(t)$.
(b) Find the optimal state-feedback control law $\mathbf{u}(t)$ which minimizes the cost functional

$$J_{\mathbf{u}}(\mathbf{x}_0) \triangleq \int_0^\infty (\|\mathbf{d}(t)\|_2^2 + \|\mathbf{u}(t)\|_2^2)dt$$

where $\mathbf{d} \triangleq (d_1, \ldots, d_4)^\top$. What are the Q and R weights you should choose?

(c) Use the Hamiltonian matrix method to solve the Riccati equation you got in part (b). Verify that you get the same result as in part (b).
(d) Using the spring coefficients $k_i = 1$ for all i, plot $\mathbf{d}(t)$ versus time t for both the open-loop dynamics ($\mathbf{u}(t) = 0$) and the feedback-stabilized closed-loop dynamics. Plot your trajectories for at least 3 different initial conditions (not equal to 0).

Stochastic Differential Equations, Stochastic Systems

Problem 12: Euler-Maruyama Algorithm. Recall that for deterministic linear systems $\dot{\mathbf{x}}(t) = A\mathbf{x}(t) + B\mathbf{u}(t)$, we discussed several methods of discretization to DT systems of the form $\mathbf{x}_{t+1} = A_d\mathbf{x}_t + B_d\mathbf{u}_t$. This allows us to approximately simulate a CT linear system on a computer using well-known numerical algorithms (e.g., Euler integration, leapfrog, etc.). Note that discretization can be applied to any deterministic system, even nonlinear ones.

An analogous technique used to discretize and simulate SDEs is the *Euler-Maruyama algorithm*, whose pseudocode is written as follows.

Algorithm 1 Euler-Maruyama

Given SDE $d\mathbf{x}(t) = f(t, \mathbf{x}(t))dt + \sigma(t, \mathbf{x}(t))dW(t)$, initial condition $\mathbf{x}_0$.
Partition time interval $[0, T]$ into $\{t_n\}_{n=0}^{N}$, where $t_{n+1} - t_n = \Delta T$ for some chosen ΔT.
Set $X_0 = \mathbf{x}_0$.
for index $n = 0, \ldots, N-1$ **do**
  Generate normal random variable $\Delta W_n \sim \mathcal{N}(0, \Delta T)$.
  Set $X_{n+1} = X_n + f(t_n, X_n)\Delta T + \sigma(t_n, X_n)\Delta W_n$
end for
Return stochastic process $\{X_n\}_{n=0}^{N}$, a sample path approximating the true $\mathbf{x}(t)$.

A few remarks:

- The basic idea is that we are discretizing the SDE $d\mathbf{x}(t) = f(t, \mathbf{x}(t))dt + \sigma(t, \mathbf{x}(t))dW(t)$ using some fixed timestep ΔT, and transforming it into a *stochastic difference equation* $X_{n+1} = X_n + f(t_n, X_n)\Delta T + \sigma(t_n, X_n)\Delta W_n$.
- Note that $\Delta W_n \triangleq W(t_{n+1}) - W(t_n)$ is just a $\mathcal{N}(0, \Delta T)$ Gaussian random variable because $t_{n+1} - t_n = \Delta T$ for all n, and because of the stationary increments property of $W(t)$.

Code and implement (via Python, MATLAB, your choice) the Euler-Maruyama algorithm for the two scalar linear SDEs we discussed in class:

(a) the mean-reverting Ornstein-Uhlenbeck process $\mathbb{R} \ni dx(t) = \theta(\mu - x(t))dt + \sigma dW(t)$ with $\theta = 0.7$, $\mu = 0.5$, $\sigma = 0.06$.
(b) the Geometric Brownian motion $\mathbb{R} \ni dx(t) = \theta x(t)dt + \sigma x(t)dW(t)$ with $\theta = 0.7$, $\sigma = 0.2$.

For each one of your plots, fix the initial condition at $\mathbf{x}_0 = 1$.
Note: in contrast to deterministic linear systems, each $\mathbf{x}_0$ does not give us a unique solution trajectory $\mathbf{x}(t)$ in stochastic systems. This is clearly due to the randomness in the system's dynamics.

Now code and implement the Euler-Maruyama algorithm for the following *nonlinear* stochastic system:

(c) $dx(t) = -\sin x(t) + \sqrt{2\sigma^2}dW(t)$ with $\sigma = 0.1$. Plot both $x(t)$ versus t **and** $\sin x(t)$ versus t on separate subplots.

Problem 13: Itô's Formula. *Itô's formula* (or *Itô's lemma*) is an important identity in stochastic calculus. It is used to find the differential of a (time-dependent) function of some $\mathbf{x}(t)$, where $\mathbf{x}(t)$ evolves according to a stochastic process or SDE. One can think of it as the stochastic calculus version of the Taylor series expansion, or the chain rule.

Given a scalar-valued SDE $dx(t) = f(t, x(t))dt + \sigma(t, x(t))dW(t)$, suppose we are interested in computing $y(t) \triangleq V(t, x(t))$, where $V \in C^{(1,2)}$ is a real-valued function. Then the formula tells us:

$$dy(t) = \partial_t V(t, x(t))dt + \partial_x V(t, x(t))dx(t) + \frac{1}{2}\partial_x^2 V(t, x(t))d[x, x](t)$$

where $[\cdot, \cdot]$ is the quadratic variation notation we discussed in class. Note that this can be written in an integral form:

$$y(t) = y(0) + \int_0^t \partial_s V(s, x(s))ds + \int_0^t \partial_x V(s, x(s))dx(s) + \frac{1}{2}\int_0^t \partial_x^2 V(s, x(s))d[x, x](s)$$

In order to complete the calculations, one must now substitute in the expressions for $dx(t)$ and $d[x, x](t)$.

Now consider the following problems.

(a) Use Itô's formula to prove that

$$\int_0^t W(s)dW(s) = \frac{1}{2}\left(W^2(t) - t\right) \tag{21.1}$$

Hint: Consider applying Itô's formula to some function f of $W(t)$:

$$f(W(t)) \triangleq f(W(0)) + \int_0^t f'(W(s))dW(s) + \frac{1}{2}\int_0^t f''(W(s))\underbrace{d[W, W](s)}_{=ds}.$$

Then choose f which gives you an expression for $\int_0^t W(s)dW(s)$ in terms of $W(t)$ and t.

Note: For more general functions $g(s)$, it is not always possible to compute stochastic integrals $\int(s)dW(s)$ explicitly, and the best we can do is to leave them in that expression.

(b) Recall the solution to the mean-reverting Ornstein-Uhlenback process $dx(t) = \theta(\mu - x(t))dt + \sigma dW(t)$ is given by

$$x(t) = e^{-\theta t}x_0 + \mu(1 - e^{-\theta t}) + \sigma\int_0^t e^{-\theta(t-s)}dW(s) \tag{21.2}$$

Apply Itô's formula to the function $V(x(t)) \triangleq x(t)e^{\theta t}$ and verify that you get the same result as (21.2).

(c) Recall we derived a solution to the Geometric Brownian motion $dx(t) = \theta x(t)dt + \sigma x(t)dW(t)$ as

$$\frac{dx(t)}{x(t)} = \theta dt + \sigma dW(t) \Longrightarrow x(t) = x_0 \exp\left(\int_0^t \theta ds + \int_0^t \sigma dW(s)\right) \tag{21.3}$$

by directly applying separation of variables. **However**, it turns out this solution is actually incorrect! Apply Itô's formula to the function $V(x(t)) \triangleq \log x(t)$ and derive the correct solution.

(d) The *Brownian bridge* is another common type of SDE with many applications in physics and statistics.

$$dx(t) = -\frac{1}{1-t}x(t)dt + dW(t) \quad \forall\, t \in [0, 1)$$

Show that the solution is given by

$$x(t) = (1-t)\int_0^t \frac{1}{1-s}dW(s) \tag{21.4}$$

Verify this solution using **both** types of methods: (1) separation of variables and (2) Itô's formula. For Itô's formula, how do you think you should choose $V(t, x(t))$?

Note: the Brownian bridge can be viewed as a linear time-varying (LTV) stochastic system, where the $a(t)$ coefficient is $-1/(1-t)$.

(e) You may have noticed that typical methods to solve deterministic ODEs (e.g., integrating factors, separation of variables) works for *some* SDEs, but not others. Why do you think this is the case?

A good textbook that is used a lot in SDE courses at various universities is Øksendal's "Stochastic Differential Equations" [3]; Chap. 4 discusses Itô's formula [4, 5].

Problem 14: Brownian Motion Processes. This problem is concerned with some miscellaneous properties of Gaussian processes and the Brownian motion.

(a) Let $\mathbf{x} \sim \mathcal{N}(\mu, \Sigma)$ be a Gaussian random vector with mean $\mu \in \mathbb{R}^n$ and covariance matrix $\Sigma \in \mathbb{R}^{n\times n}$. Let $\mathbf{y} \triangleq A\mathbf{x} + \mathbf{b}$ be an affine transformation of $\mathbf{x}$, for some $A \in \mathbb{R}^{n\times n}$ and $\mathbf{b} \in \mathbb{R}^n$. What is the mean and covariance matrix of $\mathbf{y}$?

(b) Recall that the *autocorrelation function* of a real-valued stochastic process $\{X(t), t \geq 0\}$ is given by $R_X(s, t) \triangleq \mathbb{E}[X(t)X(s)]$ for any two times $s, t \in \mathbb{R}^{\geq 0}$. Prove that $R_W(s, t) = \min(s, t)$, where W is the standard Brownian motion process.

(c) Suppose X_1 and X_2 are two Brownian motion processes with variance parameters σ_1 and σ_2, respectively. What is the autocorrelation function of $X_1 - X_2$? Also, what is its pdf, $f_{X_1-X_2}(x)$ for $x \in \mathbb{R}$?

Hint: Generalizing part b), the autocorrelation function of a Brownian motion processes with variance parameter σ is $\sigma \min(s, t)$.

(d) Let $\mathbf{x} \triangleq [x_1^\top, \ldots, x_n^\top]$ be a Gaussian random vector with mean 0 and covariance matrix Σ_X. Show there exists a unitary matrix $U \in \mathbb{R}^{n\times n}$ (i.e., $U^\top = U^{-1}$) such that $\{\mathbf{y}_1, \ldots, \mathbf{y}_n\}$, defined as the entries in random vector $\mathbf{y} \triangleq U\mathbf{x}$, are uncorrelated random variables.
Hint: Find a U such that the covariance matrix Σ_Y of $\mathbf{y}$ becomes a diagonal matrix.

$\mathcal{H}_2$ Control, and Linear Quadratic Gaussian

Problem 15: Discounted LQG. We will consider the following *discounted* LQG problem. First, as usual, the DT linear dynamics are given by

$$\mathbf{x}_{t+1} = A\mathbf{x}_t + B\mathbf{u}_t + \mathbf{w}_t, \quad \mathbf{w}_t \sim \mathcal{N}(0, W) \text{ i.i.d. } \forall t \in \mathbb{Z}^{\geq 0}$$

The infinite-horizon cost functional we are seeking to optimize is

$$J_{\mathbf{u}}(\mathbf{x}) \triangleq \min_{\mathbf{u}\in\mathcal{U}} \mathbb{E}\left[\sum_{t=0}^{\infty} \gamma^k (\mathbf{x}_t^\top Q\mathbf{x}_t + \mathbf{u}_t^\top R\mathbf{u}_t)\middle| \mathbf{x}_0 = \mathbf{x}\right]$$

where we have the usual assumptions on Q and R, and $\gamma \in (0, 1]$ is called the *discount factor*.

(a) Derive an optimal linear state-feedback control policy of the form $\mathbf{u}_t^* = -K^*\mathbf{x}_t$, with an appropriate gain K^*. What is the form of the cost-to-go function (value function) $V(\mathbf{x}) \triangleq \mathbb{E}[\sum_{t=0}^{\infty} \gamma^k (\mathbf{x}_t^\top Q\mathbf{x}_t + \mathbf{u}_t^\top R\mathbf{u}_t)|\mathbf{x}_0 = \mathbf{x}]$ that you should use?

(b) In other fields (e.g., reinforcement learning), it is common to keep track of another type of value function called the *Q-function* (or *state-action value function*), defined as $\mathcal{Q}(\mathbf{x}, \mathbf{u}) \triangleq \mathbb{E}[\sum_{t=0}^{\infty} \gamma^k (\mathbf{x}_t^\top Q\mathbf{x}_t + \mathbf{u}_t^\top R\mathbf{u}_t)|\mathbf{x}_0 = \mathbf{x}, \mathbf{u}_0 = \mathbf{u}]$ Use Bellman's principle of optimality to derive a similar recursion for the Q-value function. Write your final answer as a matrix equation in terms of $\begin{bmatrix} \mathcal{Q}_{11} & \mathcal{Q}_{12} \\ \mathcal{Q}_{21} & \mathcal{Q}_{22} \end{bmatrix}$, which comes from

$$\mathcal{Q}(\mathbf{x}, \mathbf{u}) = \begin{bmatrix} \mathbf{x}^\top & \mathbf{u}^\top \end{bmatrix} \begin{bmatrix} \mathcal{Q}_{11} & \mathcal{Q}_{12} \\ \mathcal{Q}_{21} & \mathcal{Q}_{22} \end{bmatrix} \begin{bmatrix} \mathbf{x} \\ \mathbf{u} \end{bmatrix} + F$$

where F is an extra term due to the noise (you have to find an equation for F too).
Hint: It should look somewhat similar to your recursion in part a. Note the two value functions are related via $V(\mathbf{x}) = \mathcal{Q}(\mathbf{x}, -K^*\mathbf{x})$.

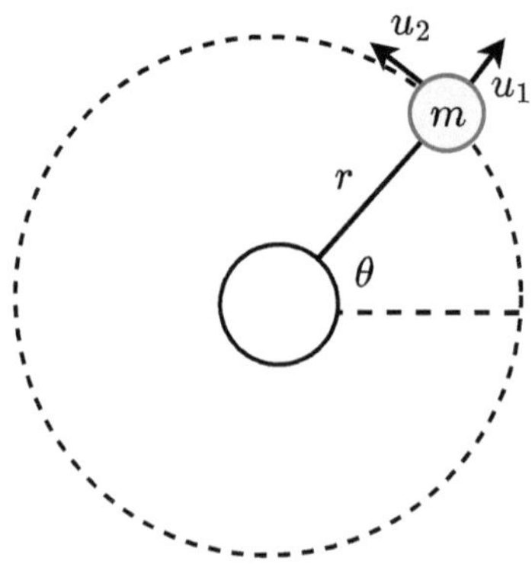

Fig. 21.3 Satellite system for Problem 5

Problem 16: Satellite Control. Consider the system of a satellite orbiting around a spherical planet (like Earth). See Fig. 21.3. Its dynamics of motion can be expressed as

$$m(\ddot{r} - r\dot{\theta}^2) = -\frac{km}{r^2} + u_1 + w_1, \quad m(2\dot{r}\dot{\theta} + r\ddot{\theta}) = u_2 + w_2$$

where m is the mass of the satellite, r is the radius of its orbit, θ is its angle with respect to the horizontal line, u_1 and u_2 are its thrust in the radial and tangential directions, respectively. Here, w_1 and w_2 are independent scalar Gaussian white noise processes with variances σ_1^2 and σ_2^2, respectively.

Defining the state as $\mathbf{x} \triangleq (r, \theta, \dot{r}, \dot{\theta})^\top$ and linearizing the dynamics around $(\bar{r}, 0, \bar{\omega}t, \bar{\omega})$, where $\bar{\omega} \triangleq \sqrt{k/\bar{r}^3}$ yields

$$\dot{\mathbf{x}} = \begin{bmatrix} 0 & 0 & 1 & 0 \\ 0 & 0 & 0 & 1 \\ 3\bar{\omega}^2 & 0 & 0 & 2\bar{r}\bar{\omega} \\ 0 & 0 & -2\bar{\omega}/\bar{r} & 0 \end{bmatrix} \mathbf{x} + \begin{bmatrix} 0 & 0 \\ 0 & 0 \\ \frac{1}{m} & 0 \\ 0 & \frac{1}{mr} \end{bmatrix} \left(\begin{bmatrix} u_1 \\ u_2 \end{bmatrix} + \begin{bmatrix} w_1 \\ w_2 \end{bmatrix} \right)$$

Use an optimization toolbox of your choice (e.g., `CVX`, `YALMIP`, etc.) to find stabilizing controls

(a) using u_2 only.
(b) using both u_1 and u_2.

For both parts, use the values $m = 100\,\text{kg}$, $\bar{r} = (R + r)\text{km}$, where $r = 300\,\text{km}$, $\sigma_1^2 = \sigma_2^2 = 0.1\,\text{N}$, and $k = GM$, where $G \approx 6.673 \times 10^{-11}\,\text{Nm}^2/\text{kg}^2$ is the universal gravitational constant, and $M \approx 5.98 \times 10^{24}\,\text{kg}$, $R \approx 6.37 \times 10^3\,\text{km}$ are the mass, radius of Earth.

Linear State Estimation and Kalman Filtering

Problem 17: Delayed-Form DT Kalman Filter. Derive an alternative formulation of the discrete-time Kalman filter (DTKF) equations which updates $\hat{\mathbf{x}}_{t|t-1}, \hat{\Sigma}_{t|t-1}$ to $\hat{\mathbf{x}}_{t+1|t}, \hat{\Sigma}_{t+1|t}$. Be sure to apply the two steps of the Bayesian filtering procedure we used in class.

Problem 18: CT Kalman Filter. In this problem, we will derive the Kalman filter in *continuous-time* (i.e., the CTKF), for the continuous-time LTI system

$$\dot{\mathbf{x}}(t) = A\mathbf{x}(t) + B_w\mathbf{w}(t), \quad \mathbf{y}(t) = C\mathbf{x}(t) + \mathbf{v}(t)$$

where $\mathbf{x}_0 \sim \mathcal{N}(0, \Sigma_0)$, $\{\mathbf{w}(t)\}$, and $\mathbf{v}(t)$ are all pairwise uncorrelated, with $\mathbb{E}[\mathbf{w}(t)] = 0$, $\mathrm{Cov}(\mathbf{w}(t), \mathbf{w}(s)) = \Sigma_w\delta(t-s)$ and $\mathbb{E}[\mathbf{v}(t)] = 0$, $\mathrm{Cov}(\mathbf{v}(t), \mathbf{v}(s)) = \Sigma_v\delta(t-s)$. As usual, the goal is to estimate $\hat{\mathbf{x}}(t)$ of $\mathbf{x}(t)$ given observations $\mathcal{A}(t) \triangleq \{\mathbf{y}(s) : 0 \le s < t\}$ such that the MSE $\mathbb{E}[\|\mathbf{x}(t) - \hat{\mathbf{x}}(t)\|_2^2]$ is minimized for each t.

(a) Similar to observer dynamics (for deterministic systems), we can use the following dynamics for the state estimate

$$\dot{\hat{\mathbf{x}}}(t) = A\hat{\mathbf{x}}(t) + L(t)(\mathbf{y}(t) - C\hat{\mathbf{x}}(t)) \tag{21.5}$$

Note that $L(t)$ is time dependent. Derive dynamics for the error $\mathbf{e}(t) \triangleq \mathbf{x}(t) - \hat{\mathbf{x}}(t)$ and the error covariance $\hat{\Sigma}(t) \triangleq \mathbb{E}[\mathbf{e}(t)\mathbf{e}^\top(t)]$. Make sure that for both equations, you get a form that looks like $\dot{\mathbf{a}} = A_{cl}\mathbf{a} +$ other terms..., with the same "A_{cl}" and $\mathbf{a}$ a placeholder for $\mathbf{e}$, Σ. (A_{cl} is something you have to find in terms of the given parameters.)

(b) Use the state transition matrix of the error equation (i.e., $\Phi(t, \tau) \triangleq e^{A_{cl}(t-\tau)}$, with the A_{cl} you derived in part a) to prove that

$$\mathbb{E}[\mathbf{e}(t)\mathbf{w}^\top(t)B_w^\top] = \mathbb{E}[B_w\mathbf{w}(t)\mathbf{e}^\top(t)] = \frac{1}{2}B_w\Sigma_w B_w^\top \text{ for all } t \tag{21.6}$$
$$\mathbb{E}[\mathbf{e}(t)\mathbf{v}^\top(t)L^\top(t)] = \mathbb{E}[L(t)\mathbf{v}(t)\mathbf{e}^\top(t)] = -\frac{1}{2}L(t)\Sigma_v L^\top(t) \text{ for all } t$$

Hint: Here, use $\int_0^t g(s)\mathbf{w}(s)ds$ instead of stochastic integral $\int_0^t g(s)dW(s)$. Also, the following formula may be helpful:

$$\int_a^b g(x)\delta(b-x)dx = \int_{b-a}^0 g(b-u)\delta(u)du = \frac{1}{2}g(b).$$

The factor $\frac{1}{2}$ can be justified by regarding the delta function $\delta(x)$ as a "limit" of $\sqrt{\frac{a}{\pi}}e^{-ax^2}$ as $a \to \infty$.

(c) Substitute the identities (21.6) into your dynamics for $\hat{\Sigma}(t)$ from part a) and simplify. Show that

$$\dot{\hat{\Sigma}}(t) = (A - L(t)C)\hat{\Sigma}(t) + \hat{\Sigma}(t)(A - L(t)C)^\top + B_w \Sigma_w B_w^\top + L(t)\Sigma_v L^\top(t) \tag{21.7}$$

Note: in the steady-state case (i.e., $\dot{\Sigma}(t) = 0$ and $L(t) \equiv L$), this is another algebraic Riccati equation!

(d) Recall that the objective is to minimize $J(t) \triangleq \mathbb{E}[\|\mathbf{x}(t) - \hat{\mathbf{x}}(t)\|_2^2]$. How would you rewrite $J(t)$ in terms of $\hat{\Sigma}(t)$? Show that the Kalman gain $L(t) \triangleq \hat{\Sigma}(t)C^\top \Sigma_v^{-1}$ yields the minimum $J(t)$ (where Σ_v is assumed to be invertible). Substitute $L(t)$ back into the dynamics for $\hat{\mathbf{x}}(t)$ and $\hat{\Sigma}(t)$ to complete the derivation of the CTKF.
Note: Some common alternative names for the CTKF are the *Kalman-Bucy filter* or the *linear quadratic estimator (LQE)*.

Problem 19: CT Kalman Filter. For the CT scalar integrator dynamics below:

$$\dot{x}(t) = w(t)$$

where $\Sigma_w = 2$, derive a Kalman filter for each of the following three sensor models.

(a) one noisy measurement of x is available: $y(t) = x(t) + v(t), \quad \Sigma_v = 1$.
(b) two independent noisy measurements of x are available:

$$\begin{bmatrix} y_1(t) \\ y_2(t) \end{bmatrix} = \begin{bmatrix} x(t) \\ x(t) \end{bmatrix} + \begin{bmatrix} v_1(t) \\ v_2(t) \end{bmatrix}, \quad \Sigma_v = \begin{bmatrix} 1 & 0 \\ 0 & 2 \end{bmatrix}$$

(c) two dependent noisy measurements of x are available:

$$\begin{bmatrix} y_1(t) \\ y_2(t) \end{bmatrix} = \begin{bmatrix} x(t) \\ x(t) \end{bmatrix} + \begin{bmatrix} v_1(t) \\ v_2(t) \end{bmatrix}, \quad \Sigma_v = \begin{bmatrix} 2 & 1 \\ 1 & 2 \end{bmatrix}$$

Problem 20: CT Kalman Filter. Here, we will derive a Kalman filter for the following 2-dimensional CT system

$$\dot{\mathbf{x}}(t) = \begin{bmatrix} 0 & 1 \\ 1 & 0 \end{bmatrix} \mathbf{x}(t) + \begin{bmatrix} 1 \\ 0 \end{bmatrix} \mathbf{u}(t) + \mathbf{w}(t), \quad \mathbf{y}(t) = \begin{bmatrix} 1 & 0 \end{bmatrix} \mathbf{x}(t) + \mathbf{v}(t)$$

where $\{\mathbf{w}(t)\}$ and $\{\mathbf{v}(t)\}$ are Gaussian white noise processes with covariance matrices $\Sigma_w = 3I_2$ and $\Sigma_v = 1$.

(a) Calculate the minimum observer error covariance $\hat{\Sigma}(t)$ and the optimal Kalman gain $L(t)$.
(b) Derive the filter equation for the state estimate $\hat{\mathbf{x}}(t)$.

(c) Use the `lqe` command to compute $\hat{\Sigma}(t)$ and $L(t)$. Compare these with your answers in part (a).

General Bayesian Filtering

Problem 21: Unscented Kalman Filter. In this problem, implement the unscented Kalman filter for the following three sample nonlinear systems:

(a) tracking a planar circle, with x-position x_1, y-position x_2, and angle position x_3.

$$\begin{bmatrix} x_1 \\ x_2 \\ x_3 \end{bmatrix} [t+1] = \begin{bmatrix} R\cos(x_3[t]) \\ R\sin(x_3[t]) \\ \mathrm{mod}(x_3[t] + \frac{\pi}{M}, 2\pi) \end{bmatrix} + \mathbf{w}[t], \quad \begin{bmatrix} y_1 \\ y_2 \end{bmatrix} [t] = \begin{bmatrix} x_1 \\ x_3 \end{bmatrix} [t] + \mathbf{v}[t]$$

where $R = 5$ (the radius of the circle), $M = 12$, $\mathbf{w}[t] \sim \mathcal{N}(0, 0.05^2 I_3)$, and $\mathbf{v} \sim \mathcal{N}(0, 0.1^2 I_2)$.
Note: we are using discrete-time notation $\mathbf{x}[t] \equiv \mathbf{x}_t$ (with brackets instead of subscript) because "$x_{i,t}$" doesn't look as good. Both types of notations are standard for DT systems anyway.

(b) tracking a planar spiral.

$$\begin{bmatrix} x_1 \\ x_2 \\ x_3 \end{bmatrix} [t+1] = \begin{bmatrix} \gamma^t R\cos(x_3[t]) \\ \gamma^t R\sin(x_3[t]) \\ \mathrm{mod}(x_3[t] + \frac{\pi}{M}, 2\pi) \end{bmatrix} + \mathbf{w}[t], \quad \begin{bmatrix} y_1 \\ y_2 \end{bmatrix} [t] = \begin{bmatrix} x_1 \\ x_3 \end{bmatrix} [t] + \mathbf{v}[t]$$

with $\gamma = 0.95$ and all other values are the same as in part a.

(c) tracking a 3D helix (i.e., tornado), with x-position x_1, y-position x_2, and z-position x_3, and angle x_4.

$$\begin{bmatrix} x_1 \\ x_2 \\ x_3 \\ x_4 \end{bmatrix} [t+1] = \begin{bmatrix} \gamma^t R\cos(x_4[t]) \\ \gamma^t R\sin(x_4[t]) \\ (P x_4[t]/2\pi) \\ x_4[t] + \frac{\pi}{M} \end{bmatrix} + \mathbf{w}[t], \quad \begin{bmatrix} y_1 \\ y_2 \\ y_3 \end{bmatrix} [t] = \begin{bmatrix} x_1 \\ x_2 \\ x_4 \end{bmatrix} [t] + \mathbf{v}[t]$$

with pitch $P = 0.5$, $\mathbf{w}[t] \sim \mathcal{N}(0, 0.05^2 I_4)$, $\mathbf{v} \sim \mathcal{N}(0, 0.1^2 I_3)$, and all other values are the same as in parts a and b.

For each part of the problem, plot at least three different sample paths for $T_{sim} = 100$ timesteps, with various different $\mathbf{x}_0 \sim \mathcal{N}(\mu, P)$ (i.e., you are free to choose μ and P), and $\lambda = n\alpha^2 - n$ with various different α and β.

Problem 22: Particle Filter. Recall the following dynamics for the uncontrolled double-pendulum:

$$2mL^2\ddot{\theta}_1 + mL^2\ddot{\theta}_2\cos(\theta_1 - \theta_2) + mL^2\dot{\theta}_2^2\sin(\theta_1 - \theta_2) + J_1\ddot{\theta}_1 - 2mgL\sin(\theta_1) = 0$$
$$mL^2\ddot{\theta}_2 + mL^2\ddot{\theta}_1\cos(\theta_1 - \theta_2) - mL^2\dot{\theta}_1^2\sin(\theta_1 - \theta_2) + J_2\ddot{\theta}_2 - mgL\sin(\theta_2) = 0$$

(These can be derived from the Euler-Lagrange equations.)

Form a state space with state $\mathbf{x} \triangleq \left[\theta_1\ \theta_2\ \dot{\theta}_1\ \dot{\theta}_2\right]^T$ and assume additive Gaussian white noise of $\mathbf{w}_t \sim \mathcal{N}(0, I_4)$. Then, implement a particle filter that will estimate the entire state given only noisy measurements of θ_2 and $\dot{\theta}_1$, perturbed by additive Gaussian white noise of $\mathbf{v}_t \sim \mathcal{N}(0, I_2)$.

In your simulations, use the following the following system parameters: $m = 1$, $L = 2$, and J_1, J_2 are inertias of the two respective pendulum bobs (i.e., $J_1 = mL^2/12$, $J_2 = 4mL^2/12$). Also, start with initial condition $x_0 = \left[0\ \frac{\pi}{5}\ 0\ 0\right]^T$. Discretize the original dynamics via Euler integration with a timestep of $\Delta t = 0.1$, and simulate your system for $T_{sim} = 50$ timesteps with $N = 1000$ particles.

References

1. Thomas Kaliath. *Linear Systems*. Prentice Hall, Inc., 1980.
2. John C. Doyle, Bruce Francis, and Allen Tannenbaum. *Feedback Control Theory*. Dover, 2009.
3. B. Øksendal. *Stochastic Differential Equations: An Introduction with Applications*. Universitext. Berlin, Germany: Springer Berlin, 2003.
4. N Ikeda and S Watanabe. *Stochastic differential equations and diffusion processes; 2nd ed.* North-Holland mathematical library. Amsterdam: North-Holland, 1989.
5. Erhan Çinlar. *Probability and Stochastics*. Springer Science & Business Media, 2011.

MIX
Papier aus verantwortungsvollen Quellen
Paper from responsible sources
FSC® C105338

If you have any concerns about our products,
you can contact us on
ProductSafety@springernature.com

In case Publisher is established outside the EU,
the EU authorized representative is:
Springer Nature Customer Service Center GmbH
Europaplatz 3, 69115 Heidelberg, Germany

Printed by Libri Plureos GmbH
in Hamburg, Germany